Wynn Kapit / Lawrence M. Elson

The **ANATOMY** COLORING BOOK

THIRD EDITION

Benjamin
Cummings

San Francisco Boston New York
Capetown Hong Kong London Madrid Mexico City
Montreal Munich Paris Singapore Sydney Tokyo Toronto

Publisher: Daryl Fox
Sponsoring Editor: Amy Teeling
Project Editor: Susan Teahan
Managing Editor: Wendy Earl
Production Supervisor: David Novak
Text Design and Art: Wynn Kapit
Manufacturing Supervisor: Stacey Weinberger
Senior Marketing Manager: Lauren Harp
Compositor: TypeStudio, Gerald Ichikawa
Copy Editor: Jill Breedon

Library of Congress Cataloging-in-Publication Data

Kapit, Wynn.
   The anatomy coloring book/Wynn Kapit, Lawrence M. Elson.—3rd ed.
     p. cm.
   Includes bibliographical references and index.
   ISBN 0-8053-5086-1
   1. Human anatomy—Atlases. 2. Coloring books. I. Elson, Lawrence M., 1935-II. Title.

QM25 .K36 2001
611—dc2l

2001017095

ISBN 0-8053-5086-1
17 18 19 20 21 22   V031   14 13 12 11
www.aw.com/bc

DEDICATION

For my wife, Lauren, and sons, Neil and Eliot.

WYNN KAPIT

My work in this book is dedicated to my incredibly talented and loving wife Ellyn, and my family: Jennifer, Chris and Gina, Amelia and Bill, Bill and Chris, Aunt Boo, Hilary and Jim, Jason, Jodi, Stephanie, and all living extensions of the remarkable Elson, Stembel, Green, Kornblau, and Gilberg families… and especially to Andrea, who made it happen.

LARRY ELSON

ABOUT THE AUTHORS

WYNN KAPIT

Wynn Kapit, the designer and illustrator of this book, has had careers in law, graphic and advertising design, painting, and teaching.

In 1955, he graduated from law school, with honors, from the University of Miami and was admitted to the Florida Bar. He practiced law both before and after military service. Four years later, he decided to pursue a childhood ambition and enrolled at what is now the Art Center College in Los Angeles, where he studied graphic design. Afterwards, he worked in the New York advertising world for six years as a designer and art director. He "dropped out" in the late 60s, returned to California, and began painting. His numerous exhibitions included a one-man show at the California Palace of the Legion of Honor in 1968. He returned to school and received a Masters in painting from the University of California at Berkeley in 1972.

Kapit was teaching figure drawing in Adult Ed in San Francisco in 1975 when he decided he needed to learn more about bones and muscles. He enrolled in Dr. Elson's anatomy class at San Francisco City College. While he was a student, he created the word and illustration coloring format that seemed to be a remarkably effective way of learning the subject. He showed some layouts to Dr. Elson and indicated his intention to do a coloring book on bones and muscles for artists. Immediately recognizing the potential of this method, Dr. Elson encouraged Kapit to do a "complete" coloring book on anatomy and offered to collaborate on the project. The first edition of *The Anatomy Coloring Book* was published in 1977, and its immediate success inspired the development of a completely new field of publishing: educational coloring books.

Kapit went on to create *The Physiology Coloring Book* with the assistance of two professors who were teaching at Berkeley: Dr. Robert A. Macey and Dr. Esmail Meisami. That book was published in 1987 and has gone through two editions. In the early '90s, Kapit wrote and designed *The Geography Coloring Book*, now in its second edition.

LAWRENCE M. ELSON

Lawrence M. Elson, Ph.D., planned the content and organization, provided sketches, and wrote the text for the book. This is his seventh text, having authored *It's Your Body* and *The Zoology Coloring Book* and co-authored *The Human Brain Coloring Book* and *The Microbiology Coloring Book*. He received his B.A. in zoology and pre-med at the University of California at Berkeley and continued there to receive his Ph.D. in human anatomy. Dr. Elson was assistant professor of anatomy at Baylor College of Medicine in Houston, participated in the development of the Physician's Assistant Program, lectured and taught dissection and anatomy at the University of California Medical School in San Francisco, and taught general anatomy at City College of San Francisco.

In his younger days, Dr. Elson trained to become a naval aviator and went on to fly dive-bombers off aircraft carriers in the Western Pacific. While attending college and graduate school, he remained in the Naval Air Reserve and flew antisubmarine patrol planes and helicopters. His last position in his 20-year Navy career was as commanding officer of a reserve antisubmarine helicopter squadron. He continues to fly his own airplane for business and pleasure.

Currently, Dr. Elson is a consultant and lecturer on the anatomic bases and mechanics of injury, a practice that has taken him throughout the United States and Canada. He has testified in hundreds of personal injury trials and arbitrations. His research interests are focused on the anatomic bases and mechanisms of injury.

To report errors or make suggestions to enhance the effectiveness of this book, Dr. Elson can be contacted at: foranat@earthlink.net.

TABLE OF CONTENTS

AUTONOMIC OR VISCERAL NERVOUS SYSTEM

SPECIAL SENSES

CARDIOVASCULAR SYSTEM

LYMPHATIC SYSTEM

IMMUNE (LYMPHOID SYSTEM)

RESPIRATORY SYSTEM

DIGESTIVE SYSTEM

URINARY SYSTEM

ENDOCRINE SYSTEM

REPRODUCTIVE SYSTEM

HUMAN DEVELOPMENT

PREFACE

You may ask why we are creating a third edition. Does anatomy change between editions? With respect to what is taught to students of anatomy, formally or informally, there is not much change. Occasionally, a new variation is noted, but substantively and as a practical matter, anatomy doesn't change. On a grander scale, the anatomic arrangement of our bodies IS subject to infinitesimal change in an evolutionary sense, but that's not why we composed a new edition. We did it because it was time to freshen the illustrations and text, to go over material and find new and better ways to illustrate and express the anatomy and its function. We also cleaned up errors and made the presentation more clear.

We worked to improve the visual appeal of the plates, struggling to resist putting 10 lbs. of information in a 5-lb. plate. We found our earlier coverage of joints to be inadequate. We produced nine new plates, five of which make up that deficit: the temporomandibular, shoulder, elbow, sacroiliac, hip, and knee joint plates. We rearranged the order of presentation of material to make it easier for anatomy teachers to integrate our material with commonly employed anatomy texts.

Our visually-based quizzes on bones, arteries, and veins have been expanded to include joints. We reorganized the lists of vessels in those quizzes to make them more digestible. We revised and gave new life to a full third of the existing plates, including expansion of the integument to two plates, updating the plate on HIV-induced immunosuppression, and vastly improving the plates on distribution of spinal nerves, the meninges, the visual system, and the renal tubules. The literature on innervation of skeletal muscle has been reviewed and updated, as has Appendix B.

To borrow from the Preface of the second edition, we are grateful to the thousands of colorers who have advised and encouraged us, including coaches, trainers, teachers, paramedics, body workers, court reporters, attorneys, insurance claims adjusters, judges, and students and practitioners of dentistry and dental hygiene, nursing, medicine/surgery, chiropractic, podiatry, massage therapy, myotherapy, physical therapy, occupational therapy, and exercise therapy. More informal seekers of self-realization and those with impairments have been drawn to *The Anatomy Coloring Book* because of its lighter, more visual approach. Truly, a picture is worth a thousand words!

ACKNOWLEDGMENTS

We had the support of many and thank them for their participation in the development of this work. Our reviewers gave us excellent advice and pointed out errors that otherwise would not have been caught. Arlene Klepatsky, R.N., J.D., carried out key research work for us, and we are grateful. Michael Loftus, M.D., cardiologist, and fellow racquetball player, kindly reviewed the electrocardiogram material. Maureen Larsen and Carolyn Scott typed much of the backmatter under rush conditions and late-night hours, and we are very grateful for their work. Dolores Espinoza was terrific in overseeing the preparation of materials for the typesetter. Thank you, Stephanie Luros, for coloring the plates and giving us insight into potential problems. Jason Luros's computer and software skills and advice were gratefully received, as was the assistance from software engineer Clifford Clark on Macintosh™ conversion. Our editor, Susan Teahan, helped us in many expert ways and perhaps was at her best as she drove us to meet the deadline—which we did. To Gerry Ichikawa and Jill Breedon, copy editor, at the TypeStudio in Santa Barbara, we can only repeat what was said after the second edition—that their superb and accurate work was most welcomed and appreciated. To all who helped us and remain unnamed in this acknowledgment, we are most grateful for your participation.

WYNN KAPIT   LARRY ELSON
Santa Barbara, California   Napa Valley, California

INTRODUCTION TO COLORING

(Important tips on how to get the most out of this book)

HOW THE BOOK IS ARRANGED

The book is divided by subject matter into sections. The sections contain groups of plates, each dealing with a separate topic within that subject heading.

A plate consists of an illustration with various parts to be colored, related titles (also to be colored), an explanatory paragraph of text, and coloring notes (**CN**).

You can begin with any section, but it is best to color that section in the order in which the plates are presented. Feel free to skip the plates that might be too complex or irrelevant to your area of interest.

HOW MANY COLORS NEEDED

It is best to have at least 10 pens or pencils (no crayons). Pencils are more versatile because one can lighten or darken each color. Felt-tipped pens, on the other hand, produce brighter colors.

The more colors you have available, the greater the pleasure. If you are able to purchase your colors individually (as opposed to a set), you should choose mostly light colors, but be sure to include gray and black.

HOW THE COLORING SYSTEM WORKS

The parts of an illustration that are meant to be colored are drawn with, or separated from each other by, dark outlines. They are also identified by small letter labels (A, B, etc.). The "titles" (names or terms referring to those parts) are printed with outlined letters, followed by the same letter labels. *Color a part and its respective title with the same color.* Do not use that color again on a different part and title on that plate, unless you run out of colors, and have to repeat some colors.

When different parts of an illustration are related to each other in some fundamental way, they will receive the same letter labels, but with different superscripts (A^1, A^2) for identification purposes. All those parts will receive the same color.

Occasionally, you will come across a title or general heading that is meant to be colored but does not refer to any specific part of the illustration. In such cases, the small letter label will be followed by a dash (A-, B-), and only the title or heading will be colored.

Areas or words meant to be colored gray are identified by an asterisk (∗); if colored black, by a black circle (•); and when not to be colored at all, by the "don't color" sign (-¦-).

HOW TO APPROACH EACH PLATE

Whether you read the explanatory material before coloring or vice versa, you should always read the coloring notes (**CN**) before starting to color. The notes (located at the top of the plate) contain recommendations about which colors to use and what to take notice of when coloring that particular plate.

Begin by coloring the first title in the list of titles to be colored. The title will be followed by a small letter label (A). Locate and color the part of the illustration to which the title refers. *It is important that you color the titles in the order that they are presented*; they are usually listed that way for specific reasons.

The titles are usually placed away from the illustrations to facilitate your review. Try covering them when testing your recall of the material.

It is recommended that you reserve your lightest colors for the largest areas to be colored. A dark color on an especially large part of the illustration would dominate the plate. Certain colors are traditionally associated with certain structures of the body: red for arteries, blue for veins, purple for capillaries, yellow for nerves, and green for lymphatics. Where you are asked to identify a diverse group of such structures (i.e., many different arteries or many veins), you will naturally have to use more than just the one representative color.

SYMBOLS USED THROUGHOUT THE BOOK

Don't color -¦-   Color gray ✳
Color black •   Not shown N.S.
A broken outline represents one form lying underneath or behind another.
The subject matter is microscopic in size.

ABBREVIATIONS

In the text and titles, the following abbreviations may precede or follow the names of the structures identified, e.g., Post. auricular m., Brachial a., Scalenus med. m.:

A. = Artery
Ant. = Anterior
Br. = Branch
Inf. = Inferior
Lat. = Lateral
Lig. = Ligament
M., Ms. = Muscle(s)
Med. (preceding term) = Medial
Med. (after term) = Medius
N. = Nerve
Post. = Posterior
Sup. = Superior, superficial
Sys. = System
Tr. = Tract
V. = Vein

ANATOMIC PLANES & SECTIONS

MEDIAN A

The median plane is the midline longitudinal plane dividing the head and torso into right and left halves. The presence of the sectioned midline of the vertebral column and spinal cord is characteristic of this plane. The median plane is the middle sagittal (mid-sagittal) plane.

SAGITTAL B

The sagittal plane is a longitudinal plane dividing the head and torso into left and right parts (not halves). It is parallel to the median (not medial) plane.

CORONAL, FRONTAL C

The coronal or frontal plane is a longitudinal plane dividing the body (head, torso, limbs) or its parts into front and back halves or parts.

TRANSVERSE, CROSS,

The transverse plane divides the body into upper and lower halves or parts (cross sectons). It is perpendicular to the longitudinal planes. Transverse planes may be horizontal planes of the upright body. Transverse planes are called "axial" or "transaxial" sections/slices by radiologists.

CN: (1) Use your lightest colors on A–D. (2) Color a body plane in the center diagram; then color its title, related sectional view, and the sectioned body example. (3) Color everything within the dark outlines of the sectional views.

Study of the human body requires visualization of internal regions and parts. Dissection (dis, apart; sect-, cut) is the term given to preparing the body for internal inspection. One method of dissection permits consistent visual orientation by cutting the body into parts, called "sections," along the lines of reference, called "planes." The viewing and study of internal human structure in these planes is possible through medical imaging, such as computerized tomography (CT) and magnetic resonance imaging (MRI).

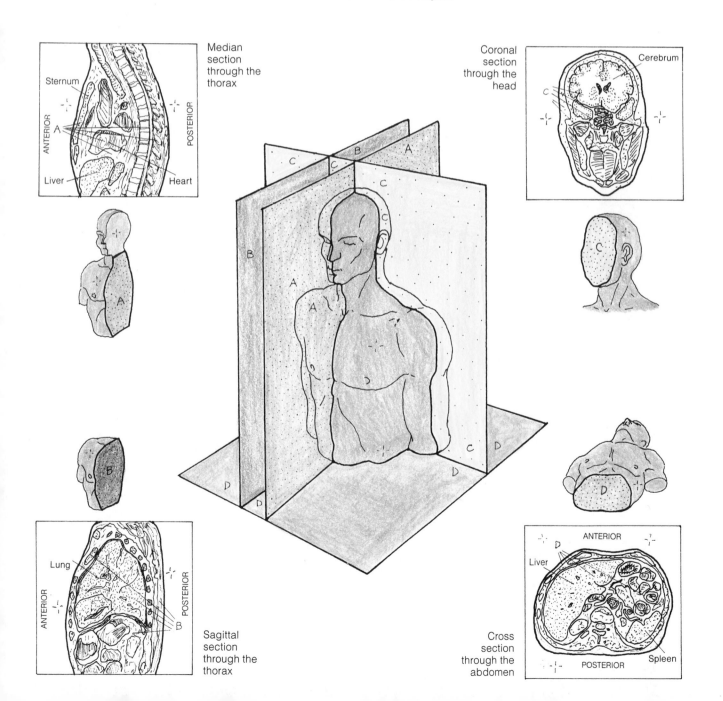

Median section through the thorax

Sternum

ANTERIOR

POSTERIOR

A

Liver

Heart

Coronal section through the head

Cerebrum

C

Sagittal section through the thorax

Lung

ANTERIOR

POSTERIOR

B

Cross section through the abdomen

ANTERIOR

D

Liver

Spleen

POSTERIOR

TERMS OF POSITION & DIRECTION

CN: Color the arrows and titles, but not the illustrations.

Terms of position and direction describe the relationship of one organ to another, usually along one of the three body planes illustrated in the previous plate. To avoid confusion, these terms are related to the standard anatomical position: body standing erect, limbs extended, palms of the hands forward.

CRANIAL, SUPERIOR, ROSTRAL A

These terms refer to a structure being closer to the head or higher than another structure of the body. These terms are not used with respect to the limbs.

ANTERIOR, VENTRAL B

These terms refer to a structure being more in front than another structure in the body. The term "anterior" is preferred.

POSTERIOR, DORSAL C

These terms refer to a structure being more in back than another structure in the body. The term "posterior" is preferred.

MEDIAL D

This term refers to a structure that is closer to the median plane than another structure in the body. "Medial" is not synonymous with "median."

LATERAL E

This term refers to a structure that is further away from the median plane than another structure in the body.

PROXIMAL F

Employed only with reference to the limbs, this term refers to a structure being closer to the median plane or root of the limb than another structure in the limb.

DISTAL G

Employed only with reference to the limbs, this term refers to a structure being further away from the median plane or the root of the limb than another structure in the limb.

CAUDAL, INFERIOR H

These terms refer to a structure being closer to the feet or the lower part of the body than another structure in the body. These terms are not used with respect to the limbs.

SUPERFICIAL I DEEP J

The term "superficial" is synonymous with external, the term "deep" with internal. Related to the reference point on the chest wall, a structure closer to the surface of the body is superficial; a structure further away from the surface is deep.

IPSILATERAL K CONTRALATERAL L

The term "ipsilateral" means "on the same side" (in this case, as the reference point); "contralateral" means "on the opposite side" (of the reference point).

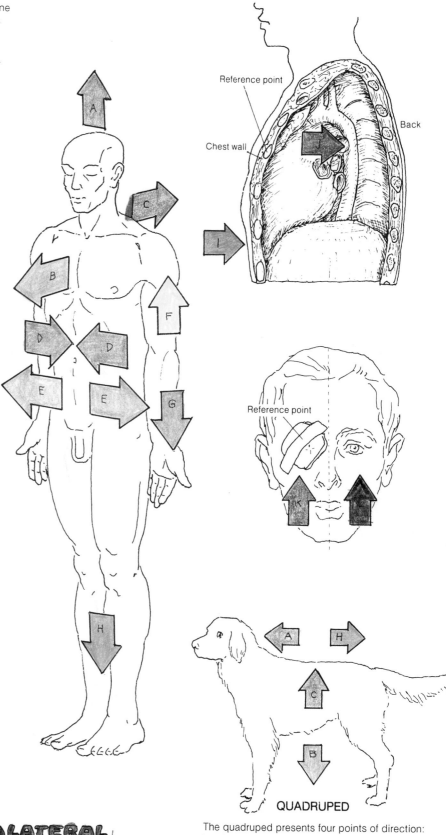

The quadruped presents four points of direction: head end (cranial), tail end (caudal), belly side (ventral), back side (dorsal). In the biped (e.g., human), the ventral side is also anterior, the dorsal side is also posterior, the cranial end is also superior, and the caudal end is inferior.

SYSTEMS OF THE BODY (1)

CN: Use light colors. Color the skeleton (A). Color the musculature (B) brown. Color the major arteries and heart red (with darker outlines), veins blue (C). Color all lymphatic vessels (D) green. Color nerves, brain and spinal cord (E), yellow. Color the insets representing the endocrine system (F). Pick a skin color for the integumentary system (G). Note that the latter two are independent systems, but are graphically combined here in one body.

Collections of similar cells constitute tissues. The four basic tissues are integrated into body wall and visceral structures/organs. A *system* is a collection of organs and structures sharing a common function. Organs and structures of a single system occupy diverse regions in the body and are not necessarily grouped together.

SKELETAL A ARTICULAR A'

The skeletal system consists of the skeleton of bones and their periosteum, and the ligaments that secure the bones at joints. By extension, this system could include the varied fasciae that ensheath the body wall/skeletal muscles and contribute to the body's structural stability. The *articular system* comprises the joints, both movable and fixed, and the related structures, including joint capsules, synovial membranes, and discs/menisci.

MUSCULAR B

The muscular system includes the skeletal muscles that move the skeleton, the face, and other structures and give form to the body; the cardiac muscle of the heart walls; and the smooth muscle of the walls of viscera and vessels and in the skin.

CARDIOVASCULAR C

The cardiovascular system consists of the four-chambered heart, arteries conducting blood to the tissues, capillaries through which nutrients, gases, and molecular material pass to and from the tissues, and veins returning blood from the tissues to the heart. Broadly interpreted, the cardiovascular system includes the lymphatic system.

LYMPHATIC D

The lymphatic system is a system of vessels assisting the veins in recovering the body's tissue fluids and returning them to the heart. The body is about 60% water, and the veins alone are generally incapable of meeting the demands of tissue drainage. Lymph nodes, which filter lymph, are located throughout the body

NERVOUS E

The nervous system consists of impulse-generating/conducting tissue organized into a central nervous system (brain and spinal cord) and a peripheral nervous system (nerves), which includes the visceral (autonomic) nervous system involved in involuntary "fight or flight" and vegetative responses.

ENDOCRINE F

The endocrine system consists of glands that secrete chemical agents (hormones) into the tissue fluids and blood, affecting the function of multiple areas of the body. Many of these glands are under some control by the brain (hypothalamus). Hormones help maintain balanced metabolic functions in many of the body's systems.

INTEGUMENTARY G

The integumentary system is the skin, replete with glands, sensory receptors, vessels, immune cells and antibodies, and layers of cells and keratin that resist environmental factors harmful to the body.

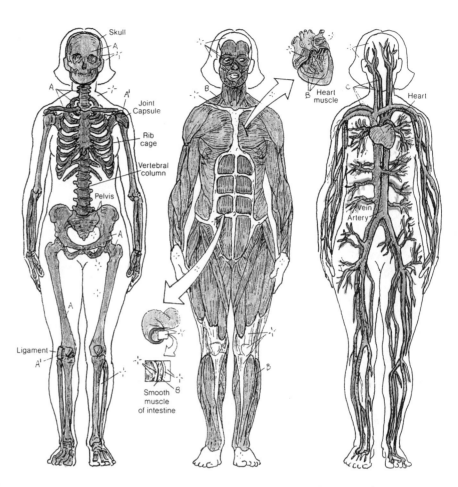

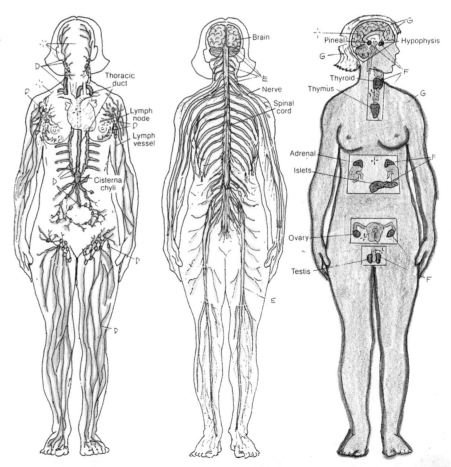

SYSTEMS OF THE BODY (2)

CN: Use different light colors from those used on the preceding plate.

RESPIRATORY H

The respiratory system consists of the upper (nose through larynx) and lower respiratory tract (trachea through the air spaces of the lungs). Most of the tract is airway; only the air spaces (alveoli) and very small bronchioles exchange gases between alveoli and the lung capillaries.

DIGESTIVE I

The digestive system is concerned with the breakdown, digestion, and assimilation of food as well as excretion of the residua. Its tract begins with the mouth and continues down to the abdomen, wherein it takes a convoluted course to open again at the anus. Associated glands include the liver, the pancreas, and the biliary system (gall bladder and related ducts).

URINARY J

The urinary system is concerned with the conservation of water and maintenance of a neutral acid-base balance in the body fluids. The kidneys are the main functionaries of this system; residual fluid (urine) is excreted through ureters to the urinary bladder for retention and discharged to the outside through the urethra.

IMMUNE / LYMPHOID K

The lymphoid system consists of organs concerned with body defense: thymus, bone marrow, spleen, lymph nodes, tonsils, and smaller aggregates of lymphoid tissue. This system, including a diffuse arrangement of immune-related cells throughout the body, is concerned with resistance to invasive microorganisms and the removal of damaged or otherwise abnormal cells.

FEMALE REPRODUCTIVE L

The female reproductive system is concerned with the secretion of sex hormones, production and transportation of germ cells (ova), receipt and transport of male germ cells to the fertilization site, maintenance of the developing embryo/fetus, and initial sustenance of the newborn.

MALE REPRODUCTIVE M

The male reproductive system is concerned with the secretion of male sex hormones, formation and maintenance of germ cells (sperm), and transport of germ cells to the female genital tract.

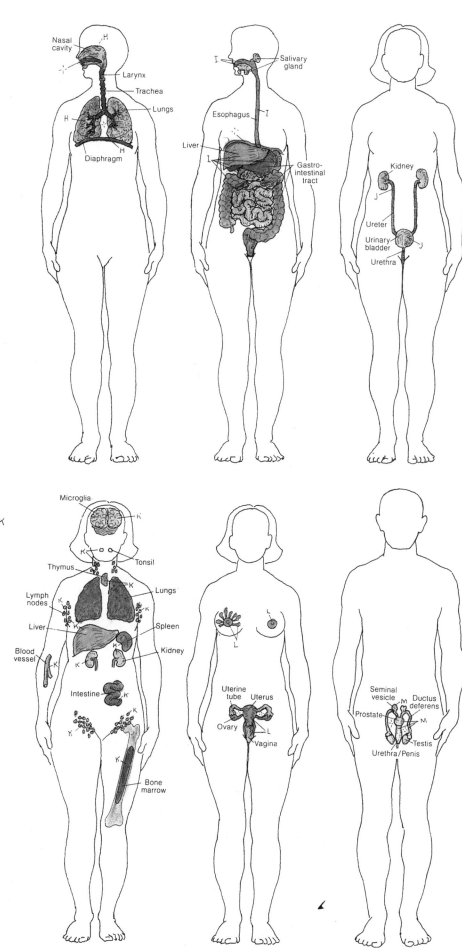

REGIONS OF THE BODY (ANTERIOR)

CN: The text for this and the next plate is located on the next plate. (1) The anterior/lateral regions have been grouped according to larger areas: e.g., head, neck. The regions of each area (A^1, A^2, etc.) all receive a single color. Color a title and the arrow pointing to its region. (2) Although the title "pudendal" (D-) is to be colored, that region, consisting of the female external genitals, is not shown (N.S.). The same is true for the perineum (D-), that region between the pubis and the coccyx, below the pelvic floor.

HEAD A-
  FRONTAL A^1 (forehead)
  TEMPORAL A^2 (temple)
  ORBITAL A^3 (eye, cavity/walls)
  NASAL A^4 (nose, cavity/walls)
  BUCCAL A^5 (cheek)
  ORAL A^6 (mouth cavity)
  MANDIBULAR A^7 (lower jaw)

NECK B-
  ANTERIOR CERVICAL B^1 (front of neck)
  LATERAL CERVICAL B^2 (side of neck)
  SUPRACLAVICULAR B^3 (above clavicle)

THORAX C-
  PECTORAL C^1 (anterior chest)

ABDOMINOPELVIC D-
  ABDOMINAL D^1 (abdomen)
  INGUINAL D^2 (groin)
  PELVIC D^3 (pelvis)
  PUBIC D^4 (genital region)
  GENITAL D^5 (reproductive organs)
    PUDENDAL D- N.S. (female genitals)
  PERINEAL D- N.S. (between pubis and coccyx)

UPPER LIMB E-
  DELTOID E^1 (shoulder/upper arm)
  AXILLARY E^2 (armpit)
  BRACHIAL E^3 (arm)
  ANTECUBITAL E^4 (front of elbow)
  ANTEBRACHIAL E^5 (forearm)
  CARPAL E^6 (wrist)
  HAND: PALMAR E^7 (palm)
  HAND: DIGITAL E^8 (fingers)

LOWER LIMB F-
  COXAL F^1 (hip)
  FEMORAL F^2 (thigh)
  PATELLAR F^3 (knee cap)
  CRURAL F^4 (leg)
  FIBULAR F^5 (lateral leg)
  TARSAL F^6 (ankle)
  FOOT: DORSUM F^7 (top)
  FOOT: DIGITAL F^8 (toes)

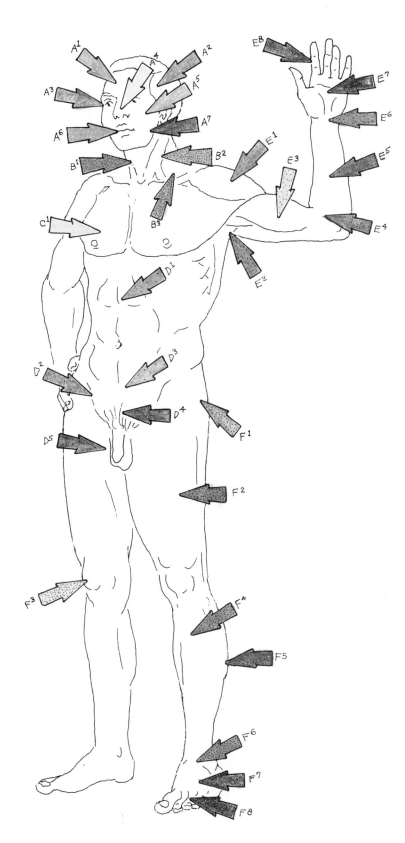

REGIONS OF THE BODY (POSTERIOR)

CN: (1) Use the same colors for divisions marked A, B, E, and F that were used for those letters on the preceding plate.

Regional anatomy is the organization of human structure by regions. Here are shown the major regions within the principal areas of the body (e.g., head, neck). There are many regions within regions, each of which includes structures from different systems, such as bone, muscles, blood vessels, and nerves. Study of the body by dissection is generally accomplished region by region. An in-depth regional awareness of human structure is fundamental for most health care providers.

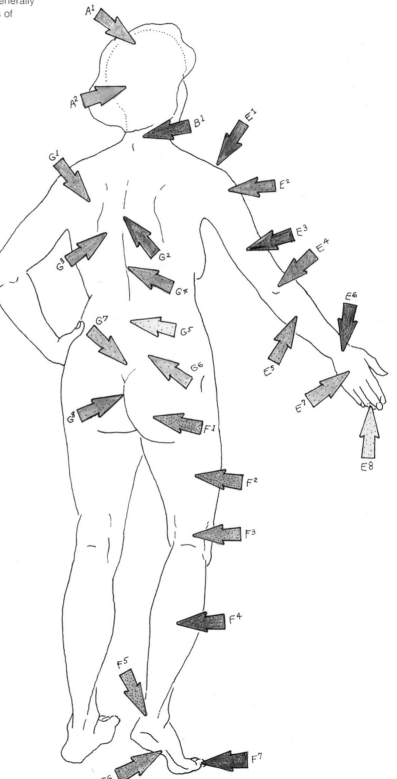

HEAD A-
PARIETAL A¹ (top and sides of head)
OCCIPITAL A² (back of head)

NECK B-
POST. CERVICAL /NUCHAL B' (back of neck)

BACK G-
SCAPULAR G¹ (shoulder blade)
VERTEBRAL G² (spinal column)
PARASPINAL G³ (along side spinal column)
THORACIC G⁴ (posterior chest)
LUMBAR G⁵ (lower back)
SACROILIAC G⁶ (vertebro-pelvic joint)
SACRAL G⁷ (posterior pelvis)
COCCYGEAL G⁸ ("tail bone")

UPPER LIMB E-
ACROMIAL E¹ (top of shoulder)
DELTOID E² (shoulder/upper arm)
BRACHIAL E³ (arm)
CUBITAL E⁴ (elbow)
ANTEBRACHIAL E⁵ (forearm)
CARPAL E⁶ (wrist)
HAND: DORSAL E⁷ (back of hand)
HAND: DIGITAL E⁸ (fingers)

LOWER LIMB F-
GLUTEAL F¹ (buttock)
FEMORAL F² (thigh)
POPLITEAL F³ (back of knee)
CRURAL F⁴ (leg)
TARSAL F⁵ (ankle)
FOOT: PLANTAR F⁶ (sole)
FOOT: DIGITAL F⁷ (toes)

CAVITIES & LININGS

CN: Except for H, use light colors. (1) Note that the linings for closed body cavities (A¹*–D¹*) are to be colored gray. (2) In the open visceral cavities shown below, the linings receive the color (H).

CLOSED BODY CAVITIES

CRANIAL A   DURA MATER A'*
VERTEBRAL B   DURA MATER B'*
THORACIC C   PLEURA C'*
ABDOMINOPELVIC D   PERITONEUM D'*

Closed body cavities (The *cranial, vertebral, thoracic,* and *abdomino-pelvic* cavities) are not open to the outside of the body. Though organs may pass through them or exist in them, the organs' cavities do not open into these closed cavities. Closed body cavities are lined with a membrane: the thick *dura mater* in the skull and vertebral cavity, the thin, watery (serous) membranes (serosa) in the thoracic and abdominopelvic cavities.

The cranial cavity is occupied by the brain and its coverings, cranial nerves, and blood vessels. The bony walls of the cranial cavity are lined by the dura mater, a tough, fibrous membrane that turns inward to form a meningeal layer that envelops the brain (Plate 81). The vertebral cavity houses the spinal cord, its coverings, related vessels, and nerve roots (Plate 77). Its dura mater is continuous with the cranial dura at the foramen magnum, and it forms a sac whose bottom is at the level of the 2nd sacral vertebra.

The thoracic cavity contains the lungs, heart, and other structures (tubular airways, blood vessels, lymphatics, nerves) in the chest. Its skeletal walls are the thoracic vertebrae and ribs posteriorly, the ribs anterolaterally, and the sternum and costal cartilages anteriorly (Plate 30). The roof of the cavity is membranous; the floor is the muscular thoracic diaphragm (Plate 50). The middle of the thoracic cavity has a partition filled with structure (e.g., heart), called the mediastinum (Plate 104). It separates the thoracic cavity into discrete left and right parts (not shown). The internal surface of each half of the thoracic cavity is completely lined with a serous membrane called pleura (Plate 133). The pleura, like all serous membranes, consists of a single layer of ells supported by a thin, vascular, connective tissue layer. These cells secrete a serous fluid that permits the pleura-lined lungs to move against the pleura-lined thoracic walls without friction.

The abdominal cavity, containing the gastrointestinal tract and related glands, the urinary tract, and great numbers of vessels and nerves, has muscular walls anterolaterally, the lower ribs and muscle laterally, and the lumbar vertebrae posteriorly. The roof of the abdominal cavity is the thoracic diaphragm. The abdominal and pelvic cavities are continuous with one another and share the muscular pelvic floor. The pelvic cavity, containing the urinary bladder, rectum, and reproductive organs, has muscular walls anteriorly, bony walls laterally, and the sacrum posteriorly. The internal surface of the abdominal wall is lined by a serous membrane, the peritoneum (Plate 140). The serous secretions enable the mobile abdominal viscera to slip and slide frictionlessly during movement. The peritoneum drapes over the pelvic viscera, does not envelop them, and does not reach the pelvic floor.

OPEN VISCERAL CAVITIES

RESPIRATORY TRACT E
URINARY TRACT F
DIGESTIVE TRACT G
MUCOSA H

Open cavities (*respiratory, digestive, urinary* tracts) are largely tubular passageways lined with a mucus-secreting layer called a *mucosa*. The mucosa is the working tissue (secretion, absorption, protection) of open cavities; it is lined with epithelial cells, is supported by vascular connective tissue, and often incorporates a smooth muscle layer. Open cavities within the thoracic and abdominopelvic cavities are open to the outside of the body. Their mucosal lining is continuous with the skin at the ends of the tubular cavities (nose, mouth, perineum).

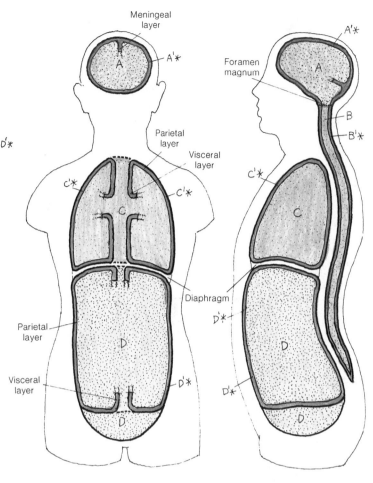

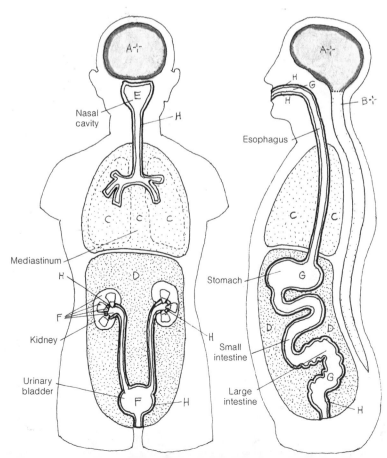

THE GENERALIZED CELL

CN: Color gray the variety of cell shapes at upper left. Use lightest colors for A, C, D, F and G. (1) Small circles representing ribosomes (H) are found throughout the cytoplasm (F) and on the rough endoplasmic reticulum (G¹); color those larger areas, including the ribosomes, first, and then color over the ribosomes again with a darker color. Each organelle shown is just one of many found in the living cell.

CELL SHAPES *

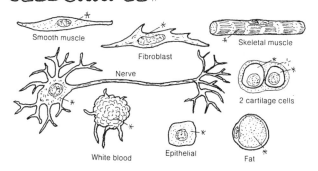

Smooth muscle
Fibroblast
Skeletal muscle
Nerve
2 cartilage cells
White blood
Epithelial
Fat

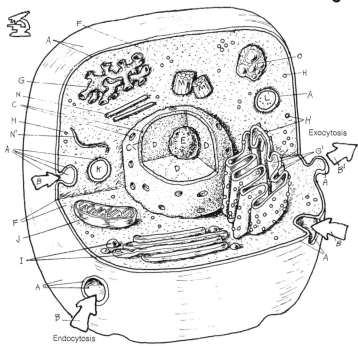

Exocytosis

Endocytosis

ORGANELLES

CELL MEMBRANE ᴀ
    ENDOCYTOSIS ʙ / EXOCYTOSIS ʙ'
NUCLEAR MEMBRANE ᴄ
NUCLEOPLASM ᴅ
NUCLEOLUS ᴇ
CYTOPLASM ꜰ
ENDOPLASMIC RETICULUM
    SMOOTH ꜰ,ɢ ROUGH ɢ'
RIBOSOME ʜ
GOLGI COMPLEX ɪ
MITOCHONDRION ᴊ
VACUOLE ᴋ
LYSOSOME ʟ
CENTRIOLE ᴍ
MICROTUBULE ɴ
MICROFILAMENT ɴ'
CELL INCLUSION ᴏ

The cell is the basic structural and functional unit of all living things. Living things are characterized by the ability to reproduce and grow, metabolize (transform or produce/consume of energy), and adapt to limited changes in their internal and external environment. Body structure lacking these characteristics, such as connective tissue fibers, is not considered to be "alive." Body structure more complex than a cell consists of a collection of cells and their products.

The activities of cells constitute the life process; they include ingestion, assimilation, and digestion of nutrients and excretion of the residue; respiration; synthesis and degradation of materials; movement; and excitability or response to stimuli. The impairment or cessation of these activities in normal cells, whether caused by trauma, infection, tumors, degeneration, or congenital defects, is the basis of a disordered or disease process.

By volume, the generalized cell is 80% water; by weight, it is composed of proteins (about 15%), lipids (3%), carbohydrates (1%), and nucleic acids and minerals (1%). These materials may be integrated into structural working units (organelles), form a more mobile functional unit (e.g., messenger RNA, globular protein-based enzymes), or form products of the cell. The basic function of a cell is to produce protein, which is essential to the acquisition and use of cell energy, formation and repair of structure, and cell activities (e.g., synthesis, secretion, absorption, contraction).

Cell membrane: the limiting lipoprotein membrane of the cell; retains internal structure; permits exportation and importation of materials. Infolding/outfolding of the cell membrane permits the introduction of material into the cell (endocytosis) or its expulsion (exocytosis) from the cell.

Nuclear membrane: porous, limiting, lipoprotein membrane; regulates passage of molecules.

Nucleoplasm: the nuclear substance containing chromatin (chromosomes during cell division) and RNA.

Nucleolus: a mass of largely RNA, it forms ribosobal RNA (RNAr) that passes into cytoplasm and becomes the site of protein synthesis.

Cytoplasm: the ground substance of the cell less the nucleus. Contains organelles and inclusions listed below.

Smooth/rough endoplasmic reticulum (ER): membrane-lined tubules to which ribosomes may be attached (rough ER; flattened tubules) or not (smooth ER; rounded tubules). Rough ER is concerned with transport of protein synthesized at the ribosomes. Smooth ER synthesizes complex molecules called steroids in some cells; stores calcium ions in muscle; breaks down toxins in liver.

Ribosome: the site of protein synthesis where amino acids are strung in sequence as directed by messenger RNA from the nucleus.

Golgi complex: flattened membrane-lined sacs that bud off small vesicles from the edges; collect secretory products and package them for export or cell use, e.g., lysomes.

Mitochondrion: membranous, oblong structure in which the inner membrane is convoluted like a maze. Energy for cell operations is generated here through a complex series of reactions between oxygen and products of digestion (oxidative reactions).

Vacuoles: membrane-lined containers that can merge with one another or other membrane-lined structure, such as the cell membrane. They function as transport vehicles.

Lysosome: membrane-lined container of enzymes with great capacity to break down structure, e.g., microorganisms, damaged cell parts, and ingested nutrients.

Centriole: bundle of microtubules in the shape of a short barrel; usually seen paired, perpendicular to one another. They give rise to spindles used by migrating chromatids during cell division.

Microtubules: formed of protein; provide structural support for the cell and/or its parts.

Microfilaments: are support structures formed of protein different from that of microtubules. In skeletal muscle, the proteins actin and myosin are examples of thin and thick microfilaments.

Cell inclusion: aggregation of material within the cell that is not a functional part (organelle) of the cell—e.g., glycogen, lipid.

CELL DIVISON / MITOSIS

CN: Use the colors you used on Plate 8 for cell membrane, nuclear membrane, nucleolus, and centriole for those titles on this plate, even though the previous letter labels may be different. Use contrasting colors for E–E² and F–F², and gray for D–D¹ to distinguish the latter from those with the contrasting colors. (1) Begin with the cell in interphase, reading the related text and completing each cell before going on to the next. (2) Color the name of each stage and its appropriate arrow of progression. Note that in interphase, the chromatin material within the nuclear membrane is in a thread-like state; color over the entire area with the appropriate color. Note that the starting chromatin (D* in interphase) is colored differently in the daughter cells (E², F²); it is the same chromatin.

CELL MEMBRANE $_A$
NUCLEAR MEMBRANE $_B$
NUCLEOLUS $_C$
CHROMATIN $_{D*}$ /CHROMOSOME $_{D^{1*}}$
CHROMATID $_E$ /CHROMOSOME $_{E^1}$
    CHROMATIN $_{E^2}$
CHROMATID $_F$ /CHROMOSOME $_{F^1}$
    CHROMATIN $_{F^2}$
CENTROMERE $_G$
CENTRIOLE $_H$
ASTER $_I$
SPINDLE $_J$

The ability to reproduce their kind is a characteristic of living things. Cells reproduce in a process of duplication and division called mitosis. Epithelial and connective cells reproduce frequently, mature muscle cells not so frequently, and mature nerve cells rarely if at all. Overactive mitosis may result in the formation of an encapsulated tumor; uncontrolled mitosis, associated with invasiveness and metastases, is called cancer.

As the main cellular changes during mitosis occur in the nucleus and surrounding area, only these parts of the cell are illustrated here. We show here how the nuclear chromatin (diffuse network of DNA and related protein), once duplicated, transforms into 46 chromosomes, which divide into paired subunits (92 chromatids); those chromatids separate and move into opposite ends of the dividing cell, forming the 46 chromosomes of each of the newly formed daughter cells. For clarity, we show only four pairs of chromatids and chromosomes. The phases of the observed nuclear changes during mitosis are as follows.

Interphase: the longest period of the reproductive cycle; the phase between successive divisions. Duplication of DNA (in chromatin) occurs during this phase. The dispersed chromatin (D*) here is a network of fine fibrils, not visible as discrete entities in the nucleoplasm. The cell membrane, nucleus, and nucleolus are intact. The centrioles are paired and adjacent to one another at one pole of the cell.

Prophase: the dispersed chromatin (D*) thickens, shortens, and coils to form condensed chromatin or chromosomes (D¹*). Each chromosome consists of two chromatids (E and F) connected by a centromere (G). Each chromatid has the equivalent amount of DNA of a chromosome. In the latter part of this phase, the nuclear membrane breaks up and dissolves, as does the nucleolus. The centrioles, having duplicated during interphase, separate, going to opposite poles of the cell. They project microtubules called asters.

Metaphase: strands of spindle fibers project across the cell center from paired centrioles. The chromatids attach to the spindle fibers at the centromere and line up in the center, half (46) on one side, half on the other.

Anaphase: the centromeres divide, each daughter centromere attached to one chromatid. Each centromere is drawn to the ipsilateral pole of the cell, along the track of the spindle fiber, taking its chromatid with it. The separated chromatids now constitute chromosomes. Anaphase ends when the daughter chromosomes arrive at their respective poles (46 on each side).

Telophase: the cell pinches off in the center, forming two daughter cells, each identical to the mother cell (assuming no mutations). The cytoplasm and organelles had duplicated earlier and are segregated into their respective newly-forming cells. As the nucleus is reconstituted, and the nuclear membrane and nucleolus reappear in each new cell, the chromosomes fade into dispersed chromatin, and the centromere disappears. Complete cleavage of the parent cell into daughter cells terminates the mitotic process. Each daughter cell enters interphase to start the process anew. The process of cell division increases cell numbers, and does not change cellular content.

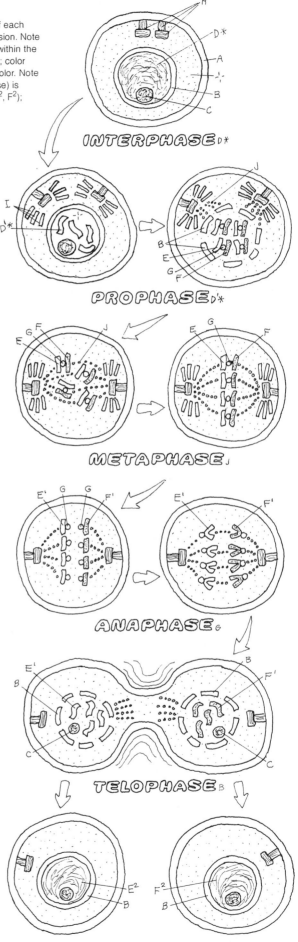

INTERPHASE $_{D*}$

PROPHASE $_{D^{1*}}$

METAPHASE $_J$

ANAPHASE $_G$

TELOPHASE $_B$

DAUGHTER CELLS $_{D*}$

TISSUES: EPITHELIUM

CN: Use very light colors throughout. (1) Color the arrows pointing to the location of the epithelial tissues in the body organs.

Epithelial tissues, one of four basic tissue types, form the working surface of skin and all body cavities, including glands, ducts, and vessels. They protect, secrete, absorb, or sense (e.g., neuroepithelia). Some even contract (myoepithelia). Epithelial tissues generally exist as one layer (simple) or more (stratified). The lowest layer of epithelia is bound to the underlying connective tissue by a basement membrane (secreted basal and reticular laminae). Epithelial cells are connected together by one or more of: adhesive glycoproteins, desmosomes, gap junctions, and circumferential bands (not shown).

SIMPLE EPITHELIUM ¦-

This surface tissue functions in filtration, diffusion, secretion, and absorption.

SQUAMOUS ₐ

Simple squamous epithelia are thin, plate-like cells. They function in diffusion. They line the heart and all blood vessels (endothelia), air cells, body cavities (mesothelia), etc.

CUBOIDAL ʙ

Simple cuboidal epithelia are generally secretory cells and make up glands throughout the body, tubules of the kidney, terminal bronchioles of the lungs, and ducts of the reproductive tracts.

COLUMNAR c

Simple columnar epithelia line the gastrointestinal tract and are concerned with secretion and absorption. Their free (apical) surface may be covered with finger-like projections of cell membrane called microvilli, increasing the cell's surface area for secretion/absorption.

PSEUDOSTRATIFIED COLUMNAR ᴅ

Columnar cells bunched together form a single layer, appearing as if stratified. Each cell is attached to the basement membrane. These cells line reproductive and respiratory tracts. Cilia on the free surface collectively move surface material by means of undulating power strokes alternating with resting strokes.

STRATIFIED EPITHELIUM -¦-

Stratified epithelia are generally resistant to damage by wear and tear because of ready replacement of cells.

STRATIFIED SQUAMOUS ᴇ

This tissue may be keratinized (skin) or not (oral cavity, pharynx, vocal folds, esophagus, vagina, anus). Basal cells are generally columnar and germinating.

TRANSITIONAL ꜰ

Multiple layers of cells line the urinary tract. In the empty bladder, the fibromuscular layer is contracted because of muscle tone; the epithelia are closely concentrated. With bladder distension, cells are stretched out; the tissue is thinner than in the contracted state. The tissue is responsive to volume changes.

GLANDULAR EPITHELIUM -¦-

Glandular cells produce and secrete/excrete materials of varying composition— e.g., sweat, milk, sebum, cerumen, hormones, enzymes. Myoepithelial cells induce discharge of the secreted material in most cases.

EXOCRINE ɢ

Exocrine glands (e.g., sweat, sebaceous, pancreatic, mammary) arise as outpocketings of epithelial lining tissue, retain a duct to the free surface of the cavity or skin, and excrete/secrete some substance. Secretory portions may have one of several shapes (tubular, coiled, alveolar/acinar) connected to one or more ducts.

ENDOCRINE ʜ

Endocrine glands arise as epithelial outgrowths but lose their connections to the surface during development. They are intimately associated with a dense capillary network and secrete their products into it.

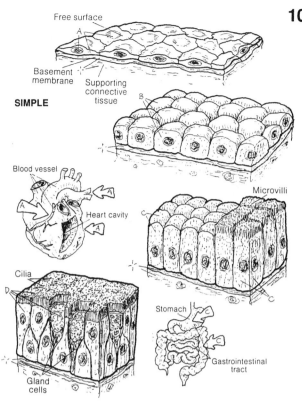

SIMPLE

Free surface · Basement membrane · Supporting connective tissue · Blood vessel · Heart cavity · Microvilli · Cilia · Stomach · Gastrointestinal tract · Gland cells

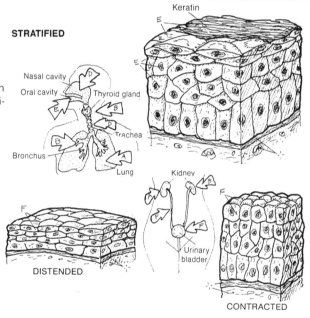

STRATIFIED

Keratin · Nasal cavity · Oral cavity · Thyroid gland · Trachea · Bronchus · Lung · Kidney · Urinary bladder · DISTENDED · CONTRACTED

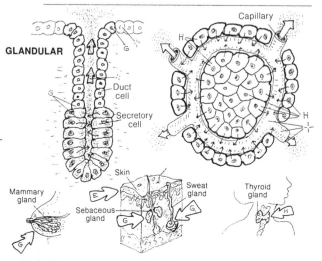

GLANDULAR

Capillary · Duct cell · Secretory cell · Skin · Mammary gland · Sebaceous gland · Sweat gland · Thyroid gland

TISSUES: FIBROUS CONNECTIVE TISSUES

CN: Use yellow for C and C¹, and red for J. (1) Begin with the illustration at middle left and the related titles (A through K). The titles and borders of the microscopic sections of dense regular/irregular c.t. (F¹, F²) receive the color of collagen (F), as that is the dominant structure in both tissues. (2) Do not color the matrix.

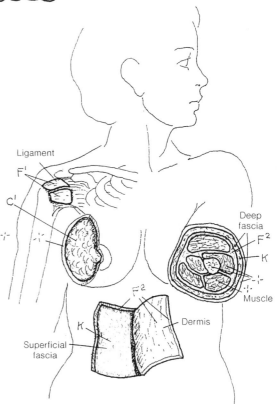

CELLS⁺
FIBROBLASTₐ
MACROPHAGEв
FAT CELLc
PLASMA CELLᴅ
MAST CELLε

FIBERS⁺
COLLAGENғ
ELASTICɢ
RETICULARн
MATRIX, GROUND SUBSTANCE ı⁺
CAPILLARYⱼ

The connective tissues (c.t.) connect, bind, and support body structure. They consist of variable numbers of cells, fibers, and ground substance (fluid, viscous sol/gel, or mineralized). At the microscopic level (here illustrated at about 600× magnification), connective tissues range from blood (cells/fluid), through the fibrous tissues (cells/fibers/variable matrix) to the more stiff supporting tissues (cells/fibers/dense matrix) of cartilage and mineralized bone. Connective tissue can be seen at visible levels of body organization as well, in fascial layers of the body wall, tendons, ligaments, bone, and so on. This plate introduces the fibrous connective tissues (c.t. proper).

LOOSE, AREOLAR C.T. ᴋ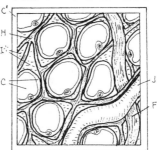

Loose, areolar connective tissue is characterized by many cells, a loose, irregular arrangement of fibers, and a moderately viscous fluid matrix. *Fibroblasts* secrete the fibers and ground substance of this tissue. Mobile *macrophages* engulf cell debris, foreign matter, and microorganisms. *Fat cells,* storing lipids, may be seen in small numbers or large (adipose tissue). *Plasma cells* secrete antibodies in response to infection. *Mast cells* contain heparin and other secretory products, some of which initiate allergic reactions when released. Numerous other cells may transit the loose fibrous tissues, including white blood cells (leukocytes). *Collagen* (linkages of protein exhibiting great tensile strength) and *elastic fibers* (made of the protein elastin) are the fibrous support elements in this tissue. *Reticular tissue* is a smaller form of collagen, forming supporting networks around cell groups of the blood-forming tissues, the lymphoid tissues, and adipose tissue. The *matrix* (consisting largely of water with glycoproteins and glycosaminoglycans in solution) is the intercellular ground substance in which all of the above function; it is fluid-like in the fibrous tissue. Numerous *capillaries* roam throughout this tissue. Loose connective tissue found deep to the skin is called superficial fascia, subcutaneous tissue, or hypodermis. It is found deep to the epithelial tissues of mucous and serous membranes of hollow organs.

ADIPOSE C.T. c¹

Adipose tissue is an aggregation of *fat cells,* supported by reticular and collagenous fibers and closely associated with both blood and lymph capillaries. The storage/release of fat in/from adipose tissue is regulated by hormones (including nutritional factors) and nervous stimuli. It serves as a source of fuel, an insulator, and mechanical padding and stores fat-soluble vitamins. Adipose tissue is located primarily in the superficial fasciae (largely breast, buttock, anterior abdominal wall, arm, and thigh), yellow marrow, and the surface of serous membranes.

DENSE REGULAR C.T. F¹

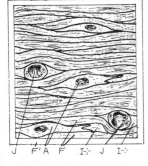

Dense, parallel-arranged, masses of collagenous/elastic fibers form ligaments and tendons that are powerfully resistant to axially loaded tension forces, yet permit some stretch. Tendons/ligaments contain few cells, largely fibroblasts. Elastic, dense regular ligaments are found in the posterior neck and between vertebrae; the tendocalcaneus is the largest elastic structure (tendon or ligament) in the body, storing energy used in gait.

DENSE IRREGULAR C.T. F²

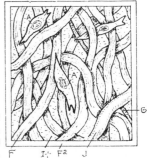

Dense, irregularly arranged masses of interwoven collagenous (and some elastic) fibers in a viscous matrix form capsules of joints, envelop muscle tissue (deep fasciae), encapsulate certain visceral organs (liver, spleen, and others), and largely make up the dermis of the skin. The tissue is impact resistant (bearing stress omnidirectionally), contains few cells, and is minimally vascularized.

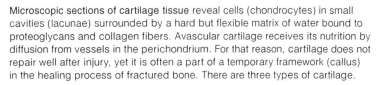

TISSUES: SUPPORTING CONNECTIVE TISSUES

CN: Use the same colors as used on the previous plate for collagen (D) and elastic (E) fibers. Use a light tan or yellow for F and red for L. Use light colors for A, B, G, I, and M. Complete the upper material before coloring the bone section.

Microscopic sections of cartilage tissue reveal cells (chondrocytes) in small cavities (lacunae) surrounded by a hard but flexible matrix of water bound to proteoglycans and collagen fibers. Avascular cartilage receives its nutrition by diffusion from vessels in the perichondrium. For that reason, cartilage does not repair well after injury, yet it is often a part of a temporary framework (callus) in the healing process of fractured bone. There are three types of cartilage.

Bone is unique for its mineralized matrix (65% mineral, 35% organic by weight). The skeleton is bone. Bone is a reservoir of calcium; it is an anchor for muscles, tendons, and ligaments; it harbors many viscera; it assists in the mechanism of respiration; its cavity in certain bones is a center of blood-forming activity (hematopoiesis); in other bones, its cavity is a storage site for lipid.

CARTILAGE
CHONDROCYTE A
LACUNA B
MATRIX C
COLLAGEN FIBER D
ELASTIC FIBER E

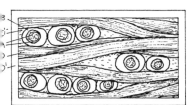

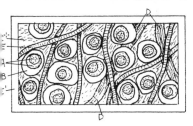

FIBROCARTILAGE D'

Fibrocartilage offers strength with flexibility, resisting both impact and tensile forces. The best example of this tissue is the intervertebral disc. It consists of dense fibrous tissue interspersed with cartilage cells and a relatively small amount of intercellular matrix.

ELASTIC CARTILAGE E'

This tissue is essentially hyaline cartilage with elastic fibers and some collagen. It supports the external ear and the epiglottis of the larynx. Feel its unique flexibility in your own external ear.

HYALINE CARTILAGE A'

Well known as the covering at bone ends (articular cartilage), hyaline cartilage is avascular, insensitive, and compressible. It is porous, enhancing absorption of nutrients and oxygen. It supports the external nose (feel and compare with the elastic cartilage of the ear). It is the main structural support of the larynx and much of the lower respiratory tract. It forms the model for most early developing bone (Plate 168).

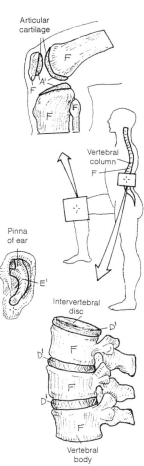

Articular cartilage

Vertebral column

Pinna of ear

Intervertebral disc

Vertebral body

BONE F
PERIOSTEUM F'
COMPACT BONE G
HAVERSIAN SYS. D
HAV. CANAL H
LAMELLAE G'
OSTEOCYTE I
OSTEOCLAST I'
LACUNA B
CANALICULI J
VOLKMANN CANAL K
BLOOD VESSEL L
SPONGY BONE G²

As you read, check Plate 20. Bone has compact and cancellous forms. Compact bone is the impact-resistant, weight-bearing shell of bone lined by a sheath of life-supporting fibrous periosteum. Compact bone consists of columns called haversian systems or osteons: concentric layers (lamellae) of mineralized, collagenous matrix around a central (haversian) canal containing blood vessels. Volkmann's canals interconnect the haversian canals. Note the interstitial lamellae between columns and the circumferential lamellae enclosing the columns. Between lamellae are small cavities (lacunae) interconnected by little canals (canaliculi). Bone cells (osteocytes) and their multiple extensions fill these spaces, which connect with the haversian canal. In areas of resorbing bone matrix, large, multinucleated, avidly phagocytic osteoclasts can be seen with multiple cytoplasmic projections facing the matrix they are destroying. Bone-forming cells (osteoblasts) can be seen in Plate 168. Cancellous bone is internal to compact bone and is especially well seen at the ends of long bones. It consists of irregularly-shaped, interwoven beams (trabeculae) of bone, lacking haversian systems.

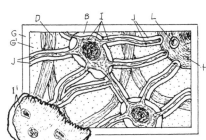

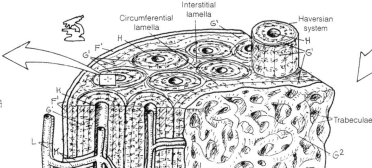

Circumferential lamella

Interstitial lamella

Haversian system

Trabeculae

TISSUES: MUSCLE

Muscle tissue, one of the four basic tissue types of the body, consists of muscle cells ("fibers") and their fibrous connective tissue coverings. There are three kinds of muscle tissues: skeletal, cardiac, and smooth. Muscle tissue shortens (contracts) in response to nerve, nerve-like, or hormonal stimulation. Depending on their attachments, skeletal muscles move bones at joints, constrict cavities, and move the skin; cardiac muscle compresses a heart cavity or orchestrates the sequence of cardiac muscle contraction; and smooth muscle moves the contents of cavities by rhythmic contractions, constricts vessels it surrounds, and moves hairs/closes pores of the skin. The surrounding *connective tissue* transfers the force of contraction from cell to cell and supports the muscle fibers and the many blood *capillaries* and nerves that supply them.

SKELETAL/STRIATED MUSCLE, E
SARCOLEMMA F   CELL E'

Skeletal muscle cells are long, striated, and *multinucleated*, formed of myofibrils, *mitochondria*, and other organelles within the cytoplasm (sarcoplasm). Each cell is enveloped in a cell membrane called *sarcolemma*. Collections of muscle cells make up the belly of a muscle. The highly vascularized skeletal muscles contribute greatly to the size and shape of the body. Skeletal muscles attach to bones or other muscles at their tendinous ends. Between bony attachments, muscles cross one or more joints, moving them. Muscles always pull; they never push. Skeletal muscle contractions consist of rapid, brief shortenings, often generating considerable force. Each contracting cell shortens maximally. Three kinds of skeletal muscle fibers are recognized: red (small, dark, long-acting, slow-contracting, postural muscle fibers with oxygen-rich myoglobin and many mitochondria), white (relatively large, pale, anaerobic, short-acting, fast-contracting muscle fibers with few mitochondria), and intermediate fibers. With exercise, fast fibers can convert to slow; slow fibers can convert to fast. Contraction of skeletal muscle requires nerves (innervation). Without a nerve supply (denervation), skeletal muscle cells cease to shorten; without reinnervation, the cells will die. A denervated portion of muscle loses its tone and becomes flaccid. In time, the entire muscle will become smaller (atrophy). Muscle contraction is generally under voluntary control, but the brain involuntarily maintains a degree of contraction among the body's skeletal muscles (muscle tone). After injury, skeletal muscle cells can regenerate from myoblasts with moderate functional significance; such regeneration may also occur in association with muscle cell hypertrophy in response to training/exercise.

CARDIAC/STRIATED MUSCLE, G
INTERCALATED DISC H   CELL G'

Cardiac muscle cells make up the heart muscle. They are branched, striated cells with one or two centrally located nuclei and a sarcolemma surrounding the sarcoplasm. They are connected to one another by junctional complexes called *intercalated discs*. Their structure is similar to skeletal muscle, but less organized. Cardiac muscle is highly vascularized; its contractions are rhythmic, strong, and well regulated by a special set of impulse-conducting muscle cells, not nerves. Rates of contraction of cardiac muscle are mediated by the autonomic (visceral) nervous system, the nerves of which increase/decrease heart rate. Cardiac muscle is probably not capable of regeneration.

VISCERAL/SMOOTH MUSCLE, I
PLASMALEMMA F'   CELL I'

Smooth muscle cells are long, tapered cells with centrally placed nuclei. Each cell is surrounded by a *plasmalemma* (cell membrane). These cells are smooth (nonstriated). Myofibrils are not seen; the myofilaments intersect with one another in a pattern less organized than that seen in skeletal muscle. Smooth muscle cells occupy the walls of organs with cavities (viscera) and serve to propel the contents along the length of those cavities by slow, sustained, often powerful rhythmic contractions (consider menstrual or intestinal cramps). Smooth muscle cells, oriented perpendicular to the flow of tubular contents, act as gates (sphincters) in specific sites, regulating the flow, as in delaying the flow of urine. Well-vascularized, smooth muscle fibers contract in response to both autonomic nerves and hormones. They are also capable of spontaneous contraction. Regeneration of smooth muscle, to some extent, is possible after injury.

CN: Use red for C and your lightest colors for B, E, G, and I. (1) The sarcolemma (F), which covers each skeletal and cardiac muscle cell, is colored only at the cut ends. The plasmalemma (F^1), which covers each smooth muscle cell, is colored only at the cut ends. (2) The nuclei of cardiac and smooth muscle cells, located deep within the cells, are to be colored only at the cut ends (A). (3) One of the intercalated discs (H) of the cardiac cells has been separated to reveal its structure (schematically). (4) The cellular views are microscopic.

NUCLEUS A
CONNECTIVE TISSUE B
CAPILLARY C
MITOCHONDRION D

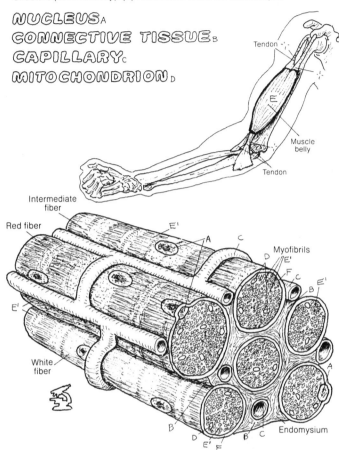

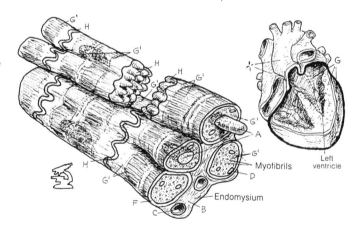

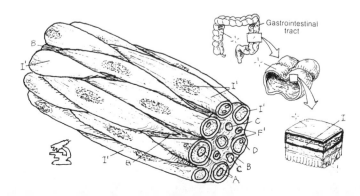

TISSUES: SKELETAL MUSCLE MICROSTRUCTURE

CN: Use the same colors used on Plate 13 for sarcolemma (A) and mitochondrion (D). Use the same color used on the skeletal muscle cell for the myofilbril (E) here. Use light colors for G and J, a dark color for H, and very dark colors for F and K. The cell nucleus is not shown here. (1) Begin with the drawing of the arm. (2) Color the parts of the muscle cell in the central illustration; note the presence of mitochondria (D) between the myofibrils. (3) Color the parts of the exposed (lowest) myofibril and the color-related letters, bands, lines, zone. Note that the cut end of this myofibril receives the color E, for identification purposes, and is part of the A band of the sarcomere adjacent to the one to be colored. (4) Color the relaxed and contracted sarcomere, the filaments, and the mechanism for contraction, noting the color relationship with the myofibril and its parts.

A part of a skeletal muscle cell is shown with the *sarcolemma* opened to reveal some cellular contents. The most visible of the contents are the *myofibrils*, the contractile units of the cell. They are enveloped by a flat tubular *sarcoplasmic reticulum* (SR) that, in part, regulates the distribution of calcium ions (Ca++) into the myofibrils. Inward tubular extensions of the sarcolemma, called the *transverse tubule system* (TTS), run transversely across the SR, at the level of the Z lines of the myofibrils. The TTS, containing stores of sodium ions (Na+) and calcium ions (Ca++), conducts electrochemical excitation to the myofibrils from the sarcolemma. *Mitochondria* provide energy for the cell work.

The myofibrils consist of myofilaments: *thick filaments* (largely myosin) with heads that project outward as *cross bridges*, and *thin filaments* (largely actin) composed of two interwoven strands. These two filament types are arranged into contractile units, each of which is called a *sarcomere*. Each myofibril consists of several radially arranged sarcomeres. At the end of each sarcomere, the thin filaments are permanently attached to the *Z line*, which separates one sarcomere from the next. The relative arrangement of the thick and thin filaments in the sarcomere creates *light* (I, H) and *dark* (A) *bands/zone* and the *M line*, all of which contribute to the appearance of cross-striations in skeletal (and cardiac) muscles.

Shortening of a myofibril occurs when the thin filaments slide toward the center (H zone), bringing the Z lines closer together in each sarcomere. The filaments do not shorten; the myosin filaments do not move. The close relationship of the TTS to the Z lines suggests that this site is the "trigger area" for induction of the sliding mechanism. This sliding motion is induced by *cross bridges* (heads of the immovable thick filaments) that are connected to the thin filaments. Activated by high-energy bonds from ATP, the paddle-like cross bridges swing in concert toward the H zone, drawing the thin filaments with them. The sarcomere shortens as the opposing thin filaments meet or even overlap at the M line.

Occurring simultaneously in all or most of the myofibrils of a muscle cell, shortening of sarcomeres translates to a variable shortening of the resting length of the muscle cell. Repeated in hundreds of thousands of conditioned muscle cells of a professional athlete, the resultant contractile force can pull a baseball bat through an arc sufficient to send a hardball a hundred meters or more through the air.

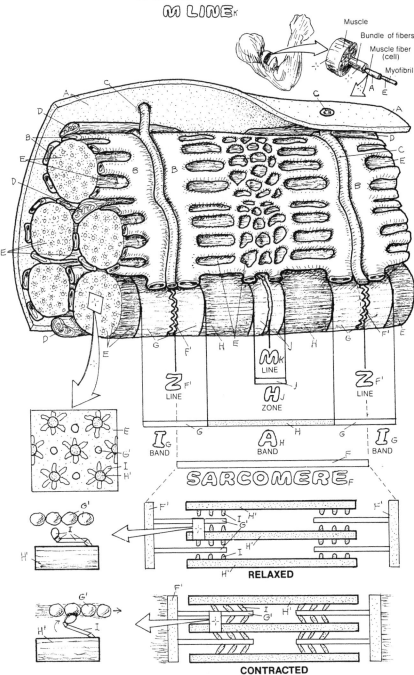

SKELETAL MUSCLE CELL
SARCOLEMMA A
SARCOPLASMIC RETICULUM B
TRANSVERSE TUBULE SYS. C
MITOCHONDRION D
MYOFIBRIL E
SARCOMERE F
I BAND G
THIN FILAMENT (ACTIN) G'
Z LINE F'
A BAND H
THICK FILAMENT (MYOSIN) H'
CROSS BRIDGE I
H ZONE J
M LINE K

Muscle
Bundle of fibers
Muscle fiber (cell)
Myofibril

THIN FILAMENT G'

THICK FILAMENT H'

RELAXED

CONTRACTED

SARCOMERE F

TISSUES: NERVOUS

CN: Use a light color for A. Note the small arrows that indicate direction of impulse conduction. The neurons of the peripheral nervous system shown at lower left are illustrated in the orientation of the left upper limb, although highly magnified.

NEURON
CELL BODY A
PROCESS(ES)
DENDRITE B
AXON C

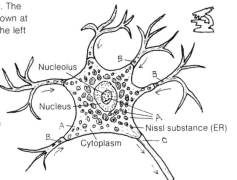

Nucleolus
Nucleus
Cytoplasm
Nissl substance (ER)

TYPES OF NEURONS

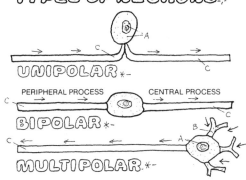

UNIPOLAR

PERIPHERAL PROCESS   CENTRAL PROCESS
BIPOLAR

MULTIPOLAR

Nervous tissue consists of *neurons* (nerve cells) and *neuroglia*. Neurons generate and conduct electrochemical impulses by way of neuronal (cellular) *processes*. Neuroglia are the supporting, non-impulse-generating/conducting cells of the nervous system. The main, nucleus-bearing part of the neuron is the *cell body*. Its cytoplasm contains the usual cell organelles. Uniquely, the endoplasmic reticulum occurs in clusters called Nissl substance. Neurons do not undergo mitosis after birth, compromising their ability to regenerate after injury. Neuronal growth consists of migration and arborization of processes. Neurons are the impulse-conducting cells of the brain and spinal cord (central nervous system, or CNS) and the spinal and cranial nerves (peripheral nervous system, or PNS).

Neurons fall into three structural categories based on numbers of processes ("poles"). Processes that are highly branched (arborized) and uncovered are called *dendrites*. Slender, long, minimally branched processes are called *axons*. Within each category, there is a great variety of shape and size of neurons. *Unipolar* neurons have or appear to have (pseudounipolar) one process that splits near its cell body into a central and peripheral process. Both processes conduct impulses in the same direction, and each is termed an axon (see the sensory neuron at lower left). *Bipolar* neurons have two (central and peripheral) processes, called axons, conducting impulses in the same direction (see Plate 71). *Multipolar* neurons have three or more processes, one of which is an axon (see PNS motor neuron at lower left and CNS neuron at lower right).

PERIPHERAL NERVOUS SYSTEM (PNS)

Nucleus

MOTOR NEURON

SENSORY NEURON

CENTRAL PROCESS

PERIPHERAL PROCESS

Nerve

Brain

Spinal cord

Skin

Receptor

Neuromuscular junction

Skeletal muscle

CENTRAL NERVOUS SYSTEM (CNS)

Blood vessel

Blood vessel

NEUROGLIA
PROTOPLASMIC ASTROCYTE G
FIBROUS ASTROCYTE H
OLIGODENDROCYTE I
MICROGLIA J

Most axons are enveloped in one or more (up to 200) layers of an insulating phospholipid (*myelin*) that enhances impulse conduction rates. Myelin is produced by *oligodendrocytes* in the CNS (lower right) and by Schwann cells in the PNS (lower left). All axons of the PNS are ensheathed by the cell membranes of Schwann cells (neurilemma) but not necessarily myelin. The gaps between Schwann cells are *nodes of Ranvier*, making possible rapid node-to-node impulse conduction. Schwann cells make possible axonal regeneration in the PNS. Significant axonal regeneration in the CNS has not been observed.

NODE OF RANVIER D
AXON COVERINGS
MYELIN E
SCHWANN CELL F

Neuroglia exist in both the CNS and PNS (Schwann cells). *Protoplasmic astrocytes* occur primarily in gray matter (dendrites, cell bodies) of the CNS, *fibrous astrocytes* in the white matter (myelinated axons). Their processes attach to both neurons and blood vessels and may offer metabolic, nutritional, and physical support. They may play a role in the blood/brain barrier. Oligodendrocytes are smaller than astrocytes, have fewer processes, and are seen near neurons. *Microglia* are the small scavenger cells of the brain and spinal cord.

NEUROMUSCULAR INTEGRATION

CN: Use very light colors for A and E, and a dark color for F.
(1) Begin with the skeletal muscle lifting the heel of the foot and complete the motor unit and the enlarged view of the neuromuscular junction. (2) Color carefully the motor units and related titles at the bottom of the plate: only the discharging motor units (in dark outline) are to be colored. Note that the word "partial" is not colored under the example of partial contraction.

SKELETAL MUSCLE A
MUSCLE FIBER A'
MOTOR END PLATE B
MOTOR NERVE C
AXON C'
AXON BRANCH D
AXON TERMINAL E

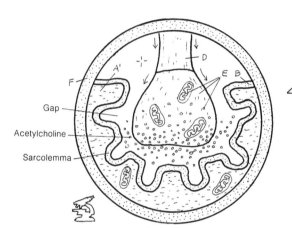

Gap

Acetylcholine

Sarcolemma

MOTOR UNIT
AXON C'
AXON BRANCH D
NEUROMUSCULAR JUNCTION F
MUSCLE FIBER A'

An axon of a single motor neuron, its axon branches, and the skeletal muscle fibers with which they form neuromuscular junctions constitute a *motor unit*. Within any given skeletal muscle, the number of muscle fibers innervated by a single motor neuron largely determines the specificity of contraction of that muscle; the fewer the number of muscle fibers in each motor unit, the more selective and refined the degree of contraction of that skeletal muscle.

NEUROMUSCULAR JUNCTION F
AXON TERMINAL E
MOTOR END PLATE B

Skeletal muscle consists of innumerable muscle fibers (cells). Skeletal muscle requires an intact nerve (innervation) to shorten (contract). Such a nerve, called a *motor nerve*, consists of numerous axons of motor neurons. A motor neuron (see Plate 15) is dedicated solely to stimulating muscle fibers to contract. Each single *muscle fiber* in a skeletal muscle is innervated by a *branch of an axon*. The microscopic site at which the axon branch attaches to the skeletal muscle fiber is called the *neuromuscular junction*. Each neuromuscular junction consists of an *axon terminal* closely applied to an area of convoluted muscle fiber sarcolemma called the *motor end plate*. There is a gap between the two surfaces. When a skeletal muscle fiber is about to be stimulated, a chemical neurotransmitter, called acetylcholine, is released by the axon terminal into the gap. The neurotransmitter induces a change in the permeability of the sarcolemma to sodium (Na^+), which initiates muscle fiber contraction. A muscle fiber can only contract maximally ("all or none" law).

GRADES OF CONTRACTION

Given the fact of "all or none" contraction by individual skeletal muscle fibers, grades of contraction of a skeletal muscle are made possible by activating a number of motor units and not activating others. A *resting muscle* activates no motor units. In a *partial contraction*, only some of the motor units are activated. In *maximal contraction* of a skeletal muscle, all motor units are activated. Gluteus maximus consists of skeletal muscle fibers having a nerve-to-muscle ratio of 1:1000 or more. There is no possibility of controlled, refined contractions from this muscle. The facial muscles, on the other hand, have a much lower nerve-to-muscle ratio, closer to 1:10. Here small numbers of muscle fibers can be contracted by implementing one or a few motor units, generating very fine control on the muscular effect (facial expression) desired.

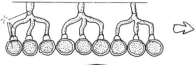

AT REST

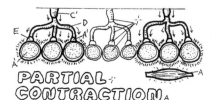

PARTIAL CONTRACTION A

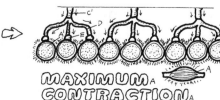

MAXIMUM CONTRACTION A

INTEGRATION OF TISSUES

This plate has one goal: to aid you in visually integrating the four basic tissues into somatic (body wall) and visceral (cavity-containing organs) structure. Concentrate on how the four tissues are arranged in each example of body structure. Consider the general function of each tissue in the overall function of the part/organ. There are an infinite number of functionally related variations in the way these four tissues form a discrete construction of the soma and viscera of the body.

SOMATIC STRUCTURE *-
EPITHELIAL TISSUE -:
SKIN (OUTER LAYER)A
CONNECTIVE TISSUE -:
SKIN (DEEP LAYER)B
SUPERFICIAL FASCIA B'
DEEP FASCIA B²
LIGAMENT B³
BONE B⁴
PERIOSTEUM B⁵
MUSCLE TISSUE -:
SKELETAL MUSCLE c
NERVOUS TISSUE -:
NERVE D

CN: Use yellow for D and light, contrasting colors for A and B, and a medium brown for C. The various vessels that are shown in these tissues—arteries and veins above, and arterioles, venules, capillaries, and lymph vessels below—are not to be colored, as they are made up of more than one basic tissue. Note that within deep fascia, arteries are generally paired with veins.

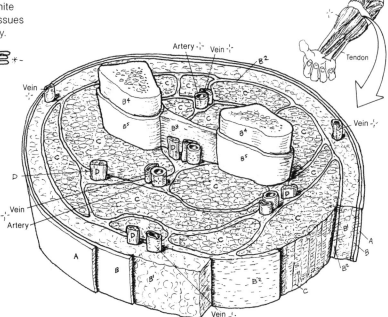

Somatic structure, making up the skin-covered musculoskeletal frame of the body, is concerned with stability, movement, and protection. Its construction reflects these functions. The outermost covering of the body wall everywhere is a protective keratinized *stratified squamous epithelial tissue*, constituting the *outer layer of skin* (epidermis). Other epithelial tissues in somatic structure are the inner layers of blood vessels, and the glands (not shown). Connective tissue layers of the body wall include the *deep layer of skin* (dermis), consisting of dense, irregular fibrous *connective tissue*; and the sub-adjacent, variously mobile, subcutaneous *superficial fascia*

(loose connective and adipose tissues), containing cutaneous nerves, small vessels, and occasional large veins. *Deep fascia*, a more vascular, sensitive, dense, irregular fibrous tissue, ensheathes skeletal muscle (myofascial tissue) as well as the supporting nerves and vessels. *Ligaments* (dense regular connective tissue) bind *bone* to bone by way of *periosteum* (vascular, cellular, dense, irregular, fibrous tissue). Skeletal *muscles* and their *nerves* are packaged in groups, separated by slippery septa of deep fascia securing neurovascular structure. The fibrous investments of skeletal muscle converge to form tendons of the muscle.

VISCERAL STRUCTURE *-
EPITHELIAL TISSUE -:
MUCOSAL LINING A'
GLAND A²
SEROSA (OUTER LAYER) A³
CONNECTIVE TISSUE -:
LAMINA PROPIA B⁶
SUBMUCOSA B⁷
SEROSA (INNER LAYER) B⁸
MUSCLE TISSUE -:
SMOOTH MUSCLE c'
NERVOUS TISSUE -:
NERVE CELLS D'

Visceral structure is generally concerned with absorbing, secreting, trapping, and/or moving food, air, secretions, and/or waste in its cavities. *Epithelial tissue* is the innermost layer (*mucosal lining*) of the thin and pliable visceral wall. It faces the lumen (cavity of the viscus); it is often a single layer of cells (esophagus, urinary tract, and reproductive tract excepted) and deals with the contents of the visceral cavity. *Glands*, unicellular or larger in the mucosa or submucosa, are epithelial, as are the inner layers of blood and lymph vessels. The mucosa includes a sub-epithelial layer of loose fibrous tissue (*lamina propria*), supporting mobile

cells, glands, vessels, and *nerves*. The deepest layer of the mucosa (when present) is a thin *smooth muscle* layer moving finger-like projections (villi) of the mucosal surface. Deep to the mucosa is a dense fibrous tissue (*submucosa*), replete with large vessels and small nerves/nerve cells (intramural ganglia) supplying the mucosa. Deeper yet, two or three layers of smooth muscle (tunica muscularis), innervated by local nerve cells, move the visceral wall in peristaltic contractions. The outermost layer of the gastrointestinal tract is the slippery serosa: an *outer* secretory simple squamous epithelial layer and an *inner* supporting layer of light fibrous tissue.

THE INTEGUMENT: EPIDERMIS

CN: Use very light colors except for E, G, and H. (1) To the right of these notes, color the entire epidermis gray. (2) Color the strata of the epidermis in the larger skin section. The thicker part of stratum corneum (A) reflects the nature of glabrous (hair-deficient) skin. The stratum lucidum (C) exists only in glabrous skin; it is too thin a layer to be shown in these views. (3) Color the strata and their constituent cells in the lower illustration, beginning with the bottom layer (F) and working upward in the direction of cell migration. (4) Color the section of the nail and its supporting elements.

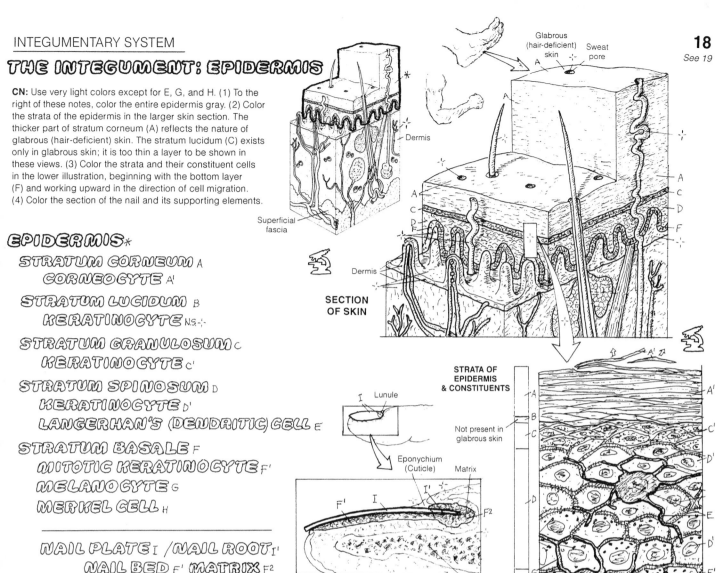

EPIDERMIS*

STRATUM CORNEUM A
CORNEOCYTE A'
STRATUM LUCIDUM B
KERATINOCYTE N.S.
STRATUM GRANULOSUM C
KERATINOCYTE C'
STRATUM SPINOSUM D
KERATINOCYTE D'
LANGERHAN'S (DENDRITIC) CELL E
STRATUM BASALE F
MITOTIC KERATINOCYTE F'
MELANOCYTE G
MERKEL CELL H

NAIL PLATE I / NAIL ROOT I'
NAIL BED F' MATRIX F²

NAIL PLATE & RELATIONS

STRATA OF EPIDERMIS & CONSTITUENTS

Not present in glabrous skin

Melanin granules
Basement membrane
Capillaries
Sensory axon
Dermis

"There is no magician's mantle to compare with the skin in its diverse roles of waterproof, overcoat, sunshade, suit of armor and refrigerator, sensitive to the touch of a feather, to temperature, and to pain, withstanding the wear and tear of three score years and ten, and executing its own running repairs."[1]

The skin is composed of an avascular, stratified squamous epithelial layer (*epidermis*) and a vascular fibrous layer (*dermis*). Within each layer, there is considerable variation. The epithelial layer consists of 4–5 levels of keratin-producing epithelial cells (keratinocytes). Absent capillaries, the layers of epithelia receive their nutrition by diffusion. The outer layers of the epidermis reflect the effects of dehydration.

Mitotic keratinocytes are columnar or cuboidal epithelia forming a single layer (*stratum basale*) separated from the dermis by a basement membrane (epidermal-dermal junction). These are the germinating cells; their progeny are pushed upward by succeeding generations. *Melanocytes* produce melanin granules that disperse along their cytoplasmic extensions (dendrites). These dendrites are woven among the cells of the strata basale and spinosum, and they disseminate melanin among the keratinocytes. Melanin protects the skin from ultraviolet (UV) radiation. Merkel cells are very sensitive to mechanical deformation (touch) of the surface of the skin. The connection with the sensory axon (nerve fiber) is probably similar to a synapse (see Plates 72, 91).

The stratum spinosum consists of several levels of cuboidal and squamous keratinocytes. The cells here have many intracellular filaments that converge on the cell membrane at desmosomes (recall Plate 10). Intercellular tonofibrils, radiating out from the cell surface, can be seen in tissue preparations where their appearance is enhanced by cellular dehydration during processing. This gives a "prickly" appearance to the cells of this stratum. Another kind of dendritic cell, the *Langerhans cell*, is seen in both strata basale and spinosum as well as the dermis. These dendritic cells are essentially phagocytic and present antigen to T lymphocytes (see Plate 124).

The stratum granulosum consists of flattened keratinocytes characterized by disintegrated nuclei and cytoplasmic keratohyalin and lamellar granules. The lipid-rich content of the lamellar granules fills the intercellular spaces, greatly contributing to the impermeability of the skin.

The thin stratum lucidum is seen only in glabrous (hair-deficient) thick skin. Its squamous keratinocytes are filled with filaments; the nuclei of these cells have largely disappeared.

The outermost stratum corneum is composed of multiple layers of squamous, lifeless, keratin-filled cells (*corneocytes*). Keratin is a scleroprotein, the polypeptides of which are intertwined with filaments within the cytoplasm. Loosening and detaching of the dead, outer layers of the stratum corneum is ongoing and involves breaking the intercellular junctional devices (desmosomes, filaments, amorphous lipid substance). The stratum corneum may be as thin as 5 layers (skin of the eyelid) and as thick as 50 layers (plantar surface of the foot).

Nails are plates of compacted, highly keratinized cells of the stratum corneum. Located on the dorsal aspect of each digit, they are translucent, revealing the vascular *nail bed* below. The nail bed consists of the strata basale and spinosum only. The proximal part of the nail plate (*nail root*) fits into a groove under the proximal nail fold. The epithelia around the root are the matrix or the source tissue for the nail plate, and they extend from the region of the nail root to the lunule (lighter, opaque area at the proximal part of the nail plates, seen best on the thumb). The nail plate is formed as the epithelia of the matrix grow distally. The nail plate is continually pushed distally by the keratinizing epithelia migrating from the matrix.

[1]Quote taken, with permission, from Lockhart, R.D., Hamilton, G.F., and Fyfe, F.W. ANATOMY THE HUMAN BODY, 2nd ed., Faber and Faber, Publishers, Ltd., London, 1965.

THE INTEGUMENT: DERMIS

DERMIS ⊹

PAPILLARY LAYER / LOOSE C.T. A
DERMAL PAPILLA A'
RETICULAR LAYER / DENSE C.T. B
HAIR SHAFT c / FOLLICLE c'
ARRECTOR PILI MUSCLE D
SEBACEOUS GLAND E
EPITHELIAL CELL E'
SECRETION F
BURST EPITH. CELL E²
SEBUM F+E²
SWEAT GLAND G
DUCT EPITHELIUM G'
GLAND EPITHELIUM G²
SWEAT H

ARTERY I   VEIN J
LYMPHATIC VESSEL K
NERVE L / RECEPTOR L'

CN: Use red for I, blue for J, green for K, yellow for L, and light colors for the rest. (1) In the skin section, color the hair shafts (C) and sweat pores (G) in the otherwise uncolored epidermis. (2) Follow the text carefully as you color the enlarged views of the sebaceous (E) and sweat (G) glands.

The dermis consists of a fibrous connective tissue supporting *arteries* and *veins, lymphatic capillaries,* nerves and sensory receptors (see Plates 18, 91), and a number of accessory structures. The dermis is separated from the epidermis by a basement membrane (epidermal-dermal junction). On the deep side, the dermis is bordered by superficial fascia (hypodermis, subcutaneous tissue), a loose connective tissue layer with variable amounts of adipose tissue. The upper or most superficial layer of dermis is the *papillary layer*, characterized by a vascular, loose connective tissue. Pegs of this layer (dermal papillae) poke up into the epidermis. These pegs have strong attachment to the basement membrane and contain vessels, nerve endings, and axons among the collagen and elastic fibers. The subjacent *reticular layer* has a more dense fibrous character.

Hair shafts rise from epidermal *follicles* pushed down into the dermis (and hypodermis in the scalp) during development. They are not found in thick skin. The follicle begins at the site where the hair leaves the epidermis; it terminates in the form of a bulb. Hair shafts are composed of layers of keratin surrounded by layers of follicular cells (root sheaths, glassy membranes). The base of the follicle (hair bulb) is turned inward (invaginated) to accommodate a vascular dermal papilla. An obliquely placed bundle of smooth muscle attaches the outer membrane of the follicle to a papillary peg under the epidermis. This is the *arrector pili muscle.* When it is contracted, the hair to which it is attached erects to become perpendicular with the skin surface. In many mammals, hair standing on end is a sign of increased vigilance.

Sebaceous glands are grape-shaped collections of cells with a common duct (acini; holocrine gland) that surround hair follicles. The base of each gland is mitotically active; the daughter cells move into the gland center and become filled with lipid. Continued lipid engorgement results in *burst cells.* The secretory product and the cell debris constitute *sebum.* The gland duct transports the sebum to the epidermal surface or into the upper hair follicle. Sebum coats the skin and hairs, providing a degree of waterproofing. Sebum may play a social role, in terms of olfactory identification.

Sweat glands are coiled tubular glands in the deep dermis. The *ducts* of these glands traverse the epidermis by spiraling around the keratinocytes and open onto the epidermal surface. The glandular cells at the base of the sweat gland are in intimate proximity to capillaries, just as the glomerulus is in relation to the visceral layer of the renal capsule in the kidney. The cells produce sweat, a filtrate of plasma, somewhat like the filtrate of the renal corpuscle (Plate 149). Sweat is largely salt water, with a dash of urea and other molecules. Sweating is a means by which the hypothalamus can induce a degree of cooling by evaporation.

EPIDERMIS
Stratum corneum
Stratum granulosum
Stratum spinosum
Stratum basale
Basement membrane
Collagen fibers
Elastic fiber (magnified)
Dermal papilla
Matrix
Bulb

SUPERFICIAL FASCIA

Gland cell
Lipid

SEBACEOUS GLAND (Section)

Capillary

SWEAT GLAND (Scematic section)

LONG BONE STRUCTURE

CN: Use light blue for C, a tan color for D, very light colors for E and F, yellow for I, and red for J. (1) The title "red marrow" is not to be colored as the red marrow in this bone is not shown, having been replaced by yellow marrow during maturity. Only part of the yellow marrow in the medullary cavity is shown. Leave the cavity (G) itself uncolored. (2) Color the vertical bar to the right which represents the epiphysis (A) and the diaphysis (B) of the long bone.

Bone is a living, vascular structure, composed of organic tissue (cells, fibers, extracellular matrix, vessels, nerves—about 35% of a bone's weight) and mineral (calcium hydroxyapatite—about 65% of a bone's weight). Bone functions as a support structure, a site of attachment for skeletal muscle, ligaments, tendons, and joint capsules, a source of calcium, and a significant site of blood cell development (hematopoiesis) for the entire body. Here we show a long bone, specifically the femur, the bone of the thigh.

EPIPHYSIS A

The epiphysis is the end of a long bone or any part of a bone separated from the main body of an immature bone by cartilage. It is formed from a secondary site of ossification. It is largely cancellous bone, and its articulating surface is lined with 3–5 mm of hyaline (articular) cartilage. The epiphysis is supplied by vessels from the joint capsule.

DIAPHYSIS B

The diaphysis is the shaft or central part of a long bone. It has a marrow-filled cavity (medullary cavity) surrounded by compact bone which is lined externally by periosteum and internally by endosteum (not shown). The diaphysis is formed from one or more primary sites of ossification and is supplied by one or more nutrient arteries.

ARTICULAR CARTILAGE C

Articular cartilage is smooth, slippery, porous, malleable, insensitive, and bloodless; it is the only remaining evidence of an adult bone's cartilaginous past. It is massaged by movement, permitting absorption of synovial fluid, oxygen, and nutrients. Articular (hyaline) cartilage is also nourished by vessels from the subchondral bone. Bones of a synovial joint make physical contact at their cartilaginous ends. The degenerative process of arthritis involves the breakdown and fibrillation of articular cartilage.

PERIOSTEUM D

Periosteum is a fibrous, cellular, vascular, and highly sensitive life support sheath for bone, providing nutrient blood for bone cells and a source of osteoprogenitor cells throughout life. It does not cover articular cartilage.

CANCELLOUS (SPONGY) BONE E

Cancellous (spongy) bone consists of interwoven beams (trabeculae) of bone in the epiphyses of long bones, the bodies of the vertebrae, and other bones without cavities. The spaces among the trabeculae are filled with red or yellow marrow and blood vessels. Cancellous bone forms a dynamic latticed truss capable of mechanical alteration (reorientation, construction, destruction) in response to the stresses of weight, postural change, and muscle tension.

COMPACT BONE F

Compact bone is dense bone characterized in long bones by microscopic hollow cylinders of bone (haversian systems) interwoven with non-cylindrical lamellae of bone. It forms the stout walls of the diaphysis of long bones and the thinner outer surface of other bones where there is no articular cartilage—e.g., the flat bones of the skull. Blood vessels reach the bone cells by a system of integrated canals.

MEDULLARY CAVITY G

The medullary cavity is the cavity of the diaphysis. It contains marrow: red in the young, turning to yellow in many long bones in maturity. It is lined by endosteal tissue (thin connective tissue with many osteoprogenitor cells).

RED MARROW

Red marrow is a red, gelatinous substance composed of red and white blood cells in a variety of developmental forms (hematopoietic tissue) and specialized capillaries (sinusoids) enmeshed in reticular tissue. In adults, red marrow is generally limited to the sternum, vertebrae, ribs, hip bones, clavicles, and cranial bones.

YELLOW MARROW H

Yellow marrow is fatty connective tissue that is not productive of blood cells. It replaces red marrow in the epiphyses and medullary cavities of long bones, and cancellous bone of other bones.

NUTRIENT ARTERY I /BRANCHES I'

The nutrient artery is the principal artery and major supplier of oxygen and nutrients to the shaft or body of a bone; its branches snake through the labyrinthine canals of the haversian systems and other tubular cavities of bones.

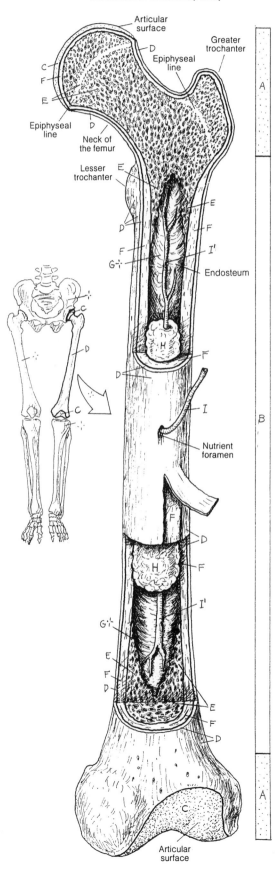

ANTERIOR VIEW
(Left femur)

Coronal section through proximal epiphysis and dissection of medullary cavity

Articular surface
Greater trochanter
Epiphyseal line
Epiphyseal line
Neck of the femur
Lesser trochanter
Endosteum
Nutrient foramen
Articular surface

AXIAL/APPENDICULAR SKELETON

CN: Use light but contrasting colors for A and B.
(1) Color the axial skeleton (A) in all three views.
Do not color the spaces between the ribs (intercostal).
(2) Color the darker, outlined appendicular skeleton (B).
(3) Color the arrows identifying bone shape/classification.

CLASSIFICATION OF BONES:

LONG c
SHORT d
FLAT e
IRREGULAR f
SESAMOID g

Bones have a variety of shapes and defy classification by shape; yet such a classification historically exists. *Long bones* are clearly longer in one axis than in another; they are characterized by a medullary cavity, a hollow diaphysis of compact bone, and at least two epiphyses—e.g., femur, phalanx. *Short bones* are roughly cube-shaped; they are predominantly cancellous bone with a thin cortex of compact bone and have no cavity—e.g., carpal and tarsal bones. *Flat bones* (cranial bones, scapulae, ribs) are generally more flat than round, and *irregular bones* (vertebrae) have two or more different shapes. Bones not specifically long or short fit this latter category. *Sesamoid bones* are developed in tendons (e.g., patellar tendon); they are mostly bone, often mixed with fibrous tissue and cartilage. They have a cartilaginous articular surface facing an articular surface of an adjacent bone; they may be part of a synovial joint ensheathed within the fibrous joint capsule. The structures are generally pea-sized and are most commonly found in certain tendons/joint capsules in hands and feet, and occasionally in other articular sites of the upper and lower limbs. The largest is the patella, integrated in the tendon of quadriceps femoris. Sesamoid bones resist friction and compression, enhance joint movement, and may assist local circulation.

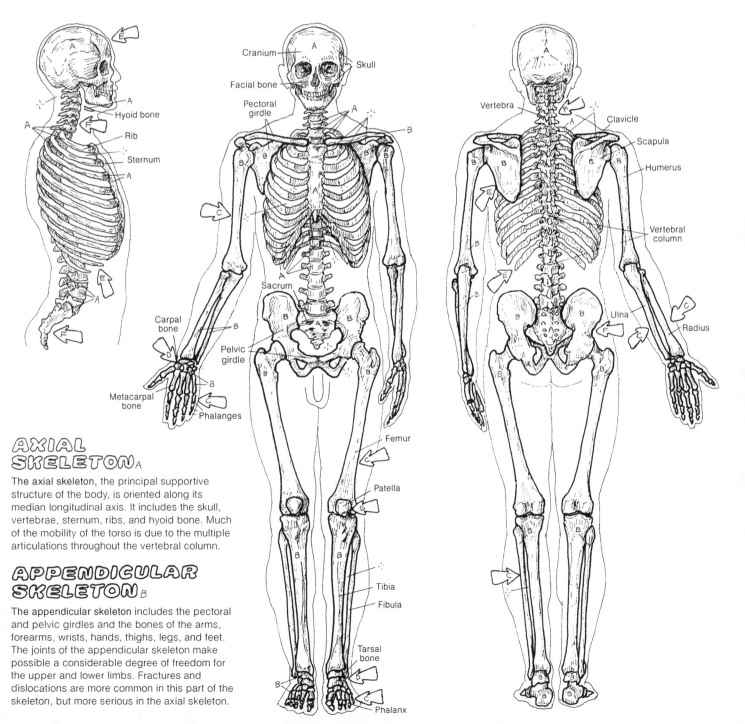

AXIAL SKELETON a

The axial skeleton, the principal supportive structure of the body, is oriented along its median longitudinal axis. It includes the skull, vertebrae, sternum, ribs, and hyoid bone. Much of the mobility of the torso is due to the multiple articulations throughout the vertebral column.

APPENDICULAR SKELETON b

The appendicular skeleton includes the pectoral and pelvic girdles and the bones of the arms, forearms, wrists, hands, thighs, legs, and feet. The joints of the appendicular skeleton make possible a considerable degree of freedom for the upper and lower limbs. Fractures and dislocations are more common in this part of the skeleton, but more serious in the axial skeleton.

CLASSIFICATION OF JOINTS

Bones are connected at joints (articulations). All bones move at joints. Joints are functionally classified as immovable (synarthroses), partly movable (amphiarthroses), or freely movable (diarthroses). The structural classification of joints is given below.

FIBROUS JOINT:
IMMOVABLE_A / PARTLY MOVABLE_A'

Fibrous joints (synarthroses) are those in which the articulating bones are connected by fibrous tissue. Sutures of the skull are essentially *immovable fibrous joints*, especially after having ossified with age. Teeth in their sockets are fixed fibrous joints (gomphoses). Syndesmoses are *partly movable fibrous joints*, such as the interosseous ligaments between bones of the forearm or the bones of the leg.

CARTILAGINOUS JOINT:
IMMOVABLE_B / PARTLY MOVABLE_B'

Cartilaginous joints (synchondroses) are essentially immovable joints seen during growth—e.g., growth (epiphyseal) plates (see Plate 168). Fibrocartilaginous joints (amphiarthroses) are partly movable—e.g., the intervertebral disc. Symphyses also are partly movable fibrocartilagious joints, as between the pubic bones (symphysis pubis) and the manubrium and body of the sternum (sternal angle).

SYNOVIAL JOINT (FREELY MOVABLE):
ARTICULATING BONES_C:
ARTICULAR CARTILAGE_D
SYNOVIAL MEMBRANE_E
SYNOVIAL CAVITY (FLUID)_F
JOINT CAPSULE_G
BURSA CAPSULE_G'
COLLATERAL LIGAMENT_H*

Synovial joints (diarthroses) are freely movable within ligamentous limits and the bony architecture. They are characterized by *articulating bones* whose ends are capped with *articular cartilage* and are enclosed in a ligament-reinforced, sensitive, fibrous (joint) *capsule* lined internally with a vascular *synovial membrane* that secretes a lubricating fluid within the *cavity*. The synovial membrane does not cover articular cartilage. A fibrous tissue–lined synovial sac of fluid (bursa) often exists between moving structures outside the joint, as between tendon and bone. Bursae facilitate friction-free movement; friction may induce painful inflammation (bursitis).

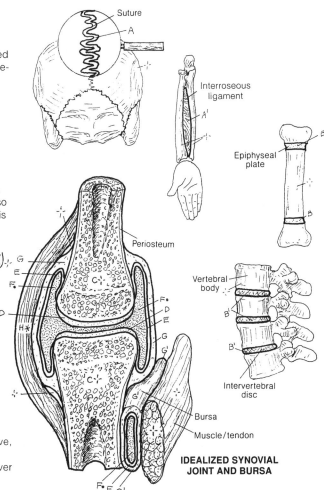

Suture
A
Interroseous ligament
A'
Epiphyseal plate
B
B
Periosteum
G
E
F
C
F
D
D
E
G
H*
G'
C
Vertebral body
B'
B'
B'
Intervertebral disc
Bursa
Muscle/tendon
F
E
G'

IDEALIZED SYNOVIAL JOINT AND BURSA

TYPES OF SYNOVIAL JOINTS:

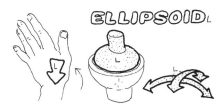

BALL & SOCKET_I

Ball-and-socket joints are best seen at the hip and shoulder. Movements in all direction are permitted —i.e., flexion, extension, adduction, abduction, internal and external rotation, and circumduction.

HINGE_J

A hinge joint permits movement in only one plane: flexion/extension. The ankle, interphalangeal, and elbow (humeroulnar) joints are hinge joints.

SADDLE_K

A saddle (sellar) joint—e.g., carpometacarpal joint at the base of the thumb—has two concave articulating surfaces, permitting all motions but rotation.

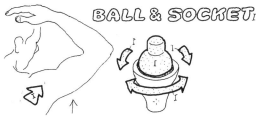

ELLIPSOID_L

The ellipsoid (condyloid, condylar) joint is a reduced ball-and-socket configuration in which significant rotation is largely excluded—e.g., the bicondylar knee, temporomandibular, and radiocarpal (wrist) joints.

PIVOT_M

A pivot joint has a ring of bone around a peg; e.g., the C1 vertebra rotates about the dens of C2, a rounded humeral capitulum on which the radial head pivots (rotates).

GLIDING_N

Gliding joints (e.g., the facet joints of the vertebrae, the acromio-clavicular, intercarpal, and intertarsal joints) has generally flat articulating surfaces.

TERMS OF MOVEMENTS

CN: Color the arrows pointing to the joints demonstrating the various movements of the body. Inversion (K) and eversion (L) movements occur among bones of the foot, not at the ankle.

EXTENSION_A
  DORSIFLEXION_B
FLEXION_C
  PLANTAR FLEXION_D
ADDUCTION_E
ABDUCTION_F
CIRCUMDUCTION_G
ROTATION_H
SUPINATION_I
PRONATION_J
INVERSION_K
EVERSION_L

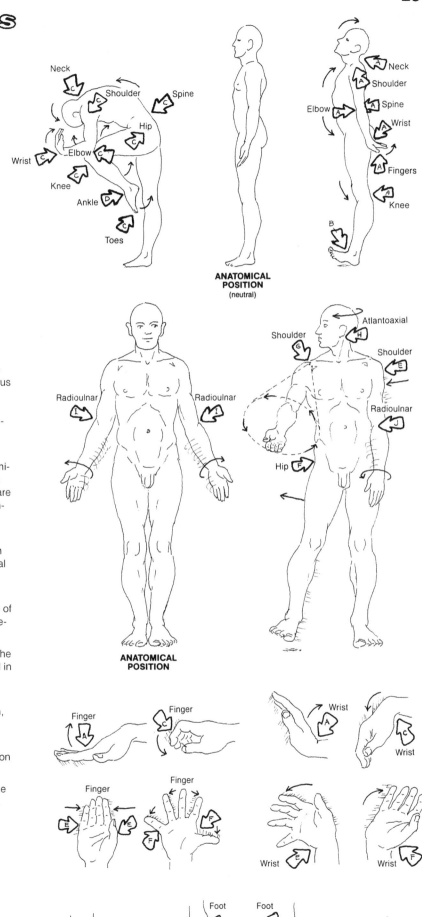

Movements of bones occur at joints. Terms of movement are therefore applicable to joints, not bones (flexion of the humerus would break it!). Ranges of motion are limited by the bony architecture of a joint, related ligaments, and the muscles crossing that joint. It is from the anatomical position that specific directions of movement can be clearly delineated and ranges of motion measured.

Extension of a joint is to generally straighten it. In the anatomical position, most joints are in relaxed extension (neutral). In relation to the anatomical position, movements of extension are directed in the sagittal plane. Extreme, even abnormal extension is called hyperextension. At the ankle and wrist joints, extension is termed dorsiflexion.

Flexion of a joint is to bend it or decrease the angle between the bones of the joint. Movements of flexion are in the sagittal plane. At the ankle joint, flexion is also called plantar flexion.

Adduction of a joint moves a bone toward the midline of the body (or in the case of the fingers or toes, toward the midline of the hand or foot). In relation to the anatomical position, movements of adduction are directed in the coronal plane.

Abduction of a joint moves a bone away from the midline of the body (or hand or foot). Movements of abduction are directed in the coronal plane.

Circumduction is a circular movement, permitted at ball and socket, condylar, and saddle joints, characterized by flexion, abduction, extension, and adduction done in sequence.

Rotation of a joint is to turn the moving bone about its axis. Rotation toward the body is internal or medial rotation; rotation away from the body is external or lateral rotation.

Supination is external rotation of the radiohumeral joint. In the foot, supination involves lifting the medial aspect of the foot.

Pronation is internal rotation of the radiohumeral joint. In the foot, pronation involves raising the lateral aspect of the foot.

Inversion turns the sole of the foot inward so that the medial border of the foot is elevated.

Eversion turns the sole of the foot outward so that its lateral border is elevated.

BONES OF THE SKULL (1)

8 CRANIAL ÷
OCCIPITAL A 2 PARIETAL B FRONTAL C
2 TEMPORAL D ETHMOID E SPHENOID F

14 FACIAL ÷
2 NASAL G VOMER H 2 LACRIMAL I
2 ZYGOMATIC J 2 PALATINE K 2 MAXILLA L
MANDIBLE M 2 INFERIOR NASAL CONCHA N

CN: Save the brightest colors for the smallest bones and the lightest colors for the largest. (1) Color one bone in as many views as it appears before going on to the next. (2) There are some very small bones to color in the orbits and in the lower part of the posterior view of the skull. Study these areas carefully before coloring to determine the color boundaries. (3) Do not color the darkened areas in the orbits and nasal cavity in the anterior view.

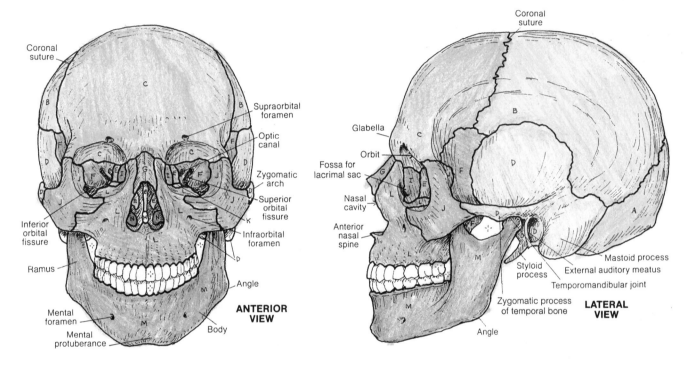

ANTERIOR VIEW

LATERAL VIEW

The skull is composed of *cranial bones* (forming a vault for the brain) and *facial bones* (giving origin to the muscles of facial expression and providing buttresses protecting the brain). Except for the temporomandibular joint (a synovial joint), all bones are connected by generally immovable fibrous sutures.

The orbit is composed of seven bones, has three significant fissures/canals, and is home to the eye and related muscles, nerves, and vessels. The most delicate of the skull bones is at the medial orbital wall. The external nose is largely cartilaginous and is therefore not part of the bony skull.

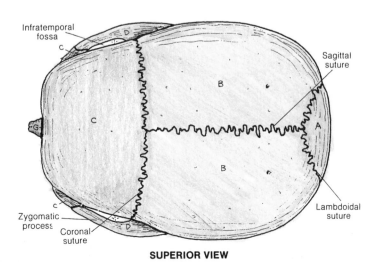

SUPERIOR VIEW

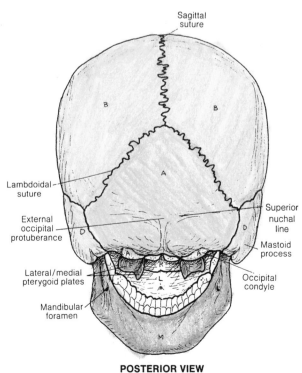

POSTERIOR VIEW

BONES OF THE SKULL (2)

CRANIAL: OCCIPITAL_A PARIETAL_B FRONTAL_C TEMPORAL_D ETHMOID_E SPHENOID_F
FACIAL: NASAL_G VOMER_H ZYGOMATIC_J PALATINE_K MAXILLA_L INFERIOR NASAL CONCHA_N

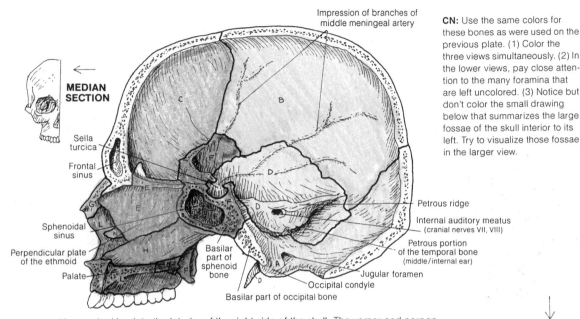

MEDIAN SECTION

Impression of branches of middle meningeal artery

Sella turcica
Frontal sinus
Sphenoidal sinus
Perpendicular plate of the ethmoid
Palate
Basilar part of sphenoid bone
Basilar part of occipital bone
Occipital condyle
Jugular foramen
Petrous portion of the temporal bone (middle/internal ear)
Internal auditory meatus (cranial nerves VII, VIII)
Petrous ridge

CN: Use the same colors for these bones as were used on the previous plate. (1) Color the three views simultaneously. (2) In the lower views, pay close attention to the many foramina that are left uncolored. (3) Notice but don't color the small drawing below that summarizes the large fossae of the skull interior to its left. Try to visualize those fossae in the larger view.

You are looking into the interior of the right side of the skull. The vomer and perpendicular plate of the ethmoid contribute significantly to the nasal septum (Plate 131). In this view, they hide the conchae on the right lateral wall of the nasal cavity.

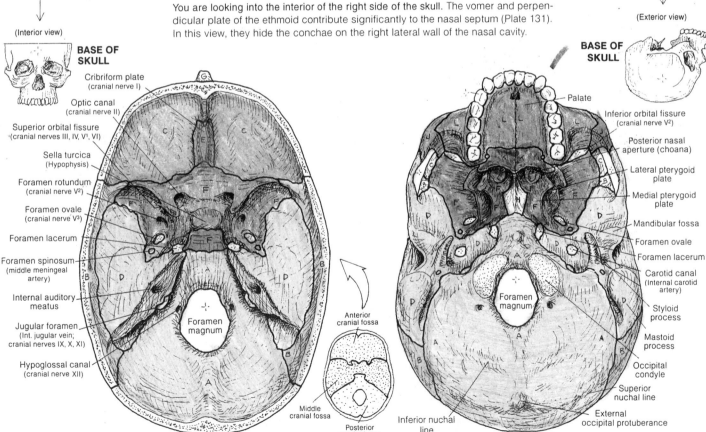

(Interior view)

BASE OF SKULL

Cribriform plate (cranial nerve I)
Optic canal (cranial nerve II)
Superior orbital fissure (cranial nerves III, IV, V¹, VI)
Sella turcica (Hypophysis)
Foramen rotundum (cranial nerve V²)
Foramen ovale (cranial nerve V³)
Foramen lacerum
Foramen spinosum (middle meningeal artery)
Internal auditory meatus
Jugular foramen (Int. jugular vein; cranial nerves IX, X, XI)
Hypoglossal canal (cranial nerve XII)
Foramen magnum

Anterior cranial fossa
Middle cranial fossa
Posterior cranial fossa

(Exterior view)

BASE OF SKULL

Palate
Inferior orbital fissure (cranial nerve V²)
Posterior nasal aperture (choana)
Lateral pterygoid plate
Medial pterygoid plate
Mandibular fossa
Foramen ovale
Foramen lacerum
Carotid canal (Internal carotid artery)
Styloid process
Mastoid process
Occipital condyle
Superior nuchal line
External occipital protuberance
Foramen magnum
Inferior nuchal line

You are looking onto the floor of the cranial cavity (base of the skull). The anterior cranial fossae support the frontal lobes of the cerebrum (Plate 73); the olfactory tracts lie over the cribriform plates, receiving the olfactory nerves (Plate 100). The middle cranial fossae embrace the temporal lobes; note the numerous foramina/canals for cranial nerves and vessels. The posterior cranial fossa retains the cerebellum and the brain stem (Plate 76) as well as related cranial nerves and vessels that enter or exit the cavity (Plate 83).

The large external surface of the occipital bone is a site of attachment of layers of posterior cervical musculature (Plate 49). The foramen magnum transmits the lower brain stem/spinal cord (Plate 76). The occipital condyles articulate with the facets of the atlas or first cervical vertebra (Plate 28). The muscular pharyngeal wall attaches around the posterior nasal apertures (Plate 139).

TEMPOROMANDIBULAR JOINT
(CRANIOMANDIBULAR)

CN: Read the text before coloring. Use light colors for A and B, light blue for C, and black or a dark color for E. (1) Begin with bones of the TM joint and the corresponding ligaments. (2) In the central illustration of the upward, lateral view of the skull, the articular fossa receives the color of its articular cartilage (C). Also color the cartilage and condyle of the mandible, which is placed here for diagrammatic purposes. (3) Finish with the sagittal views below, which describe the movement of the mandibular condyle within the TM joint.

SKULL BONES-
  TEMPORAL BONE A
  MANDIBLE B
    CONDYLAR PROCESS B¹
ARTICULAR CARTILAGE C
JOINT CAPSULE D
SYNOVIAL CAVITY E•-
  SUP. JOINT SPACE E¹•
  INF. JOINT SPACE E²•
ARTICULAR DISC F-
  ANT. BAND F¹
  POST. BAND F²
RETRODISCAL PAD G

Two temporomandibular joints form the craniomandibular joint, which consists of the heads of the left and right *condylar processes* of the *mandible* articulating with paired articular fossae of the *temporal bones*. Movement or trauma to one of these two temporomandibular joints (TMJ) always involves the contralateral joint. The TMJ is a complex synovial joint, gliding, angling, and rotating during what appears to be simple hinge movements of the lower jaw. Movements of the TMJ can be seen in Plate 47.

The TMJ is encapsulated within a fibrous (joint) capsule, the only true ligament of the joint. The *articular disc* (meniscus) is a fibrocartilaginous, oval plate between the cartilage-lined articular fossa and the articular cartilage of the condylar process. It divides the *synovial cavity* into superior and inferior *joint spaces*. The disc incorporates two avascular *bands* whose long axes lie in the coronal plane. Here we see them in cross section. These bands are connected by an intermediate zone of fibrous tissue. The disc is well connected, anteriorly to the lateral pterygoid muscle, posteriorly to the vascular, elastic *retrodiscal pad* in the bilaminar region from which it gets its nutrition, and medially/laterally to the condylar process. When the mouth is closed, the head of the condylar process abuts the larger, posterior band. As the mouth opens, the condylar head rotates forward and downward to abut the anterior band at full opening (35–50 mm between upper and lower incisors). During mouth opening, the meniscus itself is stretched as it is pulled forward with the condylar head.

The articular disc of the TMJ may fray or become dislocated or detached with aging, abuse (trauma), or misuse (clenching, grinding of teeth). This condition may be associated with bitemporal headaches (temporalis muscle overuse), clicking during jaw movement, and reduced range of motion. The disc may also be structurally incomplete (even perforated) from birth.

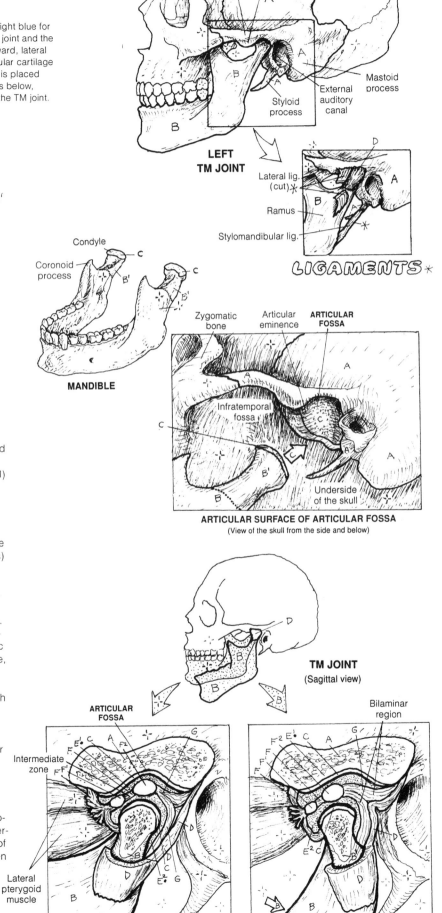

LEFT TM JOINT

LIGAMENTS *

MANDIBLE

ARTICULAR SURFACE OF ARTICULAR FOSSA
(View of the skull from the side and below)

TM JOINT
(Sagittal view)

MOUTH CLOSED

MOUTH OPEN

VERTEBRAL COLUMN

CN: Use gray for D, yellow for H, and light colors for the rest, especially C, T, L, S, and Co. L4 and L5 represent the lumbar vertebrae most involved in motion. (1) Begin with regions of the column and the three examples of vertebral disorders at lower left. (2) Color the motion segment and its role in flexion and extension. (3) Color the vertebral foramina and canal. (4) Color the example of a protruding intervertebral disc pressing on a spinal nerve.

The vertebral column has 24 individual vertebrae arranged in *cervical, thoracic,* and *lumbar* regions; the *sacral* and *coccygeal* vertebrae are fused (sacrum/coccyx). Numbers of vertebrae in each region are remarkably constant; rarely S1 may be free or L5 may be fused to the sacrum (transitional vertebrae). The seven mobile cervical vertebrae support the neck and the 3–4 kg (6–8 lb) head. The cervical spine is normally curved (*cervical lordosis*) secondary to the development of postural reflexes about three months after birth. The 12 thoracic vertebrae support the thorax, head, and neck. They articulate with 12 ribs bilaterally. The thoracic spine is congenitally curved (*kyphosis*) as shown.

The five lumbar vertebrae support the upper body,

torso, and low back. The column of these vertebrae becomes curved (*lumbar lordosis*) at the onset of walking at 1–2 years of age. The sacrum is the keystone of a weightbearing arch involving the hip bones. The sacral/coccygeal curve is congenital. The variably numbered 1–5 coccygeal vertebrae are usually fused, although the first vertebra may be movable.

Vertebral curvatures may be affected (usually exaggerated) by posture, activity, obesity, pregnancy, trauma, and/or disease; these conditions are given the same names as the normal curves. A slight lateral curvature to the spine often reflects dominant handedness; a significant, possibly disabling, lateral curve (*scoliosis*) may occur for many reasons.

REGIONS
CERVICAL c
THORACIC T
LUMBAR L
SACRAL s
COCCYGEAL Co

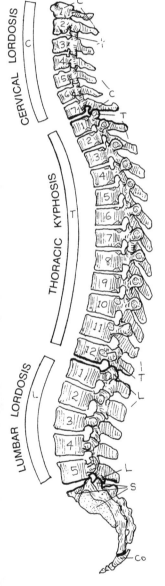

CERVICAL LORDOSIS

THORACIC KYPHOSIS

LUMBAR LORDOSIS

MOTION SEGMENT
VERTEBRA L⁴
JOINTS:
INTERVERTEBRAL DISC A
POSTERIOR (FACET) B
LIGAMENT D*
VERTEBRA L⁵
VERTEBRAL FORAMEN E
VERTEBRAL CANAL E'
INTERVERTEBRAL FORAMEN F

Each pair of individual, unfused vertebrae constitutes a *motion segment,* the basic movable unit of the back. Combined movements of motion segments underlie movement of the neck and the middle and low back. Each pair of vertebrae in a motion segment, except C1–C2, is attached by three joints: a partly movable, *intervertebral disc* anteriorly and a pair of gliding synovial *facet* (zygapophyseal) *joints* posteriorly. *Ligaments* secure the bones together and encapsulate the facet joints (joint capsules). The *vertebral* or *neural canal,* a series of *vertebral foramina,* transmits the spinal cord and related coverings, vessels, and nerve roots. Located bilaterally between each pair of vertebral pedicles are passageways, each called an *intervertebral foramen,* transmitting spinal nerves, their coverings/vessels, and some vessels to the spinal cord.

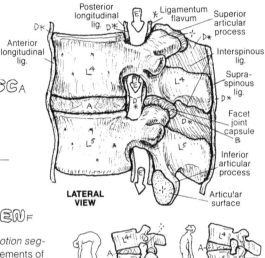

Posterior longitudinal lig. · Ligamentum flavum · Superior articular process · Anterior longitudinal lig. · Interspinous lig. · Supraspinous lig. · Facet joint capsule B · Inferior articular process · Articular surface

LATERAL VIEW

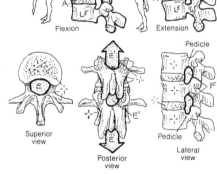

Flexion · Extension · Pedicle · Superior view · Posterior view · Lateral view

INTERVERTEBRAL DISC A
ANNULUS FIBROSUS A'
NUCLEUS PULPOSUS G
SPINAL NERVE H

The intervertebral disc consists of the *annulus fibrosus* (concentric, interwoven collagenous fibers integrated with cartilage cells) attached to the vertebral bodies above and below, and the more central *nucleus pulposus* (a mass of degenerated collagen, proteoglycans, and water). The discs make possible movement between vertebral bodies. With aging, the discs dehydrate and thin, resulting in a loss of height. The cervical and lumbar discs, particularly, are subject to early degeneration from one or more of a number of causes. Weakening and/or tearing of the annulus can result in a broad-based bulge or a localized (focal) protrusion of the nucleus and adjacent annulus; such an event can compress a *spinal nerve* root as shown.

VERTEBRAL DISORDERS

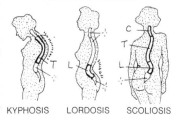

KYPHOSIS   LORDOSIS   SCOLIOSIS

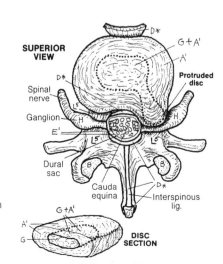

SUPERIOR VIEW

G + A' · A' · **Protruded disc** · Spinal nerve · Ganglion · Dural sac · Cauda equina · Interspinous lig.

DISC SECTION

CERVICAL & THORACIC VERTEBRAE

CN: Use red for M and use the same colors as were used on Plate 21 for C and T. Use dark colors for N, O, and R. (1) Begin with the parts of a cervical vertebra. Color the atlas and axis and note they have been given separate colors to distinguish them from other cervical vertebrae. (2) Color the parts of a thoracic vertebra and then the thoracic portion of the vertebral column. Note the three different facet/demifacet colors.

CERVICAL VERTEBRA c

BODY c'
PEDICLE b
TRANSVERSE PROCESS g
ARTICULAR PROCESS h
FACET h'
LAMINA i
SPINOUS PROCESS j

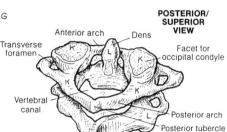

ATLAS k
AXIS l

Cranium
Mastoid process
Mandible
C 1
LATERAL VIEW
Disc — C 2
C 3 — Intervertebral foramen
C 4
Transverse foramen
C 5 — Facet joint
C 6
C 7 — Vertebra prominens

VERTEBRAL ARTERY m

POSTERIOR/ SUPERIOR VIEW

Anterior arch — Dens
Transverse foramen
Facet for occipital condyle
Vertebral canal
Posterior arch
Posterior tubercle
Bifid spine

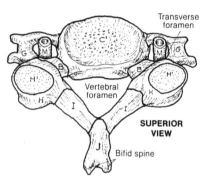

Transverse foramen
Vertebral foramen
SUPERIOR VIEW
Bifid spine

TYPICAL CERVICAL (C4) VERTEBRA

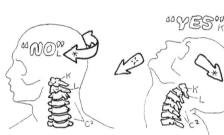

LATERAL FLEXION   ROTATION   FLEXION/EXTENSION

The small seven cervical vertebrae support and move the head and neck, supported by ligaments and strap-like paracervical (paraspinal) muscles. The ring-shaped *atlas* (C1) has no body; thus there are no weight-bearing discs between the occiput and C1, and between C1 and C2 (the *axis*). Head weight is transferred to C3 by the large *articular processes* and *facets* of C1 and C2. The atlantooccipital joints, in conjunction with the C3–C7 facet joints, permit a remarkable degree of flexion/extension ("yes" movements). The dens of C2 projects into the anterior part of the C1 ring, forming a pivot joint, enabling the head and C1 to rotate up to 80° ("no" movements). Such rotational capacity is permitted by the relatively horizontal orientation of the cervical facets. The C3–C6 vertebrae are similar; C7 is remarkable for its prominent *spinous process*, easily palpated. The anteriorly directed cervical curve and the extensive paracervical musculature preclude palpation of the other cervical spinous processes. The *vertebral arteries*, enroute to the brain stem, pass through foramina of the *transverse processes* of the upper six cervical vertebrae. These vessels are subject to stretching injuries with extreme cervical rotation of the hyperextended neck. The cervical vertebral canal conducts the cervical spinal cord and its coverings (not shown). The C4–C5 and C5–C6 motion segments are the most mobile of the cervical region and are particularly prone to disc/facet degeneration.

The twelve thoracic vertebrae—characterized by long, slender spinous processes, heart-shaped *bodies*, and nearly vertically oriented *facets*—articulate with *ribs* bilaterally. In general, each rib forms a synovial joint with two *demifacets* on the bodies of adjacent vertebrae and a single *facet* on the transverse process of the lower vertebra. Variations of these costovertebral joints are seen with T1, T11, and T12.

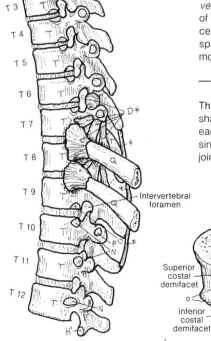

T 1
T 2 — Disc
T 3
T 4
T 5
T 6
T 7
T 8
T 9 — Intervertebral foramen
T 10
T 11
T 12

LATERAL VIEW

TYPICAL THORACIC (T5) VERTEBRA

Vertebral foramen
Superior articular process
Inferior articular process
Superior costal demifacet
LATERAL/ SUPERIOR VIEW
Inferior costal demifacet
Transverse costal facet
Outline of rib

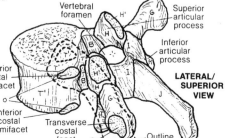

THORACIC VERTEBRA t

BODY t'
FACET n
DEMIFACET o
TRANSVERSE FACET p

RIB q
LIGAMENT r*

LUMBAR, SACRAL & COCCYGEAL VERTEBRAE

CN: Use the same colors as were used on the previous two plates for C, T, L, E, F, A, S, and Co. (1) Begin with the three large views of lumbar vertebrae. (2) Color the different planes of articular facets. (3) Color the four views of the sacrum and coccyx. Note that the central portion of the median section receives the vertebral canal color (E^1).

LUMBAR VERTEBRA L
VERTEBRAL FORAMEN E
VERTEBRAL CANAL E'
INTERVERTEBRAL FORAMEN F
INTERVERTEBRAL DISC A

The five lumbar vertebrae are the most massive of all the individual vertebrae, their thick processes securing the attachments of numerous ligaments and muscles/tendons. Significant flexion and extension of the lumbar and lumbosacral motion segments, particularly at L4-L5 and L5–S1, are possible. At about L1, the spinal cord terminates and the cauda equina (bundle of lumbar, sacral, and coccygeal nerve roots; see Plate 70) begins. The lumbar *intervertebral foramina* are large. Transiting nerve roots/sheaths take up only about 50% of the volume of these foramina. Disc and facet degeneration is common in the L4–L5 and L5–S1 segments; reduction of space for the nerve roots increases the risk of nerve root irritation/compression. Occasionally, the L5 vertebra is partially or completely fused to the sacrum (sacralized L5). The S1 vertebra may be partially or wholly non-fused (lumbarized S1), resulting in essentially six lumbar vertebrae and a sacrum of four fused vertebrae.

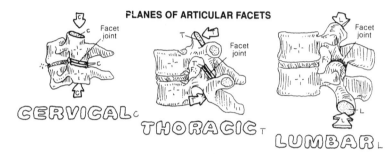

PLANES OF ARTICULAR FACETS

CERVICAL C   **THORACIC** T   **LUMBAR** L

The planes (orientation) of the articular facets determine the direction and influence the degree of motion segment movement. The plane of the *cervical facets* is angled coronally off the horizontal plane about 30°. Considerable freedom of movement of the cervical spine is permitted in all planes (sagittal, coronal, horizontal). The *thoracic facets* lie more vertically in the coronal plane and are virtually non-weightbearing. The range of motion here is significantly limited in all planes, less so in rotation. The plane of the *lumbar facets* is largely sagittal, resisting rotation of the lumbar spine, transitioning to a more coronal orientation at L5–S1. The L4–L5 facet joints permit the greatest degree of lumbar motion in all planes.

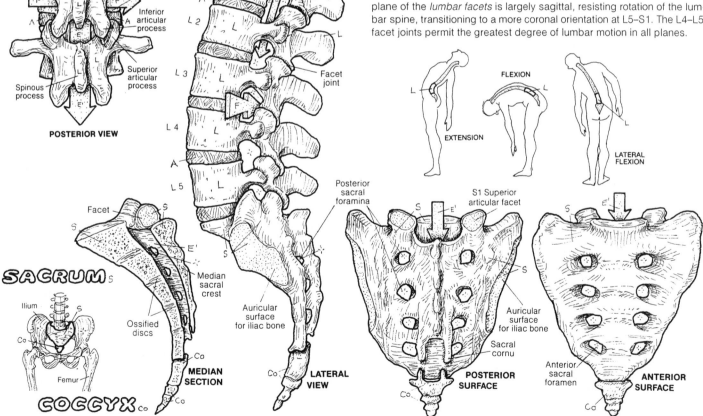

The sacrum consists of five fused vertebrae; the intervertebral discs are largely replaced by bone. The sacral (vertebral) canal contains the terminal sac of the dura mater (dural sac, thecal sac) to S2 and the sacral nerve roots, which transit the sacral foramina. The sacrum joins with the ilium of the hip bone at the auricular surface, forming the sacroiliac joint.

The sacrum and the ilia of the hip bones form an arch for the transmission and distribution of weightbearing forces to the heads of the femora. It is a strong arch, and the sacrum is its keystone. The coccyx consists of 2–4 tiny individual or partly fused, rudimentary vertebrae. The first coccygeal vertebra is the most completely developed.

BONY THORAX

CN: Use the same colors as were used on Plate 22 for true ribs, thoracic vertebrae, demifacets, and transverse process facets. Use bright colors for A–C. (1) Color the anterior view of the bony thorax. Color each rib completely before going on to the next. (2) Color the posterior view in the same manner. (3) Color the lateral view of the bony thorax. (4) When coloring the drawings of a rib and the sites of articulation, note that the rib facets (drawn with dotted lines) are to be colored even though they are on the underside of the rib.

STERNUM
MANUBRIUM A
BODY B
XIPHOID PROCESS C

12 RIBS
7 TRUE D
5 FALSE E
(2 FLOATING) E'

COSTAL CARTILAGE (10) F
THORACIC VERTEBRA (12) G

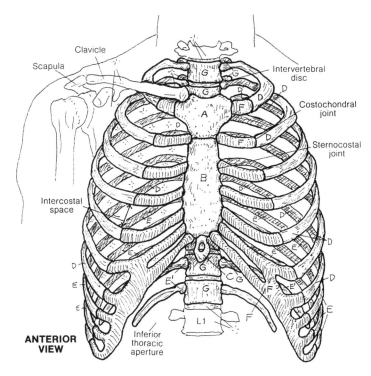

ANTERIOR VIEW

Clavicle · Scapula · Intervertebral disc · Costochondral joint · Sternocostal joint · Intercostal space · Inferior thoracic aperture

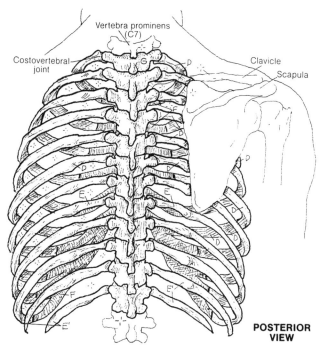

POSTERIOR VIEW

Vertebra prominens (C7) · Costovertebral joint · Clavicle · Scapula

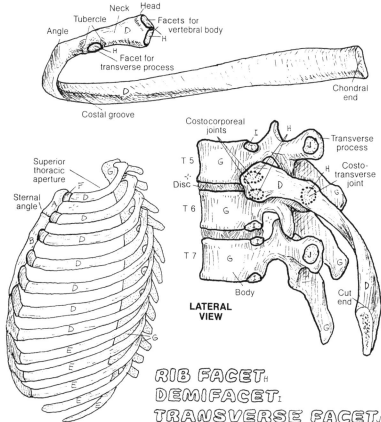

Neck · Head · Tubercle · Angle · Facets for vertebral body · Facet for transverse process · Costal groove · Chondral end

LATERAL VIEW

Superior thoracic aperture · Sternal angle · Costocorporeal joints · Transverse process · Costo-transverse joint · Disc · T 5 · T 6 · T 7 · Body · Cut end

RIB FACET H
DEMIFACET I
TRANSVERSE FACET J

The bony thorax is the skeleton of the chest, harboring the heart, lungs, and other significant organs. The superior thoracic aperture or thoracic inlet (often called thoracic outlet by surgeons) transmits the esophagus, trachea, nerves, and important ducts and vessels (Plate 104). The inferior thoracic aperture is virtually sealed by the thoracic diaphragm (muscle), through which pass the aorta, inferior vena cava, and esophagus (Plate 50). The region between each pair of ribs is the intercostal space, containing muscle, fasciae, vessels, and nerves (Plate 50). Collective rib movement is responsible for about 25% of the respiratory effort (inhalation, exhalation); the diaphragm does the rest (Plate 135).

The fibrocartilaginous joint between the *manubrium* and the body of the *sternum* (sternal angle, sternomanubrial joint) makes subtle hinge-like movements during respiration. The xiphoid makes a fibrocartilaginous (xiphisternal) joint with the body of the sternum. The sternum is largely cancellous bone containing red marrow. The *costal cartilages*, representing unossified cartilage models of the anterior ribs, articulate with the sternum by gliding-type synovial joints (sternocostal joints, except for the first joint, which is not synovial). All ribs form synovial joints with the thoracic vertebrae (costovertebral joints). Within each of these joints, the rib (2 through 9) forms a synovial joint with a demifacet of the upper vertebral body and with a *demifacet* of the lower body (costocorporeal joints). In addition, the tubercle of the rib articulates with a cartilaginous facet at the tip of the transverse process of the lower vertebra (costotransverse joint). Ribs 1, 10, 11, and 12 each join with one vertebra instead of two; ribs 11 and 12 have no costotransverse joints. *True ribs* (1–7) articulate directly with the sternum. Ribs 8–12 are called *false ribs*; ribs 8–10 articulate indirectly with the sternum (via cartilages connecting to the 7th costal cartilage) and ribs 11 and 12 (*floating ribs*) end in the muscular abdominal wall.

PECTORAL GIRDLE & ARM BONE

CLAVICLE_A

SCAPULA_B

HUMERUS_C

CN: Use very light colors in order to see surface detail. (1) Color each view completely before going on to the next.

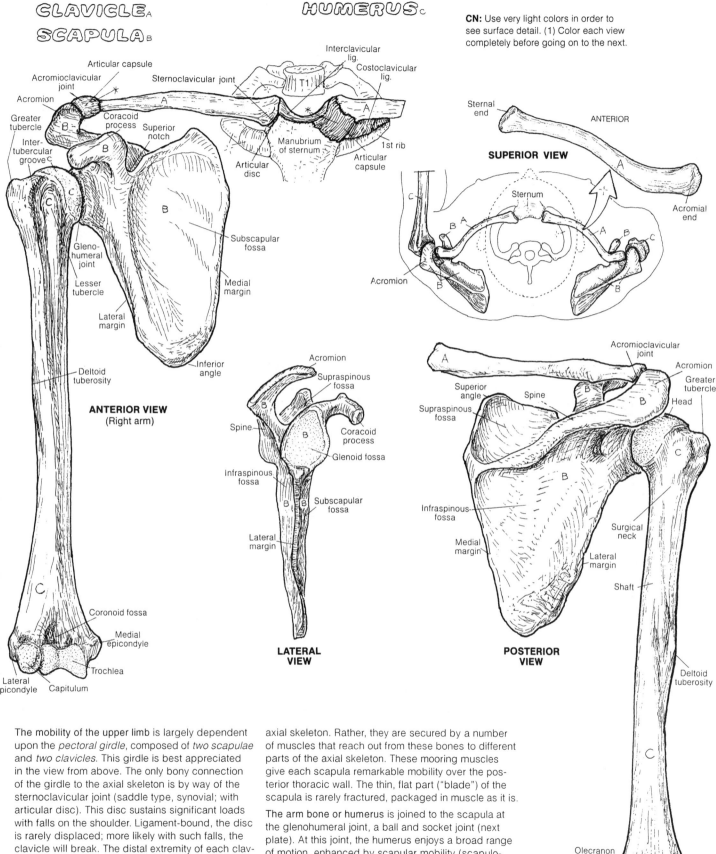

ANTERIOR VIEW
(Right arm)

SUPERIOR VIEW

LATERAL VIEW

POSTERIOR VIEW

The mobility of the upper limb is largely dependent upon the *pectoral girdle*, composed of *two scapulae* and *two clavicles*. This girdle is best appreciated in the view from above. The only bony connection of the girdle to the axial skeleton is by way of the sternoclavicular joint (saddle type, synovial; with articular disc). This disc sustains significant loads with falls on the shoulder. Ligament-bound, the disc is rarely displaced; more likely with such falls, the clavicle will break. The distal extremity of each clavicle articulates with the acromion of the scapula in a gliding type synovial acromioclavicular or AC joint. This joint is commonly disjointed (separated shoulder) with certain activities, an event not to be confused with shoulder joint dislocation.

The scapulae have no direct bony connection to the axial skeleton. Rather, they are secured by a number of muscles that reach out from these bones to different parts of the axial skeleton. These mooring muscles give each scapula remarkable mobility over the posterior thoracic wall. The thin, flat part ("blade") of the scapula is rarely fractured, packaged in muscle as it is.

The arm bone or humerus is joined to the scapula at the glenohumeral joint, a ball and socket joint (next plate). At this joint, the humerus enjoys a broad range of motion, enhanced by scapular mobility (scapulothoracic motion). Fractures of the humerus generally occur at the surgical neck, mid-shaft, or the distal extremity. The sharp sensation generated by striking the ulnar nerve under the medial epicondyle gives rise to the name "crazy bone." It is humorous that the humerus also is known as the "funny bone."

GLENOHUMERAL (SHOULDER) JOINT

CN: Use the same color that the scapula (A) and humerus (B) received on the preceding plate. Use light blue for (C). (1) The scapula (A) and humerus (B) are only to be colored in the upper right corner. (2) Color all ligaments gray. Because the glenohumeral ligaments are thickenings of the joint capsule (E), they are colored both gray and the color assigned to E. (3) The subdeltoid bursa and the subacromial bursa (F) are one continuous structure. (4) In the coronal section, part of the synovial membrane (G) has been removed to reveal the biceps brachii tendon where it attaches to the supraglenoid tubercle of the scapula.

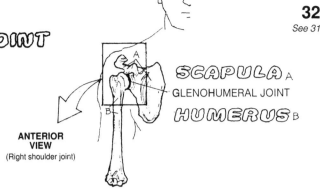

SCAPULA A
GLENOHUMERAL JOINT
HUMERUS B

ANTERIOR VIEW
(Right shoulder joint)

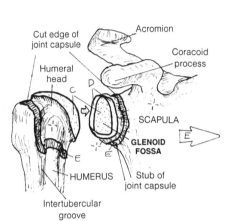

Cut edge of joint capsule
Acromion
Coracoid process
Humeral head
SCAPULA
GLENOID FOSSA
HUMERUS
Stub of joint capsule
Intertubercular groove

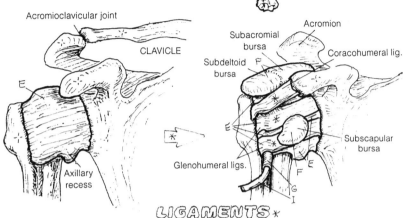

Acromioclavicular joint
CLAVICLE
Axillary recess

Acromion
Subacromial bursa
Subdeltoid bursa
Coracohumeral lig.
Subscapular bursa
Glenohumeral ligs.

LIGAMENTS *

ARTICULAR CARTILAGE c
GLENOID LABRUM D
JOINT CAPSULE E
BURSA F
SYNOVIAL MEMBRANE G
SYNOVIAL CAVITY H·
TENDON OF BICEPS BRACHII M. I

The glenohumeral (shoulder) joint is a synovial, ball-and-socket-type, multiaxial articulation between the glenoid fossa of the *scapula* and the head of the *humerus*. The shallow fossa is deepened by the *labrum* around its rim. The fossa and the humeral head are covered with a thin layer of articular (hyaline) cartilage. The connecting ends of the bones are encapsulated with a fibrous *joint capsule* lined internally with a synovial membrane and containing a small amount of synovial fluid. Usually isolated from the joint capsule and its cavity are numerous fibrous sacs of synovial fluid (*bursae*) intervening between muscle tendons crossing bones, tendons, and muscles. One of these, the subacromial bursa, has great clinical significance (see Plate 55).

The tendon of the long head of biceps brachii arises from the scapula's supraglenoid tubercle just above the 12 o'clock point of the glenoid labrum. Ensheathed by synovium, the tendon passes over the head of the humerus within the fibrous capsule and emerges below the capsule in the intertubercular groove to join the long head (muscle) of biceps brachii.

The fibrous joint capsule is lax (note the pouch of the axillary recess), yet incorporates thickened bands of glenohumeral ligaments. The capsule/ligament complex is reinforced by a musculotendinous cuff that offers great flexibility to shoulder motion. Movements of the glenohumeral joint can be see in Plates 55 and 56.

The joint suffers from overuse. Capsules become abnormally lax, the labrum becomes torn from its attachments, the tendon of biceps becomes frayed and torn, and the cavity of the joint may communicate with various bursae. Repetitive dislocations of the humeral head may induce articular cartilage damage.

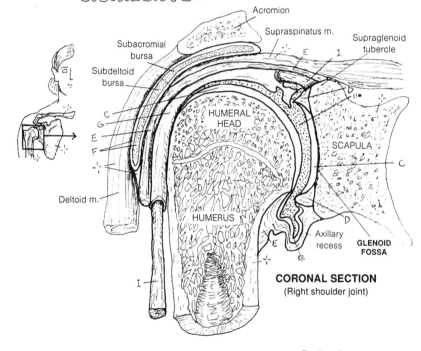

Acromion
Subacromial bursa
Supraspinatus m.
Supraglenoid tubercle
Subdeltoid bursa
HUMERAL HEAD
SCAPULA
Deltoid m.
HUMERUS
Axillary recess
GLENOID FOSSA

CORONAL SECTION
(Right shoulder joint)

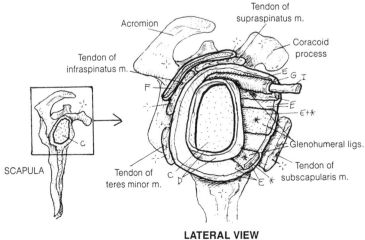

Acromion
Tendon of supraspinatus m.
Coracoid process
Tendon of infraspinatus m.
Glenohumeral ligs.
Tendon of subscapularis m.
SCAPULA
Tendon of teres minor m.

LATERAL VIEW
(Opened joint with humerus removed)

FOREARM BONES

ULNA A
RADIUS B

HUMERUS C

CN: Use very light colors for A and B, and the same color for the humerus (C) that was used on the preceding plate. Note that the distal humerus and carpal bones are left uncolored in the large illustrations. (1) Color the forearm bones in the three views, taking careful note of the callouts referring to surface details. (2) In the supination/pronation diagrams, the thumb and little finger of the hand receive the same colors as the forearm bones to which they relate, regardless of hand position.

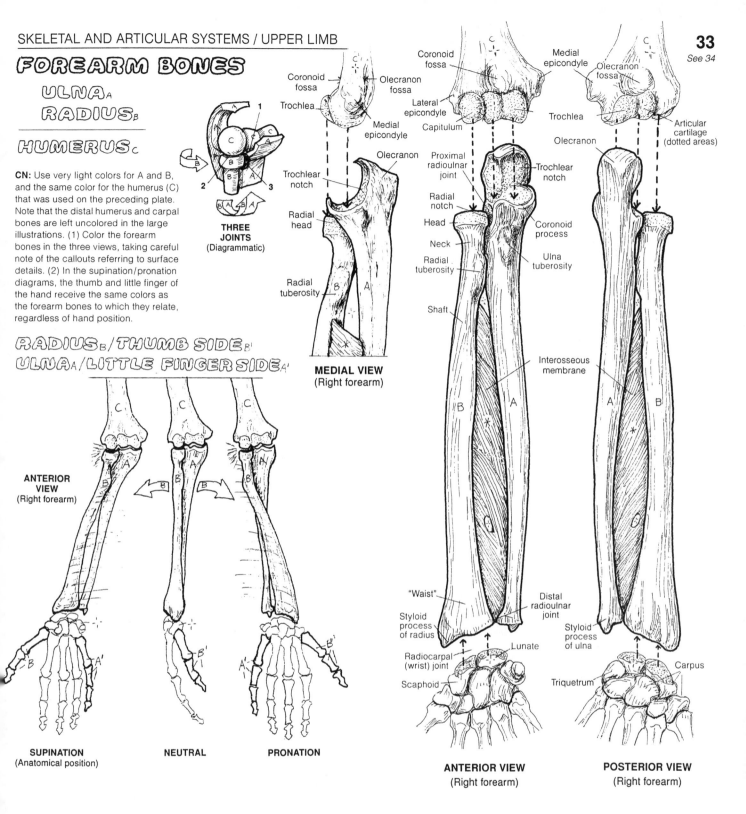

THREE JOINTS
(Diagrammatic)

MEDIAL VIEW
(Right forearm)

Coronoid fossa
Olecranon fossa
Trochlea
Medial epicondyle
Trochlear notch
Radial head
Radial tuberosity

Coronoid fossa
Medial epicondyle
Lateral epicondyle
Capitulum
Olecranon fossa
Trochlea
Olecranon
Proximal radioulnar joint
Radial notch
Head
Neck
Radial tuberosity
Shaft
Trochlear notch
Coronoid process
Ulna tuberosity
Interosseous membrane
"Waist"
Styloid process of radius
Radiocarpal (wrist) joint
Scaphoid
Lunate
Distal radioulnar joint

Medial epicondyle
Olecranon fossa
Articular cartilage (dotted areas)
Olecranon
Styloid process of ulna
Triquetrum
Carpus

ANTERIOR VIEW
(Right forearm)

POSTERIOR VIEW
(Right forearm)

RADIUS B / THUMB SIDE B'
ULNA A / LITTLE FINGER SIDE A'

ANTERIOR VIEW
(Right forearm)

SUPINATION
(Anatomical position)

NEUTRAL

PRONATION

The two bones of the forearm are quite different from one another. The posterior aspect of the proximal extremity of the *ulna* is characterized by a rather massive bone mass called the *olecranon*. You can feel it easily at the back of your elbow. On the anterior side of the olecranon is the *trochlear notch*, which articulates with the *trochlea of the humerus* at the *humeroulnar joint* (synovial; hinge). A part of this surface turns to face the *radius* (the radial head); this is the radial notch, which contributes to the *proximal radioulnar joint* (synovial; pivot). The ulnar shaft narrows distally to terminate as the head of the ulna. The head forms a pivot-type, synovial joint with the radius (*distal radioulnar joint*). This joint shares an articular disc that fits between the ulnar head and the lunate and trequetral bones of the wrist. This disc contributes to the radiocarpal (wrist) joint, but the ulnar head does not. The shaft of the ulna forms a movable, fibrous joint (syndesmosis) with the shaft of the radius by means of the interosseous membrane.

The radius has a small rounded head proximally, articulating with both the capitulum of the humerus (*radiohumeral joint*; synovial; pivot) and the radial notch of the ulna (proximal radioulnar joint). The shaft of the radius flares distally to form a broad wrist joint with the scaphoid and lunate bones of the carpus. Falls on the hands load the wrist joint and can cause a fracture of the radius at the relatively weak "waist" between the shaft and the flared distal extremity (Colles fracture, Smith fracture).

After coloring and studying the supination/pronation movements, put the palm of your right hand out in front of you, palm down (prone). In this position, the radius and ulna are in parallel. Place the fingers of the left hand on your right olecranon. Now supinate your right hand (to palm up). Notice the olecranon did not move. Thus, the ulna does not move during supination/pronation of the hand. Now find and observe the styloid process of the radius at the right wrist (on the thumb side) as you supinate/pronate the right hand. Note that the styloid process moves with the thumb. You have now demonstrated how the radius moves around the ulna during pronation and supination of the hand.

ELBOW JOINTS

CN: Use the same colors for the three bones as were used on 32 and 33. Use light blue for H. (1) Begin with the three joints of the elbow region. Note that each articulating surface (dotted) receives the color of its bone—in the lower, boxed-in illustration and in the sagittal view, those surfaces (H) are colored light blue. Color K yellow. (2) Color the remaining views of the joint capsule and ligaments.

ELBOW JOINT·
HUMERO-ULNAR ᴮ
RADIO-HUMERAL ᴀ
RADIO-ULNAR ᴮ

ANNULAR LIG. ᴅ ꜰ

ULNA ᴮ
ULNAR COLLATERAL LIG. ᴅ
HUMERUS ᴀ
RADIAL COLLATERAL LIG. ᴇ
RADIUS ᴄ

JOINT CAPSULE ɢ
ARTICULAR CARTILAGE ʜ
SYNOVIAL MEMBRANE ɪ
SYNOVIAL CAVITY ᴊ·
FAT PAD ᴋ
BURSA ʟ

The elbow joint consists of two separate articulations with the humerus: the *humeroulnar* and *radiohumeral* joints (synovial; hinge type). Movements of this joint are limited to flexion and extension. Note that the C-shaped, articular cartilage–lined trochlear notch of the ulna rotates around the pulley-shaped trochlea of the humerus during these movements. In extension, the upper part of the trochlear notch fits into the olecranon fossa of the humerus. In flexion, the coronoid process of the ulna fits into the coronoid fossa of the humerus (see Plate 33). The ligaments of the elbow joint—essentially, the radial and ulnar collateral ligaments—reinforce the fibrous joint capsule.

The joint between the radius and the ulna (*proximal radioulnar joint*) permits the radial head to pivot within the radial notch of the ulna. The ulna cannot pivot around anything because of the shape of the humeroulnar joint. Though the proximal radioulnar joint is not considered part of the elbow joint, its synovial cavity and fibrous joint capsule is continuous with that of the elbow joint, and it is secured by both radial and ulnar collateral ligaments. The annular ligament is attached at both ends to the sides of the radial notch of the ulna. It is more narrow below than above (i.e., it is beveled). It surrounds and secures the head (above) and the neck (below) of the radius and resists its displacement when the hand is pulled away from the shoulder. The lower part of the annular ligament is lined with synovial membrane; the upper part is fibrocartilaginous and is associated with the rotation of the radius at the proximal radioulnar joint. The joint capsule and the radial collateral ligament reinforce the retaining function of the annular ligament.

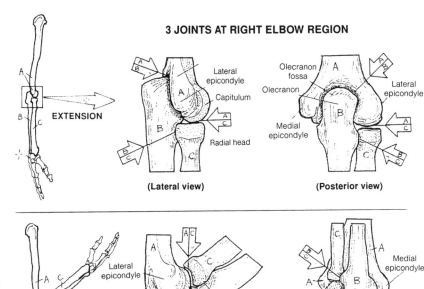

3 JOINTS AT RIGHT ELBOW REGION

EXTENSION

Lateral epicondyle

Capitulum

Radial head

Olecranon fossa

Olecranon

Lateral epicondyle

Medial epicondyle

(Lateral view)

(Posterior view)

Lateral epicondyle

FLEXION

Medial epicondyle

Olecranon

(Lateral view)

(Anterior view)

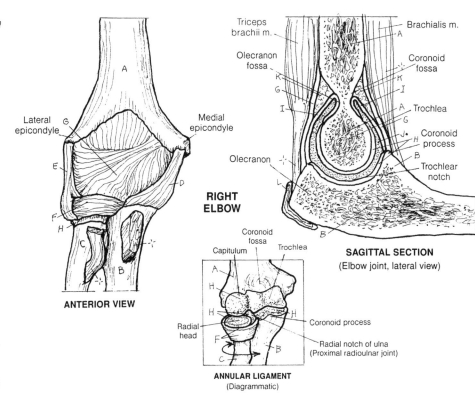

Lateral epicondyle

Medial epicondyle

ANTERIOR VIEW

RIGHT ELBOW

Triceps brachii m.

Olecranon fossa

Olecranon

Brachialis m.

Coronoid fossa

Trochlea

Coronoid process

Trochlear notch

SAGITTAL SECTION
(Elbow joint, lateral view)

Coronoid fossa

Capitulum

Trochlea

Radial head

Coronoid process

Radial notch of ulna
(Proximal radioulnar joint)

ANNULAR LIGAMENT
(Diagrammatic)

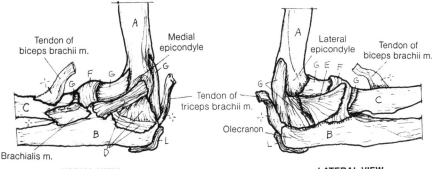

Tendon of biceps brachii m.

Medial epicondyle

Lateral epicondyle

Tendon of biceps brachii m.

Tendon of triceps brachii m.

Olecranon

Brachialis m.

MEDIAL VIEW

LATERAL VIEW

WRIST AND HAND BONES & JOINTS

CN: Use light colors other than those used for the three arm bones on the previous plates for I and J, light blue for K. (1) Color the three views of the hand and wrist: note the callouts identifying the joints that contribute to the movements shown in the satellite sketches. (2) Color the major ligaments of the wrist joints gray. Numerous carpal and phalangeal ligaments are not shown. (3) In the sectional view, color the bones and their articular cartilage (L). Color the synovial cavities (L with dark outlines) of the wrist black, but not the intercarpal joint cavities.

8 CARPALS:
SCAPHOID_A LUNATE_B TRIQUETRUM_C PISIFORM_D
TRAPEZIUM_E TRAPEZOID_F CAPITATE_G HAMATE_H
5 METACARPALS_I 14 PHALANGES_J

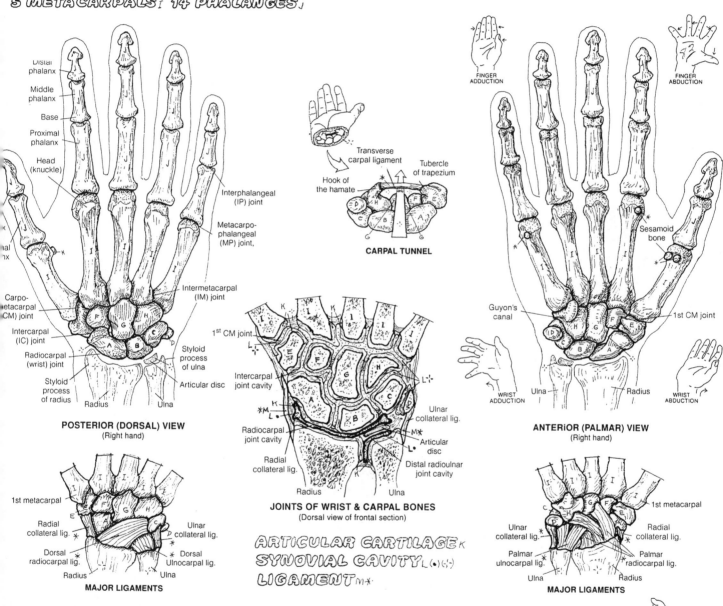

Distal phalanx
Middle phalanx
Base
Proximal phalanx
Head (knuckle)

Interphalangeal (IP) joint
Metacarpophalangeal (MP) joint

Intermetacarpal (IM) joint

Carpometacarpal (CM) joint
Intercarpal (IC) joint
Radiocarpal (wrist) joint
Styloid process of radius
Styloid process of ulna
Articular disc
Radius
Ulna

POSTERIOR (DORSAL) VIEW
(Right hand)

Transverse carpal ligament
Tubercle of trapezium
Hook of the hamate

CARPAL TUNNEL

1st CM joint
Intercarpal joint cavity
Radiocarpal joint cavity
Radial collateral lig.
Radius
Ulna
Ulnar collateral lig.
Articular disc
Distal radioulnar joint cavity

JOINTS OF WRIST & CARPAL BONES
(Dorsal view of frontal section)

ARTICULAR CARTILAGE_K
SYNOVIAL CAVITY_L (•)(-)
LIGAMENT_M *

FINGER ADDUCTION
FINGER ABDUCTION
Sesamoid bone
Guyon's canal
1st CM joint

ANTERIOR (PALMAR) VIEW
(Right hand)

WRIST ADDUCTION
WRIST ABDUCTION
Ulna
Radius

1st metacarpal
Radial collateral lig.
Dorsal radiocarpal lig.
Radius
Ulnar collateral lig.
Dorsal Ulnocarpal lig.
Ulna

MAJOR LIGAMENTS

1st metacarpal
Ulnar collateral lig.
Palmar ulnocarpal lig.
Ulna
Radial collateral lig.
Palmar radiocarpal lig.
Radius

MAJOR LIGAMENTS

The wrist joint (synovial; biaxial) is formed by the distal articular surface of the radius with the articular surfaces of the *scaphoid* and *lunate bones* primarily, and between the articular disc and the *triquetrum* secondarily. Movements here are flexion, extension, adduction, and abduction. The *wrist joint* and *carpal joints* are secured by palmar and dorsal radiocarpal and ulnocarpal ligaments and by radial and ulnar collateral ligaments. The *intercarpal joints*, between the proximal and distal rows of carpal bones, contribute to wrist movement. The trough between the *hamate* and *trapezium* bones anteriorly provides a carpal tunnel for the passage of the long flexor tendons to the thumb and fingers

as well as the median nerve. Compression by the transverse carpal ligament can irritate or depress the function of the median nerve (numbness to three radial fingers; thumb weakness). Guyon's canal transmits the ulnar artery and nerve.

Hand movement involves movements of the metacarpophalangeal (MP) and interphalangeal (IP) joints primarily, and among the carpometacarpal and intermetacarpal joints secondarily—with one exception: the unique first carpometacarpal (CM) joint (synovial; saddle). Notice the mobility it gives the thumb, as in opposing thumb and little finger, and circumduction of the thumb.

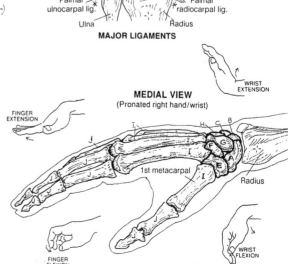

MEDIAL VIEW
(Pronated right hand/wrist)

WRIST EXTENSION
FINGER EXTENSION
1st metacarpal
Radius
FINGER FLEXION
WRIST FLEXION

BONES/JOINTS IN REVIEW

(See appendix A in the back of the book for answers)

CN: In doing this review you are asked to *write* in the name of the bone with the same color used for each bone in the two large illustrations. Since this is a review and not a test, you may wish to use the colors that were used for these bones in Plates 31, 33, and 35? Use a new color for carpals (F). (1) Color the arrows pointing to places where bones can be seen or palpated. (3) Write as many names in black pencil of the numbered joints you can recall.

A _____
B _____
C _____
D _____
E _____
F _____
G _____
H _____

BONE
SURFACE
MARKINGS

**ANTERIOR
VIEW**
(Right upper limb)

**POSTERIOR
VIEW**
(Right upper limb)

BONE
SURFACE
MARKINGS

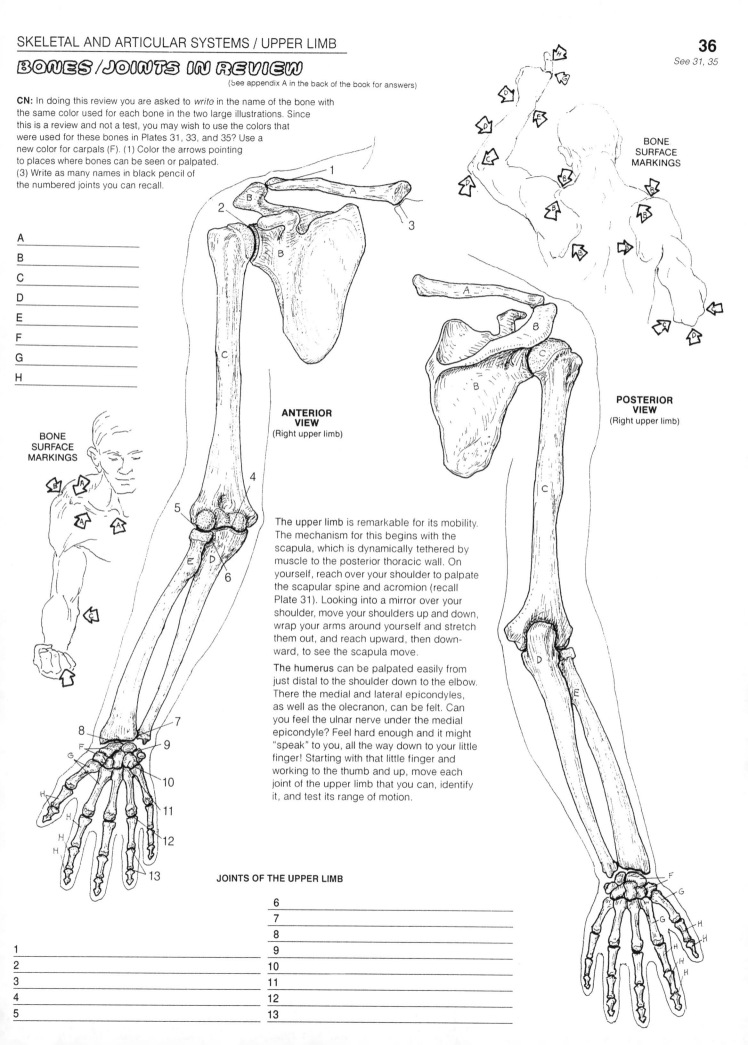

The upper limb is remarkable for its mobility. The mechanism for this begins with the scapula, which is dynamically tethered by muscle to the posterior thoracic wall. On yourself, reach over your shoulder to palpate the scapular spine and acromion (recall Plate 31). Looking into a mirror over your shoulder, move your shoulders up and down, wrap your arms around yourself and stretch them out, and reach upward, then downward, to see the scapula move.

The humerus can be palpated easily from just distal to the shoulder down to the elbow. There the medial and lateral epicondyles, as well as the olecranon, can be felt. Can you feel the ulnar nerve under the medial epicondyle? Feel hard enough and it might "speak" to you, all the way down to your little finger! Starting with that little finger and working to the thumb and up, move each joint of the upper limb that you can, identify it, and test its range of motion.

JOINTS OF THE UPPER LIMB

6 _____
7 _____
8 _____
9 _____
10 _____
11 _____
12 _____
13 _____

1 _____
2 _____
3 _____
4 _____
5 _____

HIP BONE, PELVIC GIRDLE & PELVIS

CN: Use very light colors for bones A–D. The sacrum (D) combines with A–C (below) to form the pelvis. (1) Color the diagrammatic representations of the basins of the false and true pelves. They are shown together in the lower right diagram.

HIP BONE
   ILIUM A
   ISCHIUM B
   PUBIS C
SACRUM D COCCYX D'
FALSE PELVIS E
   PELVIC INLET F
TRUE PELVIS G
   PELVIC OUTLET H

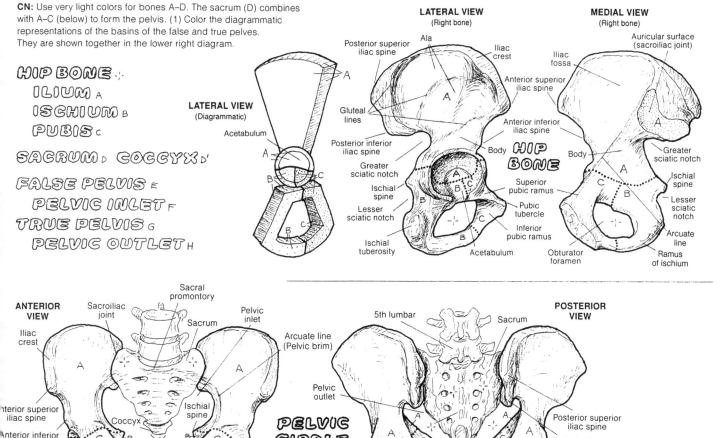

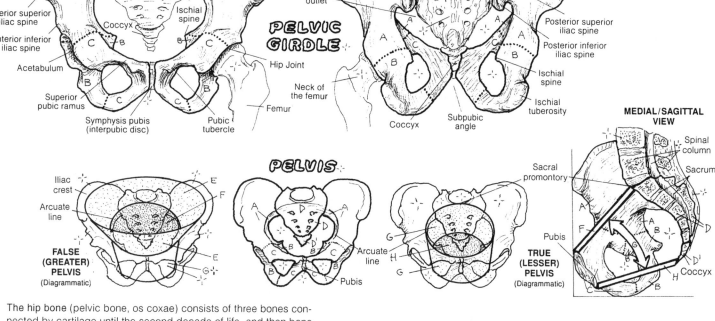

The hip bone (pelvic bone, os coxae) consists of three bones connected by cartilage until the second decade of life, and then bone thereafter: the *ilium*, *ischium*, and *pubis*. The hip bone has been likened to a propeller: the *acetabulum*, the socket for the hip joint where all three bones are fused together, is the hub. The flattened wing (ala) of the ilium would be one blade of the propeller, and the ischiopubic bone would be the other blade. The weight of the torso and upper limbs is transmitted from the sacroiliac joint to the acetabulum through the body of the ilium. The posterior and inferior ischium and the anterior and inferior pubis form a ring of bone with the obturator foramen in the center. The ischium is significant for its ischial tuberosity, upon which one sits. The pubis is easily palpable centrally at the level of the groin.

The two hip bones are connected anteriorly by the *symphysis pubis* (interpubic joint; cartilage/fibrocartilage, with cartilaginous disc). These two bones constitute the pelvic girdle. With respect to the concept of "girdle," the ischiopubic bones are somewhat similar in shape and function to the clavicle, and the iliac bones to the two scapulae. Because of its weight-bearing function, the pelvic girdle is considerably less mobile than its pectoral counterpart, which had a mobility function.

The two hip bones and the sacrum constitute the pelvis. The cavity of the pelvis (basin) consists of a false (greater) and a true (lesser) pelvis. The orientation of the pelvis can be appreciated by placing a bony pelvis in the laboratory/classroom against a vertical wall such that the anterior superior iliac spine and the pubic tubercle are in contact with the wall simultaneously. That part of the pelvis below an oblique line from the *sacral promontory*, forward and downward along the *arcuate lines* of the ilium, to the *pubic crest* (floor of the pubic tubercle) is the true pelvis. The line just described demarcates the *pelvic inlet* (superior pelvic aperture). The pelvic inlet is continuous above with the abdominal cavity, which includes the greater pelvis. The anterior wall of the greater pelvis is entirely muscular; confirm this on yourself. The true pelvic cavity has both bony and muscular walls and contains numerous structures (Plates 157, 160). The plane of the inferior pelvic aperture (*pelvic outlet*), along a line from the inferior aspect of the pubis to the tip of the coccyx, is much more horizontal than that of the inlet; the floor of the outlet is muscular (Plate 52). The pelvic cavity is continuous below with the perineum (Plate 53).

MALE & FEMALE PELVES

CN: Use very light colors for A and B. Use the same colors on structures C–F as were used on the preceding page where they were labeled E–H. (1) Carefully color the diagrammatic basins representing the false and true pelves. (2) Color the female pelvis (dotted) superimposed on the dark outline of the male pelvis (not colored). Color the two examples of subpubic angles (G). (3) In the lateral view, note the slightly forward tilt of the female pelvis, which accentuates the curve of sacrum/coccyx and lifts it away from the pelvic outlet (shown from below).

MALE PELVIS A
SACRUM / COCCYX A'
FEMALE PELVIS B
SACRUM / COCCYX B'

FALSE PELVIS C
PELVIC INLET D
TRUE PELVIS E
PELVIC OUTLET F

SUB-PUBIC ANGLE G

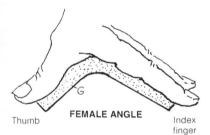

Thumb   **FEMALE ANGLE**   Index finger

Middle finger   **MALE ANGLE**   Index finger

MALE A
Iliac crest
Arcuate line

ANTERIOR VIEW
(Diagrammatic)

FEMALE B

ANTERIOR VIEW
(Diagrammatic)

MALE/FEMALE PELVES

Female and male bony pelves often differ. These differences have been investigated and analyzed for many reasons, including forensic identification of bodies, gynecologic evaluations, and anthropologic research. A primary interest of pelvic dimensions and physical characteristics is in the clinical prenatal examination. It is obviously important that the birth canal be unimpeded for the fetus at birth. Obstetric-related measurements of pelvic diameters are accomplished radiologically (pelvimetry).

In general, female pelves are wider than male pelves in all dimensions. Females have a wider *subpubic angle* than males. This angle can be measured easily on a laboratory skeleton by placing the hand against the pubis such that the thumb covers one inferior pubic ramus and the index finger covers the other. If the angle created by these two digits is superimposed rather precisely over the subpubic angle of the pelvis being measured, it is probably a female pelvis. If the subpubic angle fits between index and middle fingers, it is probably a male pelvis.

When two different pelves are compared side by side, female pelves tend to have broader *true* and *false pelves* than male pelves. The pelvic inlet and outlet is generally larger in women. The space between the ischial tuberosities is greater in females, as is the space between ischial spines and the ischial spine and the sacrum. There tends to be a larger *sacral curvature* in females, as well as a larger sciatic notch.

Posture, bone conditions such as osteomalacia, and a number of other factors can influence pelvic shape and pelvic capacities.

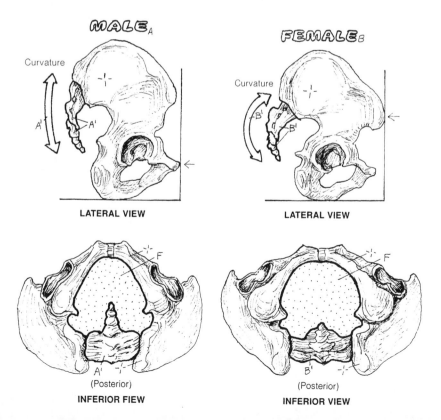

MALE A
Curvature

LATERAL VIEW

FEMALE B
Curvature

LATERAL VIEW

(Posterior)
INFERIOR FIEW

(Posterior)
INFERIOR VIEW

SACROILIAC & HIP JOINTS

CN: Use light colors for A, C, and L and light blue for B. The hip bone (C) is a fusion of the illium, ischium, and pubic bones (studied separately on Plate 37). All contribute to the hip joint. (1) The upper left inset shows only one of the two auricular surfaces of the sacrum; the partial arrow (B^1) points to the unseen surface. The synovial membrane (H) of the hip joint is shown only in the large view (where the femur is displaced). (2) In the lower two views, use color for the relevant ligaments that appear among the titles, while coloring the remaining ligaments gray.

SACROILIAC JOINT

SACRUM A
AURICULAR SURFACE B
HIP BONE C
AURICULAR SURFACE B'
SYNOVIAL CAVITY J.
INTEROSSEOUS SACROILIAC LIG. D
POSTERIOR SACROILIAC LIG. E
ANTERIOR SACROILIAC LIG. F

HIP JOINT

HIP BONE C
ACETABULUM
ACETABULAR LABRUM G
ARTICULAR CARTILAGE B²
SYNOVIAL MEMBRANE H
LIGAMENTUM TERES I
SYNOVIAL CAVITY J.
FEMUR K
ARTICULAR CARTILAGE B³
JOINT CAPSULE L
ILIOFEMORAL LIG. M
ISCHIOFEMORAL LIG. N
PUBOFEMORAL LIG. O

ADDITIONAL PELVIC LIGS. *

The sacroiliac joint is a significant load-bearing articulation. The auricular surfaces of the illium and sacrum are roughened and cartilage-lined: the sacral surface is hyaline; the iliac surface is fibrocartilaginous and thinner. Only the lower half of the joint is synovial with a cavity; the upper half is ligamentous. A fibrous capsule surrounds the entire joint. The cavity becomes smaller in later life, and the joint surfaces may fuse with advanced age. Movement of the joint is controversial; some anterior and posterior movement, with rotation, has been described. This motion may be increased during pregnancy. Movement is sharply limited by the irregularity of the articular surfaces and by the dense, thick posterior sacroiliac and the thinner anterior sacroiliac ligaments. The joint, its ligaments, and crossing muscles are implicated in the painful "sacroiliac syndrome". Inflammation of the synovial part of the joint (sacroiliitis) is well recognized in many auto-immune-related diseases (e.g., ankylosing spondylitis, rheumatoid arthritis, and inflammatory bowel disease).

The hip joint is a ball and socket, synovial joint between the acetabulum of the hip bone and the head of the femur. The joint permits flexion, extension, adduction, abduction, medial and lateral rotation, and circumduction. Each joint surface is lined with articular cartilage; that of the acetabulum is C-shaped. The incomplete bony socket of the acetabulum is completed by the transverse acetabular ligament and is enhanced by a 360° fibrocartilaginous labrum. The joint is encapsulated; the three strong iliofemoral, ischiofemoral, and pubofemoral ligaments reinforce this fibrous capsule. Arising within the acetabulum between the arms of the acetabular cartilage is the ligament of the head of the femur (lig. teres). It offers little resistance to forced distraction, but it does transmit vessels to the femoral head. An adequate blood supply to the joint requires both femoral circumflex vessels in addition to the vessels in the ligamentum teres.

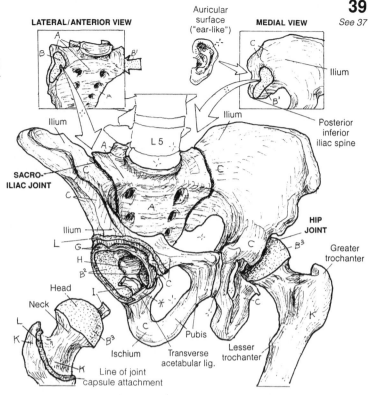

LATERAL/ANTERIOR VIEW OF JOINT LOCATIONS

FRONTAL SECTION OF SACROILIAC JOINT

LATERAL VIEW

FRONTAL SECTION OF HIP JOINT

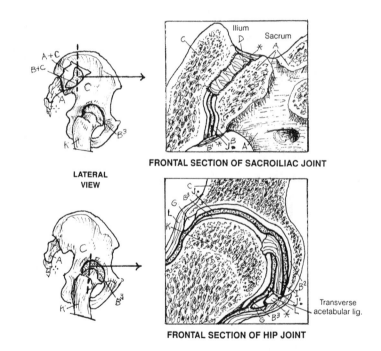

PELVIC LIGAMENTS

THIGH & LEG BONES
FEMUR_A TIBIA_B
FIBULA_C
PATELLA_D

CN: Use light colors on the four bones in order to study surface detail. (1) After coloring the two main views, color gray many of the more superficial ligaments, tendons, and muscle attachments that stablize the region of the knee. Although not distinguishable in the illustrations, the ligaments tend to be less thick and well defined compared to the tendons and muscles—important underlying structures that are introduced in the knee joint plate that follows.

ANTERIOR VIEW
(Right limb)

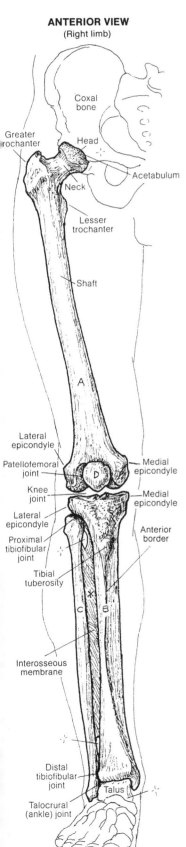

POSTERIOR VIEW
(Right limb)

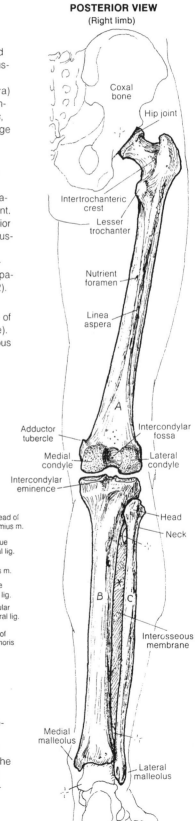

The bone of the thigh is the femur; the bones of the leg are the *tibia* and *fibula*. The *greater* and *lesser trochanters* are the site of insertion of muscles of the hip. The shaft, gently curved anteriorly along its length, is rounded circumferentially, except posteriorly where a ridge (linea aspera) along the long axis of the bone forms the origin and insertions of a number of muscles. Distally, the shaft widens to form the massive *condyles*, which contribute to the knee joint. The patella articulates with the cartilage of the femur between the two condyles. It is a sesamoid bone that is located within the tendon of quadriceps femoris (see next plate).

The major weight-bearing bone of the leg is the tibia. It is the only bone of the leg that contributes to the knee joint. This stout bone has large condyles proximally that articulate with the femoral condyles. The palpable tibial tubercle just distal to the condyles receives the patellar ligament. The tibial shaft is triangular in cross section; the apex is the sharp anterior border (shin), easily palpated. The anteromedial surface is barren of muscle; the anterolateral surface is muscle-covered. The expanded, distal extent of the tibia forms an inverted L (¬); the horizontal surface articulates with the talus of the ankle, and the vertical portion is the quite palpable medial malleolus, which also articulates with the talus (see Plate 42).

Not directly weight bearing, the fibula is a site of muscular attachment along the upper two-thirds of its shaft. Its head joins with the underside of the lateral tibial condyle (proximal tibiofibular joint; synovial, plane type). The shaft of the fibula forms an intermediate tibiofibular joint (interosseous membrane; syndesmosis) with the shaft of the tibia. Distally, the fibula joins with the tibia (distal tibiofibular joint; syndesmosis). The lateral aspect of the fibula is the palpable lateral malleolus, which articulates with the talus. The distal extremities of the fibula and tibia form a joint with the talus (ankle or talocrural joint); see Plate 42.

LIGAMENTS/TENDONS/MUSCLES AROUND RIGHT KNEE

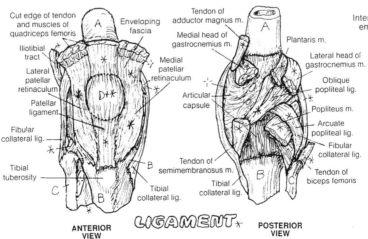

The bony parts of the knee joint provide little security during knee movement (see next plate). Tendons and muscles crossing and moving the joint also have the function of reinforcing the ligamentous stabilizers of the knee. Fibrous expansions from the medial and lateral members of the quadriceps muscle merge with the fibrous capsule on each side of the patella to form the medial and lateral retinacula. Muscles/tendons reinforcing knee stability can be seen on this plate and Plates 62–66.

KNEE JOINT

CN: The femur, tibia, fibula, and patella bones are not to be colored. (1) In the sagittal section, color (A) blue and (B) black. The synovial membrane that lines the cavity is not shown. (3) In the anterior view, color the facets on the posterior surface of the patella. (4) Color relationship between attachments and function of cruciate ligaments (E, E^1).

ARTICULAR CARTILAGE A
SYNOVIAL CAVITY B •
JOINT CAPSULE C
BURSA D
CRUCIATE LIG., ANT. E / POST. E'
MENISCUS, LAT. F / MED. F'
PATELLAR LIG. G
COLLATERAL LIG., TIBIAL H / FIBULAR H'

The knee joint consists of *two condylar synovial joints* between the femoral and tibial condyles, and a *gliding synovial joint* between the patella and the femur. Note that the fibula and the tibiofibular joint are not part of the knee joint. The movements of the knee joint, involving essentially flexion and extension with varying degrees of rotation and gliding, can be seen in Plates 62 and 64.

In the sagittal view of the joint, note the *articular cartilage*–lined patellofemoral articulation. The patella is a sesamoid bone that developed in the tendon of the quadriceps femoris muscle. It resists wear-and-tear stresses on the tendon during knee flexion and extension. Note the two facets of the patella in the anterior view and the corresponding patellar articular surface on the femur. The various bursae shown are variable in size. The *suprapatellar bursa* is an extension of the synovial joint cavity.

The fibrous (joint) capsule is incomplete around the joint, reinforced by ligaments where absent or deficient and replaced by the patella anteriorly. The synovial membrane (not shown) lines the internal surface of the fibrous capsule; it does not cover the *menisci* and the joint surfaces or the posterior fibrous capsule.

The menisci can be seen from the side in the sagittal view and from above in the superior view of the joint. They are semilunar-shaped fibrocartilaginous discs attached to the tibial condyles by ligaments; they enhance the depth of fit of the femoral condyles. The ends of the menisci (horns) are attached in the tibial intercondylar region. These horns are richly innervated, a fact reinforced to one experiencing a painful tear of the posterior horn of the medial meniscus. The medial meniscus is more fixed on the tibia than is the lateral. Thus, it is less flexible and more easily torn in the face of excessive rotation and forced abduction of the knee joint while bearing weight.

The knee joint is without bony security. It is secured by ligaments and the tendons of the muscles that cross it, including the tendon of quadriceps femoris anteriorly and the iliotibial tract and the tendon of biceps femoris laterally (Plate 62), the muscles popliteus and semimembranosus posteriorly (Plate 66), and the tendons of sartorius, gracilis, and semitendinosus (pes anserinus) medially (Plate 66). See also Plate 40.

The ligaments are particularly important in limiting ranges of motion of the knee and securing the menisci. The *collateral ligaments* resist unwanted sideward movements at the knee. The *anterior cruciate* is named for its anterior tibial attachment, the *posterior cruciate* for its posterior tibial attachment. In their ascent proximally, they cross (crux, cross). The anterior cruciate passes posteriorly and laterally to end on the posteromedial aspect of the lateral femoral condyle; the posterior cruciate passes anteriorly and medially to end on the medial aspect of the medial femoral condyle. The cruciates essentially resist forward/backward displacement of the tibia/femur; indeed, a torn cruciate ligament generally results in excessive anteroposterior movements of the tibia on the femur.

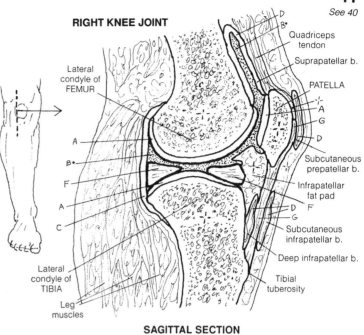

RIGHT KNEE JOINT

SAGITTAL SECTION

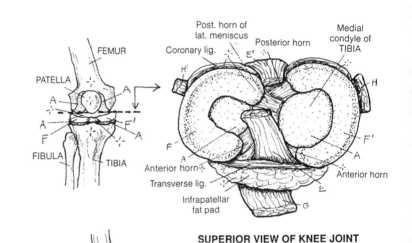

SUPERIOR VIEW OF KNEE JOINT

Posterior cru. lig. resists posterior gliding of tibia

Anterior cru. lig. resists anterior gliding of tibia

CRUCIATE "CROSSED" LIGAMENTS

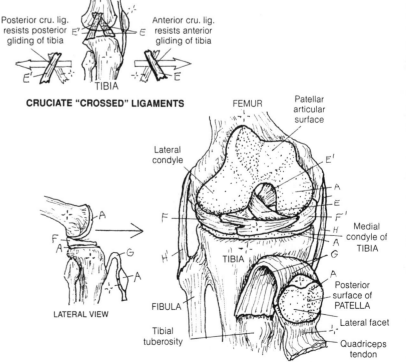

LATERAL VIEW

ANTERIOR VIEW OF EXPOSED JOINT

ANKLE & FOOT BONES

7 TARSALS:
TALUS_A CALCANEUS_B CUBOID_C
NAVICULAR_D CUNEIFORMS (3)_E
5 METATARSALS_F
14 PHALANGES_G

CN: Use different colors from those used for the hip bone on Plate 38 and for the femur, tibia, fibula, and patella on Plate 40. (1) Color the illustration and diagram of the ankle joint. (2) Then start with the talus (A); color that bone wherever it appears on the plate. Follow that procedure with each of the other bones. (3) Color gray all of the ligaments.

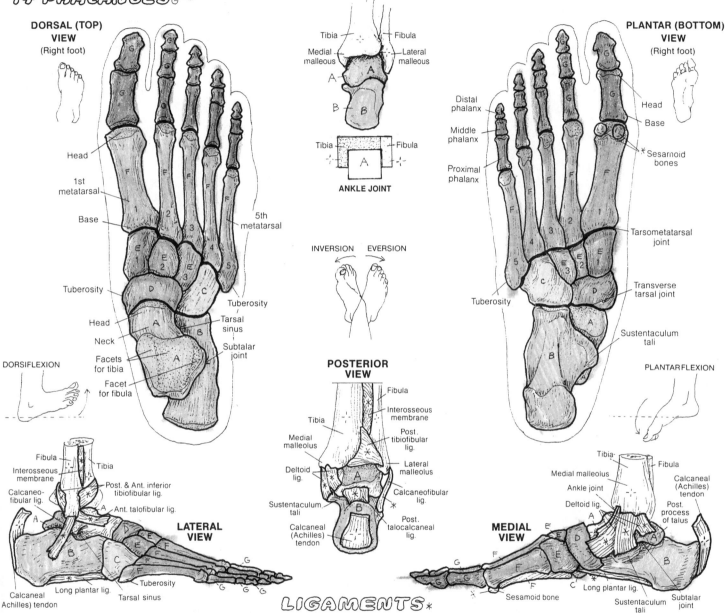

DORSAL (TOP) VIEW (Right foot)

Head
1st metatarsal
Base
5th metatarsal
Tuberosity
Head
Neck
Facets for tibia
Facet for fibula
Tuberosity
Tarsal sinus
Subtalar joint

DORSIFLEXION

Tibia — Fibula
Medial malleolus — Lateral malleolus
ANKLE JOINT

Tibia — Fibula

INVERSION   EVERSION

POSTERIOR VIEW

Fibula
Interosseous membrane
Tibia
Post. tibiofibular lig.
Medial malleolus
Lateral malleolus
Deltoid lig.
Calcaneofibular lig.
Sustentaculum tali
Post. talocalcaneal lig.
Calcaneal (Achilles) tendon

PLANTAR (BOTTOM) VIEW (Right foot)

Head
Base
Distal phalanx
Middle phalanx
Proximal phalanx
* Sesamoid bones
Tarsometatarsal joint
Transverse tarsal joint
Sustentaculum tali
Tuberosity

PLANTAR FLEXION

LATERAL VIEW

Fibula
Tibia
Interosseous membrane
Calcaneofibular lig.
Post. & Ant. inferior tibiofibular lig.
Ant. talofibular lig.
Tuberosity
Long plantar lig.
Tarsal sinus
Calcaneal (Achilles) tendon

MEDIAL VIEW

Tibia
Fibula
Medial malleolus
Calcaneal (Achilles) tendon
Ankle joint
Deltoid lig.
Post. process of talus
Long plantar lig.
Subtalar joint
Sesamoid bone
Sustentaculum tali

LIGAMENTS *

The foot is a mobile, weight-bearing structure. The *ankle joint* (hinge-type synovial joint) between the *tibia, fibula,* and *talus* forms a mortise, permitting only flexion (plantar flexion) and extension (dorsiflexion) here. With excessive rotation of this joint, characteristic fractures and torn ligaments occur. The foot can adjust to walking/running on tilted surfaces by virtue of the *subtalar* (talocalcaneal) and transverse *tarsal* (talocalcaneonavicular and calcaneocuboid) joints. Here inversion and eversion movements occur. The ankle has strong medial ligamentous (deltoid ligaments) and weaker lateral

ligamentous support. The relatively high frequency of inversion sprains (tearing the lateral ligaments) over eversion sprains seems to reflect this relative weakness. The bony architecture of the foot includes a number of arches that are reinforced and maintained by ligaments and influenced by muscles. The *medial longitudinal arch* transmits the force of body weight to the ground when standing and to the great toe in locomotion, creating a giant lever that gives spring to the gait. *Both longitudinal arches* function in absorbing shock loads and balancing the body.

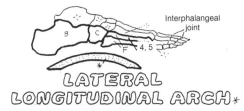

Interphalangeal joint

LATERAL LONGITUDINAL ARCH *

TRANSVERSE ARCH *

Metatarsophalangeal joint

MEDIAL LONGITUDINAL ARCH *

BONES / JOINTS IN REVIEW

(See appendix A in the back of the book for answers)

CN: Use light colors throughout. (1) Begin with the bones of the lower limb. Write the name, on the appropriate line, with the same color that you use to color the bone. Also color and name the corresponding bone of the upper limb. (2 Color the arrows pointing to the surface markings of the bones of the lower limb. (3) In black pencil, write-in the names of the joints of the lower limb. (4) Use the same colors on the fore- and hindlimbs of a quadruped. Keep in mind that these bones have the same names as their human counterparts.

See 39, 40, 42

43

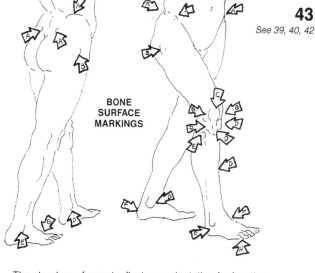

BONE SURFACE MARKINGS

LOWER LIMB

A _____
B _____
C _____
D _____
E _____
F _____
G _____
H _____

UPPER LIMB

A¹ _____
B¹ _____
D¹ _____
E¹ _____
F¹ _____
G¹ _____
H¹ _____

The structure of a part reflects an adaptation for function. This statement is borne out by comparing the bones of the upper and lower limbs in a biped (human) with those of a quadruped. The pectoral girdle provides a basis for mobility; the more sturdy pelvic girdle provides stability in both loco-motion and weight bearing. The limb bones of the lower limb are large and solid, consistent with weight-bearing function; the related joints are structurally secure, except the knee, which gives up stability for flexibility. In the upper limb, the bones are lighter and the joints are more flexible and capa-ble of greater ranges of motion (compare shoulder with hip, elbow with knee, wrist with ankle). Although the forearm and leg have two bones each, there is little functional correlation between those pairs of bones. The foot is clearly adapted for locomotion and weight bearing, the hand (especially the thumb) for mobility and dexterity.

JOINTS OF THE LOWER LIMB

1 _____
2 _____
3 _____
4 _____
5 _____
6 _____
7 _____
8 _____
9 _____
10 _____
11 _____
12 _____

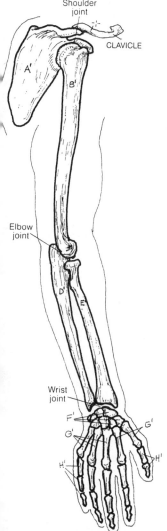

Shoulder joint

CLAVICLE

Elbow joint

Wrist joint

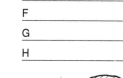

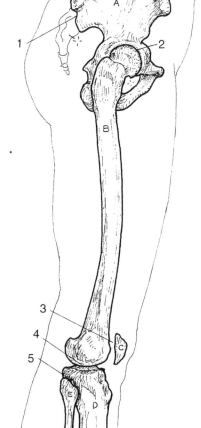

QUADRUPED (Dog)

Most mammals walk on their "fingers" and "toes."

HIND-LIMB

FORE-LIMB

INTRODUCTION TO SKELETAL MUSCLE

CN: Use light colors for A–E. (1) Begin with the muscle belly and tendons in the upper illustration. (2) When coloring the narrow borders of the endomysium (C) in the enlarged section, it is recommended that you also color over the muscle fiber ends (D) with the very light endomysium color, and then go back over the fiber ends with a darker color (D). Do not color the neurovascular bundle, or the cut ends of blood vessels and capillaries. (3) Color the lower illustration.

SKELETAL MUSCLE
BELLY
FASCIA:
EPIMYSIUM
PERIMYSIUM
ENDOMYSIUM
MUSCLE FIBER (CELL)
TENDON

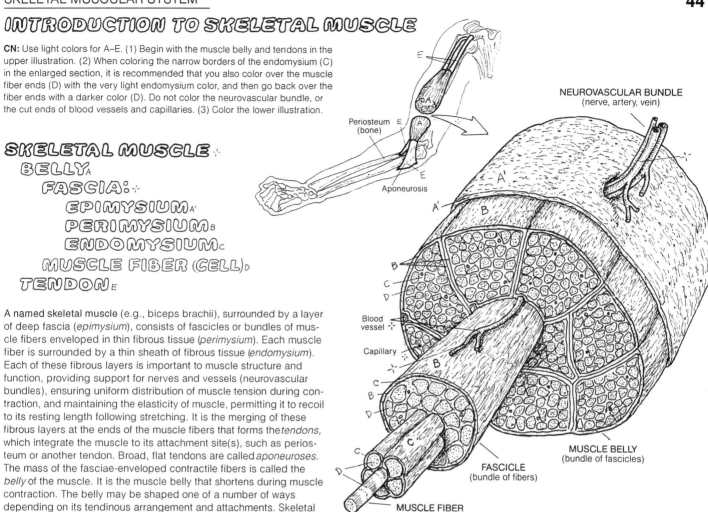

NEUROVASCULAR BUNDLE
(nerve, artery, vein)

Periosteum
(bone)

Aponeurosis

Blood
vessel

Capillary

MUSCLE BELLY
(bundle of fascicles)

FASCICLE
(bundle of fibers)

MUSCLE FIBER
(single cell)

A named skeletal muscle (e.g., biceps brachii), surrounded by a layer of deep fascia (*epimysium*), consists of fascicles or bundles of muscle fibers enveloped in thin fibrous tissue (*perimysium*). Each muscle fiber is surrounded by a thin sheath of fibrous tissue (*endomysium*). Each of these fibrous layers is important to muscle structure and function, providing support for nerves and vessels (neurovascular bundles), ensuring uniform distribution of muscle tension during contraction, and maintaining the elasticity of muscle, permitting it to recoil to its resting length following stretching. It is the merging of these fibrous layers at the ends of the muscle fibers that forms the *tendons,* which integrate the muscle to its attachment site(s), such as periosteum or another tendon. Broad, flat tendons are called *aponeuroses.* The mass of the fasciae-enveloped contractile fibers is called the *belly* of the muscle. It is the muscle belly that shortens during muscle contraction. The belly may be shaped one of a number of ways depending on its tendinous arrangement and attachments. Skeletal muscles are named in relation to their attachments (e.g., hyoglossus), shape (e.g., trapezius), number of heads (e.g., quadriceps), function (e.g., adductor magnus), or position (e.g., brachialis).

MECHANICS OF MOVEMENT
FULCRUM (JOINT)
EFFORT (MUSCLE)
RESISTANCE (WEIGHT)

Skeletal muscles employ simple machines, such as levers, to increase the efficiency of their contractile work about a joint. Mechanically, the degree of *muscular effort* required to overcome resistance to movement at a *joint* (*fulcrum*) depends upon the force of that resistance (*weight*); the relative distances from the anatomical fulcrum to the anatomical sites of *muscular effort*; and the anatomical sites of *resistance* (joints). The position of the joint relative to the site of muscle pull and the site of imposed load determines the class of the lever system in use.

1ST CLASS LEVER

In a 1st class lever, the joint lies between the muscle and the load. This is the most efficient class of lever. By flexing the neck and posturing the head forward and downward, the load (G^1) is appreciably increased (due to gravity), and the muscular effort (A) to hold that posture may induce muscle pain and stiffness/tightness (overuse).

2ND CLASS LEVER

In a 2nd class lever, the load lies between the joint and the pulling muscle. This lever system operates in lifting a wheelbarrow (the wheel is the fulcrum) as well as lifting a 75 kg (165 lb) body onto the metatarsal heads at the metatarsophalangeal joints. This is a relatively easy task for the strong calf (triceps surae) muscles; but try standing on the heads of your middle phalanges (increasing the distance F^1-G^1)!

3RD CLASS LEVER

In a 3rd class lever, the muscle lies between the joint and the load and has a poor mechanical advantage here. Consider the difference in muscular effort required to carry a 45 kg (100 lb) bag of cement in your hands with flexed elbows (elbow joint: 3rd class lever) and carrying your 75 kg (165 lb) body on the heads of your metatarsals (2nd class lever at the metatarsophalangeal joints). It is all a matter of leverage.

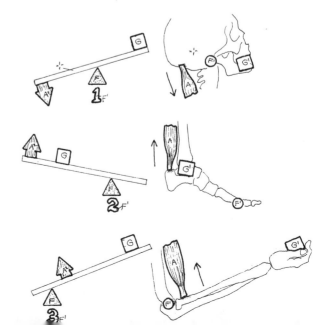

INTEGRATION OF MUSCLE ACTION

CN: Use a bright color for A and a light one for E. (1) Color the small arrows and the large letters of origin (O) and insertion (I) adjacent to the examples of contracted and stretched muscles. (2) In the lower illustration, color the portions of pronator teres and pronator quadratus that are outlined by dotted lines. These parts of the muscles are normally concealed by the radius in this lateral view.

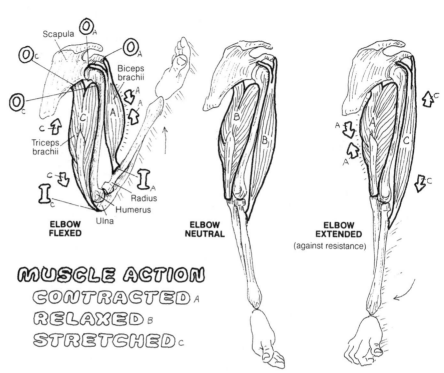

MUSCLE ACTION
CONTRACTED A
RELAXED B
STRETCHED C

Skeletal muscle generally connects two bones and crosses the joint between those two bones. When the muscle shortens (contracts), the two bones come closer together, isometric contraction excepted. Muscles never push; they always pull. In any given movement between two bones, one bone is generally fixed, and the other moves. The muscle attachment at the fixed bone is the *origin*; the attachment at the moving bone is the *insertion*. In complex movements where it is difficult to identify a "fixed" bone, the origin of the muscle is the more proximal attachment.

When a muscle contracts across a joint, other muscles crossing that joint are affected. No one muscle acts alone in joint movement. In flexion of the elbow joint, for example, biceps brachii (and brachialis, not shown) *contract*s, while triceps brachii is *stretched*. Conversely, in elbow extension, triceps is contracted, and the biceps/brachialis muscles are stretched. In neutral, all three are *relaxed* (at rest). Tense (contracted) muscles can often be relaxed by gentle stretching.

ACTORS IN ELBOW FLEXION

No muscle acts alone in the movement of a joint. In the movements shown at right, various muscles are functionally integrated in the simple act of lifting an object, with the forearm supinated in the first case and pronated in the second case.

PRIME MOVER (AGONIST) A'

The primary muscle effecting a desired joint movement is called the *prime mover* (agonist). There may be more than one; in elbow flexion with the forearm supinated, brachialis and biceps brachii are both prime movers; biceps adds significantly to the lifting power because of the added work in supinating the radius during elbow flexion. With the forearm pronated and supination resisted, the biceps loses that supinating power, and brachialis, unaffected by a pronated forearm, becomes the prime mover.

ANTAGONIST C'

Muscles that potentially or actually oppose or resist a certain movement are called *antagonists*. In the illustrations at right, triceps is the antagonist in the act of elbow flexion, even though it is being stretched and is not contracted in the case illustrated.

FIXATOR D

Fixator muscles stabilize the more proximal joints during weightbearing functions of the more distal joints. Here the trapezius muscle contracts to stabilize (immobilize) the scapula, creating a rigid platform (the scapula) for operation of the weightbearing, ipsilateral limb.

NEUTRALIZER (SYNERGIST) E

In undertaking a desired and specific movement, undesired movements are resisted by *neutralizers* (*synergists*). During flexion of the elbow with a pronated forearm, pronators of the forearm (pronator quadratus, pronator teres) contract to resist or neutralize supination of the forearm. In this action, the pronators are synergistic with the desired movement.

Globally integrated and harmonious muscle functioning makes possible painless, rhythmic, and dynamic movements, best revealed in such activities as dance, sports, and exercise. Joints affected by tense or weak interacting muscles, induced by mechanically disadvantaged posture/gait, can be subject to painful and limited movements.

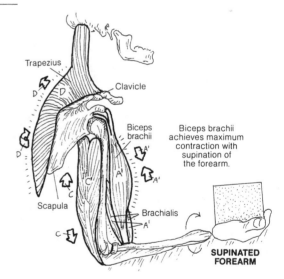

Biceps brachii achieves maximum contraction with supination of the forearm.

ACTORS IN ELBOW FLEXION
(With supination vs. pronation)

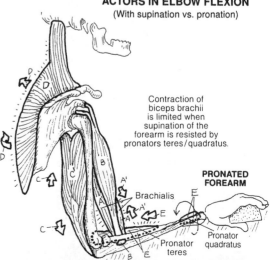

Contraction of biceps brachii is limited when supination of the forearm is resisted by pronators teres/quadratus.

MUSCLES OF FACIAL EXPRESSION

CN: Use your lightest colors for O and Q. Use warm and cheerful colors for the muscles producing a smile (A–H). Color the muscles reflecting sadness (I–O) with greens, blues, and grays. (1) Begin with the smiling side, and color only the muscles identified by titles A–H. Color those muscles in the profile view below. (2) Repeat the process with the sad side. Note that a portion of frontalis (I) has been cut away to reveal corrugator supercilii (J). (3) Color the titles at the bottom and the related muscles on the lower view. Include the portions of the auricular muscles that disappear beneath the ear.

ORBICULARIS OCULI ᴀ

NASALIS ʙ

LEVATOR LABII SUPERIORIS ALAEQUE NASI ᴄ

LEVATOR LABII SUPERIORIS ᴅ

LEVATOR ANGULI ORIS ᴇ

ZYGOMATICUS MAJOR ꜰ

ZYGOMATICUS MINOR ɢ

RISORIUS ʜ

FRONTALIS ɪ

CORRUGATOR SUPERCILII ᴊ

ORBICULARIS ORIS ᴋ

DEPRESSOR ANGULI ORIS ʟ

DEPRESSOR LABII INFERIORIS ᴍ

MENTALIS ɴ

PLATYSMA ᴏ

The muscles of facial expression are generally thin, flat bands arising from a facial bone or cartilage and inserting into the dermis of the skin or the fibrous tissue enveloping the sphincter muscles of the orbit or mouth. These muscles are generally arranged into the following regional groups: (1) the epicranial group (*occipitofrontalis* moving the scalp); (2) the orbital group (*orbicularis oculi, corrugator supercilii*); (3) the nasal group (*nasalis, procerus*); (4) the oral group (*orbicularis oris, zygomaticus major* and *minor*, the *levators* and the *depressors* of the lips and angles of the mouth, *risorius, buccinator*, and part of *platysma*); and (5) the group moving the ears (*auricular muscles*). The general function of each of these muscles is to move the skin wherever they insert. As you color each muscle, try contracting it on yourself while looking into a mirror, and see what develops.

Orbicularis oculi and oris are sphincter muscles, tending to close the skin over the eyelids and tighten the lips, respectively. Contractions of the cheek muscle buccinator makes possible rapid changes in volume of the oral cavity, as in playing a trumpet or squirting water. The nasalis muscle has both compressor and dilator parts, which influence the size of the anterior nasal openings.

BUCCINATOR ᴘ
GALEA APONEUROTICA ᴏ
OCCIPITALIS ʀ
AURICULAR MUSCLES ꜱ
PROCERUS ᴛ

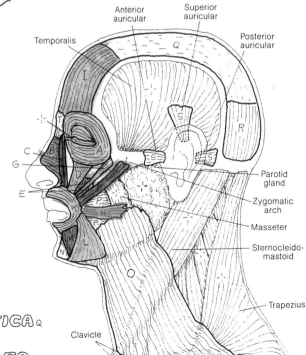

MUSCLES OF MASTICATION

CN: Use a yellowish "bone" color for the mandible (E). (1) Begin with the upper illustration and proceed to the two cut-away views exposing the muscles of mastication. On the smaller skull, two colors (A + E) are needed to indicate the insertion of the temporalis on the inner side of the mandible. Three colors (A + B + E) are needed to color the part of the external surface where the broad insertion of the masseter also covers part of the representation of the temporalis on the underside. (2) Color the directional arrows and muscles involved in moving the mandible.

MUSCLES:
TEMPORALIS A
MASSETER B
MEDIAL PTERYGOID C
LATERAL PTERYGOID L
MANDIBLE E

The act of chewing is called mastication. The muscles of mastication move the temporomandibular joint and are largely responsible for elevation, depression, protrusion, retraction, and lateral motion of the mandible. These muscles function bilaterally to effect movements of the single bone (mandible) at two joints. Chewing motions are a product of the action of the elevator muscles on one side combined with the contraction of the lateral pterygoid muscle on the opposite side.

The temporalis and masseter muscles are often contracted unconsciously (clenching teeth) during stress, giving rise to potentially severe bitemporal and preauricular headaches. The muscles can easily be palpated when contracted. Masseter, on the external surface of the ramus of the mandible, is easily palpated there. Place your fingers there and then contract the muscle (clench the teeth). Temporalis, on the other hand, inserts on the internal surface of the coronoid process and can be best palpated at the side of the head. Its dense fascia prevents the bulging you experienced with masseter.

The medial and lateral pterygoids are in the infratemporal fossa and cannot be palpated.

The muscles of mastication are all innervated by branches of the 5th cranial nerve (trigeminal), mandibular division. The muscles of facial expression (recall previous plate), on the other hand, are all supplied by the 7th cranial nerve (facial).

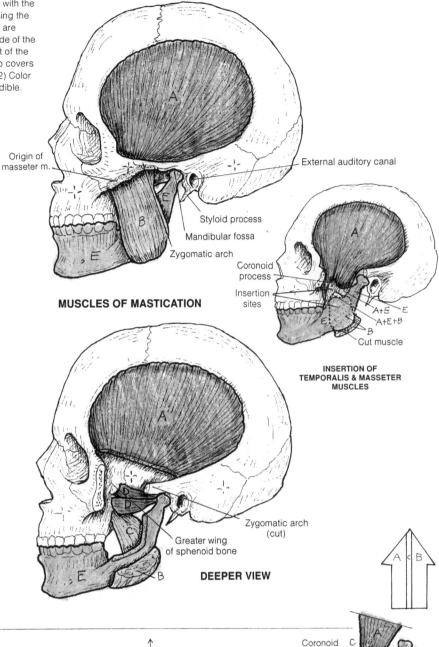

Origin of masseter m.

External auditory canal

Styloid process

Mandibular fossa

Zygomatic arch

MUSCLES OF MASTICATION

Coronoid process

Insertion sites

A+E E
A+E+B
B
Cut muscle

INSERTION OF TEMPORALIS & MASSETER MUSCLES

Zygomatic arch (cut)

Greater wing of sphenoid bone

DEEPER VIEW

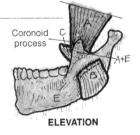

ACTION OF MUSCLES ON THE MANDIBLE

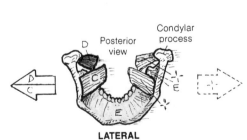

Posterior view

Condylar process

LATERAL

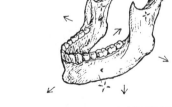

Coronoid process

A+E

ELEVATION

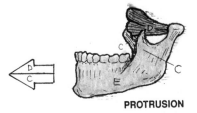

PROTRUSION

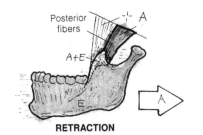

Posterior fibers

A+E

RETRACTION

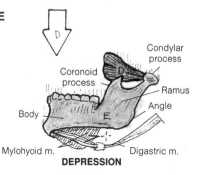

Condylar process

Coronoid process

Ramus

Angle

Body

Mylohyoid m.

Digastric m.

DEPRESSION

ANTERIOR & LATERAL MUSCLES

CN: Except for Band E, use your lightest colors throughout the plate. (1) Begin with the diagrams of the triangles of the neck and the sternocleidomastoid (A, B, C). Color over all the muscles within the triangles. (2) Then work top and bottom illustrations simultaneously, coloring each muscle in as many views as you can find it. Note the relationship between muscle name and attachment.

The neck is a complex tubular region of muscles, viscera, vessels, and nerves surrounding the cervical vertebrae. The muscles of the neck are arranged in superficial and deep groups. Here we concentrate on superficial muscles. The most superficial posterior and posterolateral muscle of the neck is trapezius (Plate 54). The deep posterior muscles are covered in Plate 49. The most superficial anterior muscle of the neck is platysma (Plate 46). The anterior and lateral muscle groups are divided into triangular areas by the *sternocleidomastoid* muscle.

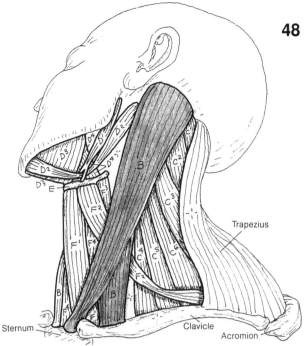

ANTERIOR TRIANGLE A

SUPRAHYOID MUSCLES D-

STYLOHYOID D' DIGASTRIC D²
MYLOHYOID D³ HYOGLOSSUS D⁴
GENIOHYOID D⁵

HYOID BONE E

INFRAHYOID MUSCLES F-

STERNOHYOID F' OMOHYOID F²
THYROHYOID F³ STERNOTHYROID F⁴

STERNOCLEIDOMASTOID B

The sternocleidomastoid muscle, acting unilaterally, tilts the head laterally on the same side while simultaneously rotating the head and pulling the back of the head downward, lifting the chin, and rotating the front of the head to the opposite side. Both muscle bellies acting together move the head forward (anteriorly) while extending the upper cervical vertebrae, lifting the chin upward.

The anterior region of the neck is divided in the midline; each half forms an *anterior triangle*. The borders of the anterior triangle of superficial neck muscles are clearly illustrated. The *hyoid bone*, suspended from the styloid processes of the skull by the stylohyoid ligaments, divides each anterior triangle into upper *suprahyoid* and lower *infrahyoid* regions.

The suprahyoid muscles arise from the tongue (glossus), mandible (mylo-, genio-, anterior digastric), and skull (stylo-, posterior digastric) and insert on the hyoid bone. They elevate the hyoid bone, influencing the movements of the floor of the mouth and the tongue, especially during swallowing. With a fixed hyoid, the suprahyoid muscles, especially the digastrics, depress the mandible.

The infrahyoid muscles generally arise from the sternum, thyroid cartilage of the larynx, or the scapula (omo-) and insert on the hyoid bone. These muscles partially resist elevation of the hyoid bone during swallowing. *Thyrohyoid* elevates the larynx during production of high-pitched sounds; sternohyoid depresses the larynx to assist in production of low-pitched sounds.

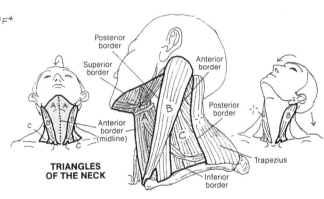

TRIANGLES OF THE NECK

POSTERIOR TRIANGLE C

SEMISPINALIS CAPITIS C'
SPLENIUS CAPITIS C²
LEVATOR SCAPULAE C³
SCALENUS: ANT. C⁴ MED. C⁵ POST. C⁶

The posterior triangle consists of an array of muscles covered by a layer of deep (investing) cervical fascia just under the skin between sternocleidomastoid and trapezius. The borders of the triangle are clearly illustrated. Muscles of this region arise from the skull and cervical vertebrae; they descend to and insert upon the upper two ribs (*scalenes*), the upper scapula (*omohyoid, levator scapulae*), and the cervical/thoracic vertebral spines (*splenius capitis, semispinalis capitis*). These muscles' function becomes clear when you visualize their attachments.

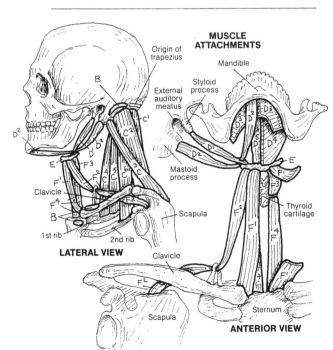

MUSCLE ATTACHMENTS

LATERAL VIEW

ANTERIOR VIEW

DEEP MUSCLES OF BACK & POSTERIOR NECK

CN: Use your lightest colors on the B and C groups. Note that splenius (A) and semispinalis (C¹) represent more than one muscle; the muscle subsets are identified. (1) After coloring the muscles of the back and posterior neck, color the lower right diagram, which describes the location and function of the deep movers of the spine.

The deep muscles of the back and posterior neck extend, rotate, or laterally flex one or more of the 24 paired facet joints and the 22 intervertebral disc joints of the vertebral column. The long muscles move several motion segments (recall Plate 27) with one contraction, while the short muscles can move one or two motion segments at a time (see intrinsic movers).

COVERING MUSCLE
SPLENIUS A

The splenius muscles extend and rotate the neck and head in concert with the opposite sternocleidomastoid muscle. Splenius capitis covers the deeper muscles of the upper spine.

VERTICAL MUSCLES
ERECTOR SPINAE B
SPINALIS B¹
LONGISSIMUS B²
ILIOCOSTALIS B³

The erector spinae group comprises the principal extensors of the vertebral motion segments. Oriented vertically along the longitudinal axis of the back, they are thick, quadrilateral muscles in the lumbar region, splitting into smaller, thinner separate bundles attaching to the ribs (iliocostalis), and upper vertebrae and head (longissimus, spinalis). Erector spinae arises from the lower thoracic and lumbar spines, the sacrum, ilium, and intervening ligaments.

OBLIQUE MUSCLES
TRANSVERSOSPINALIS GROUP C
SEMISPINALIS C¹
MULTIFIDUS C²
ROTATORES C³

The transversospinalis group extends the motion segments of the back, and rotates the thoracic and cervical vertebral joints. These muscles generally run from the transverse processes of one vertebra to the spine of the vertebra above, spanning three or more vertebrae. The semispinales are the largest muscles of this group, reaching from mid-thorax to the posterior skull; the multifidi consist of deep fasciculi spanning 1–3 motion segments from sacrum to C2; the rotatores are well defined only in the thoracic region.

DEEPEST MUSCLES
INTERTRANSVERSARII D
INTERSPINALIS E
SUBOCCIPITAL MUSCLES F

These small, deep-lying muscles cross the joints of only one motion segment. They are collectively major postural muscles. Electromyographic evidence has shown that these short muscles remain in sustained contraction for long periods of time during movement and standing/sitting postures. They are most prominent in the cervical and lumbar regions. The small muscles set deep in the posterior, suboccipital region (deep to semispinalis and erector spinae) rotate and extend the joints between the skull and C1 and C2 vertebrae.

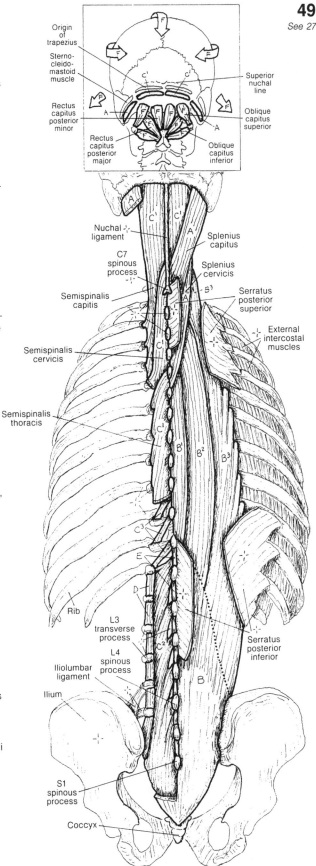

Origin of trapezius
Sterno-cleido-mastoid muscle
Rectus capitus posterior minor
Rectus capitus posterior major
Superior nuchal line
Oblique capitus superior
Oblique capitus inferior

Nuchal ligament
C7 spinous process
Semispinalis capitis
Semispinalis cervicis
Semispinalis thoracis
Splenius capitus
Splenius cervicis
Serratus posterior superior
External intercostal muscles
Rib
L3 transverse process
L4 spinous process
Iliolumbar ligament
Ilium
Serratus posterior inferior
S1 spinous process
Coccyx

INTRINSIC MOVERS
EXTENSOR E
ROTATOR C³
LATERAL FLEXOR D

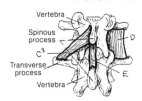

Vertebra
Spinous process
Transverse process
Vertebra

MUSCLES OF THORAX & POSTERIOR ABDOMINAL WALL

CN: Use blue for E and red for G. (1) You may wish to darken the underside of the diaphragm (A) in the anterior view. Do not confuse the arcuate ligaments with the 12th rib. (2) In the cross-sectional view at upper right, color the broken lines that represent transparent, membranous portions of the intercostal muscles.

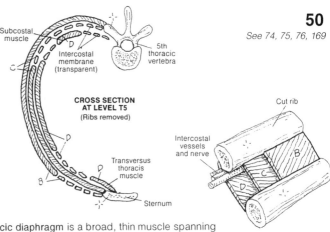

CROSS SECTION
AT LEVEL T5
(Ribs removed)

Subcostal muscle

Intercostal membrane (transparent)

5th thoracic vertebra

Transversus thoracis muscle

Sternum

Cut rib

Intercostal vessels and nerve

INTERCOSTAL MUSCLE FIBER ORIENTATION

THORAX MUSCLES:

THORACIC DIAPHRAGM ₐ
EXTERNAL INTERCOSTAL ʙ
INTERNAL INTERCOSTAL c
INNERMOST INTERCOSTAL ᴅ

The thoracic diaphragm is a broad, thin muscle spanning the thoracoabdominal cavity; the illustration shows much of its origin (all except the lower six ribs).

The left and right halves of the diaphragm insert into each other (central tendon). The diaphragm is responsible for 75% of the respiratory air flow. Openings (hiatuses) in the diaphragm provide passage for major transiting structures.

The intercostal muscles alter the dimensions of the thoracic cavity by collectively moving the ribs, resulting in 25% of the total respiratory effort. The specific function of each of these muscles, with respect to fiber orientation, is not understood. The innermost intercostals are an inconstant layer, and here include the transversus thoracis and subcostal muscles.

INFERIOR VENA CAVA ᴇ
ESOPHAGUS ꜰ
AORTA ɢ

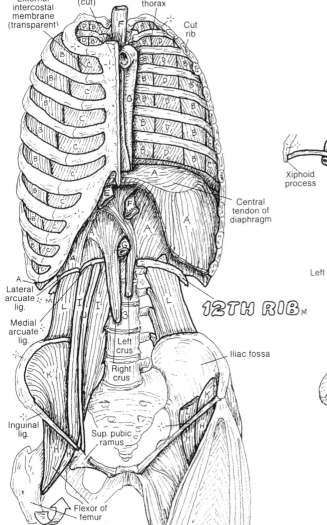

External intercostal membrane (transparent)

Sternum (cut)

Inner view of posterior thorax

Cut rib

Central tendon of diaphragm

Lateral arcuate lig.

Medial arcuate lig.

Left crus

Right crus

12TH RIB ᴍ

Iliac fossa

Inguinal lig.

Sup. pubic ramus

Flexor of femur

ANTERIOR VIEW

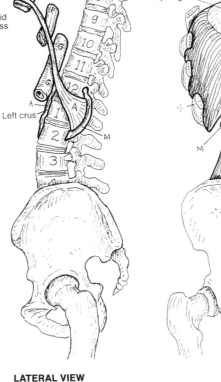

Xiphoid process

Left crus

LATERAL VIEW

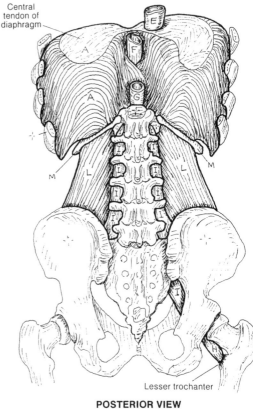

Central tendon of diaphragm

Lesser trochanter

POSTERIOR VIEW

POSTERIOR ABDOMINAL WALL MUSCLES:

ILIOPSOAS н
PSOAS MAJOR ɪ MINOR ᴊ
ILIACUS ᴋ
QUADRATUS LUMBORUM ʟ

The tendons of iliacus and psoas major converge to a single insertion (*iliopsoas*). Iliopsoas, a strong flexor of the hip joint, is a powerful flexor of the lumbar vertebrae; a weak psoas may contribute to low back pain. *Quadratus lumborum* is an extensor of the lumbar vertebrae (bilaterally) and a lateral flexor unilaterally. It functions in respiration by securing the 12th rib. Immediately anterior to these muscles is the retroperitoneum (see Plate 147).

MUSCLES OF ANTERIOR ABDOMINAL WALL & INGUINAL REGION

CN: Use a dark color for J and bright ones for B and I. (1) Color the 3 layers of the abdominal wall. (2) Color the sheath of the rectus abdominis in the lower left illustration gray. Color the two locator arrows gray in this and the upper illustration. (3) Beginning with J and K, and followed by H, color the coverings of the spermatic cord.

The anterior abdominal wall consists of three layers of flat muscles, the tendons (aponeuroses) of which interlace in the midline, and a vertically oriented pair of segmented muscles that are ensheathed incompletely by the aponeuroses of the three flat muscles (*sheath of the rectus abdominis*). The flat muscles arise from the lateral aspect of the torso (inguinal ligament, iliac crest, thoracolumbar fascia, lower costal cartilages, ribs). The lowest fibers of *external oblique* roll inwardly to form the *inguinal ligament*. These three muscles act to compress the abdominal contents during expiration, urination, and defecation. They assist in maintaining pressure on the curve of the low back, resisting "sway back" (excess lumbar lordosis) and extension of the low back.

ANTERIOR ABDOMINAL WALL
TRANSVERSUS ABDOMINIS A
RECTUS ABDOMINIS B
INTERNAL OBLIQUE C
EXTERNAL OBLIQUE D

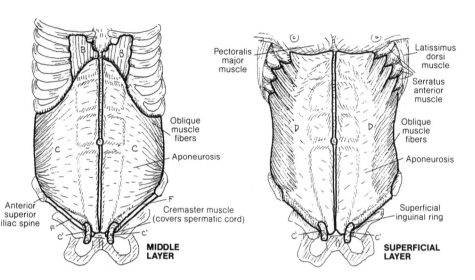

Each segmented rectus abdominis muscle arises from the pubic crest and tubercles and inserts on the lower costal cartilages and xiphoid process (sternum). They are flexors of the vertebral column. The *sheath of the rectus* varies in its extent, running from deep to superficial from below upward, as illustrated. Below the arcuate line, no muscle contributes to its posterior layer (E^{2*}); in the middle, all three flat aponeuroses contribute equally to the sheath (E^{1*}); above, the anterior sheath is formed from external oblique; posteriorly, the rectus contacts the costal cartilages.

The inguinal region is the lower medial part of the abdominal wall, characterized by a canal with inner (deep) and outer (superficial) openings or rings. This canal carries the *spermatic cord* (ductus deferens and its vessels, testicular vessels, lymphatics) in the male and the round ligament of the uterus in the female. The testes and spermatic cords "descend" (by differential growth) into outpocketings of the anterior abdominal wall, collectively called the scrotum. In their descent, they push in front of them layers of fibers from the three flat muscles of the abdominal wall and their aponeuroses, much as a finger might push against four layers of latex to form a four-layered finger glove. These are the coverings of the cord: internal, cremasteric, and external spermatic fasciae. The lower fibers of internal oblique are unique in that they continue in loops around the spermatic cord as the cremaster muscle; the two are connected by cremasteric fascia. The canal area is a weak point, subject to protrusions of fat or intestine (hernias) from within the abdominal cavity, either directly through the wall (direct inguinal hernia) or indirectly through the canal (indirect inguinal hernia).

SHEATH OF RECTUS ABDOMINIS E*

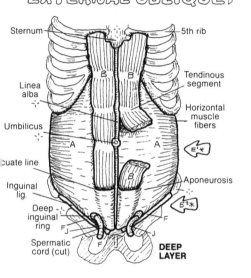

Below arcuate line there is no aponeurotic sheath behind rectus

INGUINAL REGION
INGUINAL LIG. F
CREMASTER MUS. C'
PYRAMIDALIS MUS. G
PERITONEUM H
TRANSVERSALIS FASCIA I
SPERMATIC CORD J
TESTIS / EPIDIDYMIS K

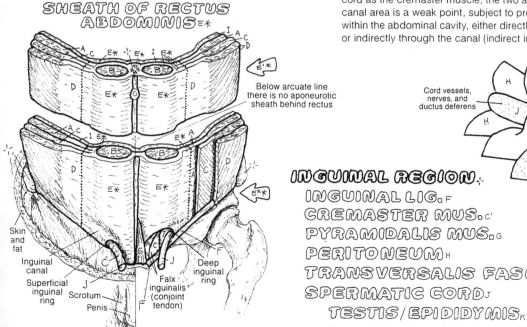

MUSCLES OF THE PELVIS

PELVIC DIAPHRAGM (FLOOR)*¹
LEVATOR ANI :-
LEVATOR PROSTATAE / VAGINAE ᴀ
PUBORECTALIS ʙ
PUBOCOCCYGEUS ᴄ
ILIOCOCCYGEUS ᴅ
COCCYGEUS ᴇ

PELVIC WALL *²
OBTURATOR INTERNUS ꜰ
PIRIFORMIS ɢ
SACROTUBEROUS LIGAMENT ʜ*
SACROSPINOUS LIGAMENT ɪ*
TENDINOUS ARCH ᴊ

CN: Use bright colors for A and J, and gray for H and I.
(1) Begin with the illustration of the pelvic floor muscles, labeled "pelvic diaphragm," just below the large illustration at mid-right. Then go to that large illustration and color the same muscles. Continue with all drawings showing pelvic floor muscles. (2) Color the pelvic wall muscles and the ligaments in the "pelvic wall" diagram. Then color these muscles/ligaments in the large illustration at right, followed by the rest of the drawings showing pelvic wall muscles/ligaments.

The muscles of the pelvis form the pelvic floor in the pelvic outlet (*coccygeus* and the *levator ani*) and the pelvic wall (*obturator internus* and *piriformis*). The fascia-covered pelvic floor muscles constitute the pelvic diaphragm, separating pelvic viscera from the perineal structures interiorly. The pelvic wall includes the *sacrotuberous* and *sacrospinous ligaments*.

The levator ani on each side arises from the pubic bone and ischial spine and the intervening *tendinous arch*, droops downward as it passes toward the midline, and inserts on the anococcygeal ligament and the coccyx with the contralateral levator ani. The muscle essentially has four parts (A, B, C, and D). Coccygeus is the posterior muscle of the pelvic floor, on the same plane as iliococcygeus and immediately posterior to it. The pelvic diaphragm counters abdominal pressure, and with the thoracic diaphragm, assists in micturition, defecation, and childbirth. It is an important support mechanism for the uterus, resisting prolapse.

The obturator internus, a lateral rotator of the hip joint, arises, in part, from the margins of the obturator foramen on the pelvic side. It passes downward and posterolaterally past the obturator foramen to and through the lesser sciatic foramen, inserting on the medial surface of the greater trochanter of the femur. Its covering fascia forms the tendinous arch from which levator ani arises in part. Piriformis, a lateral rotator of the hip joint, takes a course similar to obturator internus to arrive at the greater trochanter.

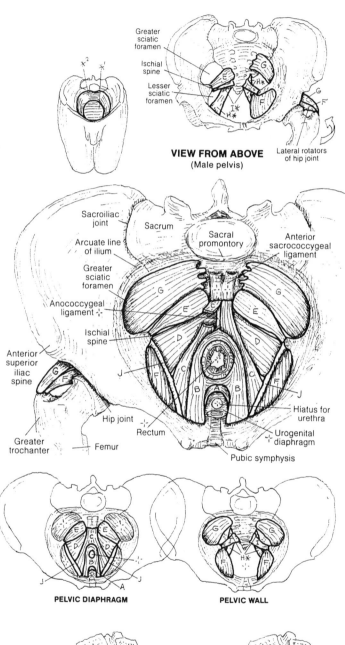

VIEW FROM ABOVE
(Male pelvis)

PELVIC DIAPHRAGM   **PELVIC WALL**

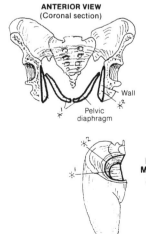

ANTERIOR VIEW
(Coronal section)

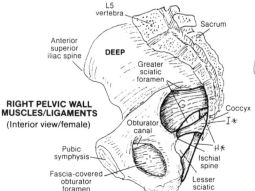

RIGHT PELVIC WALL MUSCLES/LIGAMENTS
(Interior view/female)

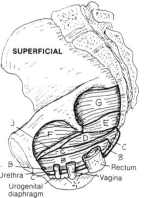

DEEP

INTERMEDIATE

SUPERFICIAL

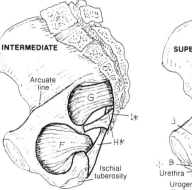

MUSCLES OF THE PERINEUM

CN: (1) Color gray the title "perineum" and the two related diamond-shaped areas at the top of the plate. Color the elements forming the boundaries of the perineum, as seen from below, (2) Color gray the titles *"urogenital triangle"* and *"anal triangle,"* and their respective triangles. (3) Color the lower views simultaneously.

PERINEUM*¹ (BOUNDARIES):
SYMPHYSIS PUBIS_A
COCCYX_B
ISCHIAL TUBEROSITY_C
SACROTUBEROUS LIGAMENT_D
ISCHIOPUBIC RAMUS_E

The 3-dimensional perineum is the region below the pelvic cavity situated within the pelvic outlet. It is bordered by structures A–E. Its floor (inferior border) is skin and fascia; its roof is the pelvic diaphragm. The diamond-shaped perineum is bisected at the *ischial tuberosities* into the urogenital and anal triangles.

UROGENITAL TRIANGLE*²
ISCHIOCAVERNOSUS M._F
BULBOSPONGIOSUS M._G
SUP. TRANSVERSE PERINEAL M._H
UROGENITAL DIAPHRAGM_I

The urogrenital triangle contains the penis, scrotum, and related structures in the male, and the clitoris, urethra and orifice, vagina and orifice, and related structures in the female. It includes: (1) the anterior recess of the ischiorectal fossa, (2) the *urogenital diaphragm* composed of the deep transverse perineal muscle and sphincter urethrae, and (3) the superficial perineal space, containing the *ischiocavernosus* and *bulbospongiosus muscles.* These muscles ensheathe the roots of the erectile bodies of the penis/clitoris, aiding in their erection. *Bulbospongiosus* also contracts rhythmically during the ejaculation of semen. The *superficial transverse perineal* muscles stabilize the fibrous perineal body, which helps anchor the perineal structures.

ANAL TRIANGLE *³
LEVATOR ANI M._J
EXTERNAL SPHINCTER ANI M._K
ANOCOCCYGEAL LIGAMENT_L

The anal triangle includes the fat-filled left and right ischiorectal fossae, the anus and its orifice, and the *external sphincter ani muscle.* This muscle is secured anteriorly by the perineal body and posteriorly by the anococcygeal ligament.

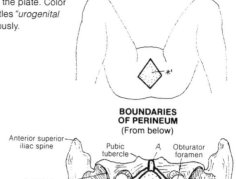

BOUNDARIES OF PERINEUM
(From below)

Anterior superior iliac spine
Pubic tubercle
Obturator foramen
Acetabulum

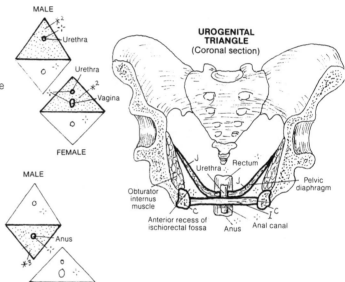

MALE
Urethra
Urethra
Vagina
FEMALE

UROGENITAL TRIANGLE
(Coronal section)

Urethra
Rectum
Obturator internus muscle
Anterior recess of ischiorectal fossa
Anus
Anal canal
Pelvic diaphragm

MALE
Anus
FEMALE
Anus

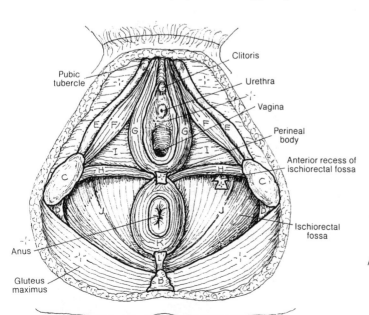

Clitoris
Pubic tubercle
Urethra
Vagina
Perineal body
Anterior recess of ischiorectal fossa
Ischiorectal fossa
Anus
Gluteus maximus

FEMALE PERINEUM

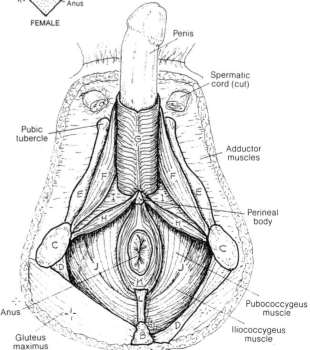

Penis
Spermatic cord (cut)
Pubic tubercle
Adductor muscles
Perineal body
Anus
Gluteus maximus
Pubococcygeus muscle
Iliococcygeus muscle

MALE PERINEUM

MUSCLES OF SCAPULAR STABILIZATION

TRAPEZIUS_A
RHOMBOID MAJOR_B, MINOR_{B'}
LEVATOR SCAPULAE_C
SERRATUS ANTERIOR_D
PECTORALIS MINOR_E

CN: (1) Color the six muscles of scapular stabilization. Note that the two rhomboids receive the same color (B). In the two main views, color gray the nuchal ligament and its title. (2) Color the attachment site diagrams at upper right. (3) In the illustrations below describing scapular movement, note that the three regions of trapezius (A) play different roles. Color gray the scapulae, the arrows, and the movement titles.

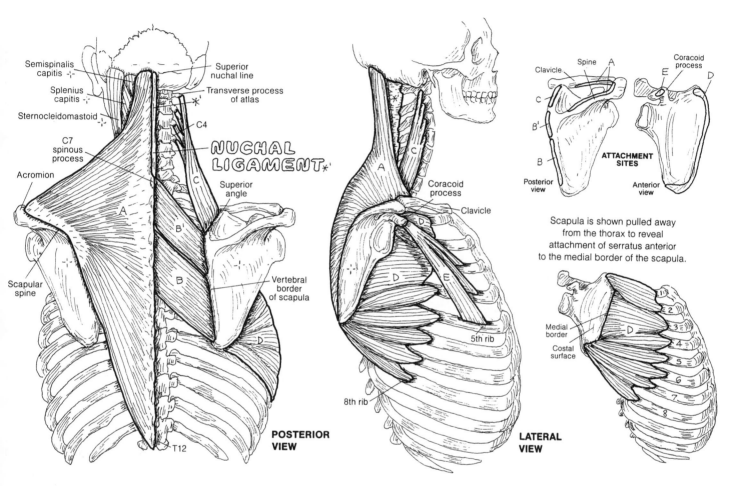

The scapula lies on the posterior thorax, roughly from T2 to T8. It has no direct bony attachment with the axial skeleton. Enveloped by muscle, it glides over the fascia-covered thorax during upper limb movement (scapulothoracic motion). Bursae have been reported between the thorax and the scapula; so has bursitis. The scapula is dynamically moored to the axial skeleton by muscles attaching the scapula to the axial skeleton. These *muscles of scapular stabilization* make possible considerable scapular mobility and, therefore, shoulder/arm mobility.

Note the roles of these six muscles in scapular movement, and note how the shoulder and arm are affected. *Pectoralis minor* assists *serratus anterior* in protraction of the scapula such as in pushing against a wall; it also helps in depression of the shoulder and downward rotation of the scapula. Consider the power resident in serratus anterior and trapezius in pushing or swinging a bat. Note the especially broad sites of attachment of the *trapezius* muscle. Trapezius commonly manifests significant tension with hard work—mental or physical. A brief massage of this muscle often brings quick relief.

MOVEMENTS OF THE SCAPULA*

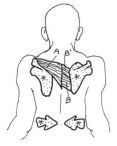

RETRACTION*
Military posture ("squaring the shoulders")

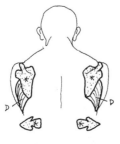

PROTRACTION*
Pushing forward with outstretched arms and hands.

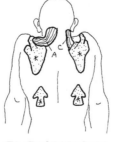

ELEVATION*
Shrugging the shoulders or protecting the head.

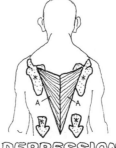

DEPRESSION*
Straight arms on parallel bars, holding weight.

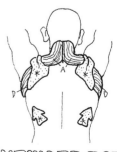

UPWARD ROT.*
Lifting or reaching over head.

MUSCLES OF MUSCULOTENDINOUS CUFF

CN: (1) In addition to the four muscles, color the arrows and titles describing their actions. (2) Color the muscular attachment sites and the diagram of the function of the cuff muscles at mid-right. (3) Do not color the problem spot numerals in the lower illustration. They are there to identify locations discussed in the text.

SUPRASPINATUS_A
INFRASPINATUS_B
TERES MINOR_C
SUBSCAPULARIS_D

Clavicle — **Trapezius** — **Acromion**
Coracoid process
Deltoid
Levator scapulae
Rhomboid minor

Greater tubercle

Rhomboid major
Teres major
Latissimus dorsi

POSTERIOR VIEW
(Scapula)

ABDUCTION A'

LATERAL ROTATION C'

Deltoid tuberosity
Deltoid
Humerus

Acromion — **Trapezius** — **Clavicle**
Coracoid process
Deltoid — **Pectoralis minor**
Coracobrachialis
Biceps brachii (short head)
Greater tubercle
Intertubercular groove
Lesser tubercle
Infraglenoid tubercle
Triceps brachii (long head)
Humerus
Teres major
Latissimus dorsi
Pectoralis major
Deltoid
Coracobrachialis
Pectoralis major
Medial margin
Serratus anterior

ANTERIOR VIEW
(Scapula)

MEDIAL ROTATION D

Shoulder joint
Head of humerus
Scapula
Insertions
Origins
Shaft

ATTACHMENT SITES
(Posterior/lateral view)

Shoulder joint
Head of humerus
Scapula
Shaft

MUSCULOTENDINOUS CUFF
(Diagrammatic/functional)

The socket at the glenohumeral joint (glenoid fossa) is too shallow to offer any bony security for the head of the humerus. As ligaments would severely limit joint movement, muscle tension must be employed to pull the humeral head in to the shallow scapular socket during shoulder movements. Four muscles fulfill this function: *supraspinatus*, *infraspinatus*, *teres minor*, and *subscapularis* ("SITS muscles"). These muscles form a musculotendinous ("rotator") cuff around the head of the humerus, enforcing joint security. Especially effective during robust shoulder movements, they permit the major movers of the joint to work without risking joint dislocation.

The SITS muscles have come to be known as the *"rotator cuff"* muscles, even though one of them, supraspinatus, is an abductor of the shoulder joint and not a rotator. Indeed, among some health care providers, supraspinatus is known as the "rotator cuff" in the context of a "rotator cuff tear."

The shoulder joint and the supraspinatus muscle/tendon are subject to early degeneration from overuse. The problem is generally one of impingement (chronic physical contact and friction) between the acromion (1), the coraco-acromial ligament (2), and the distal clavicle (3) above, and the tendon of supraspinatus (4) and the subacromial bursa (5) below. Those with a down-turned acromion or a previously dislocated, offset acromioclavicular joint are especially vulnerable to impingement (supraspinatus tendinitis and subsequent tearing, subacromial bursitis, limitation of shoulder motion, and pain). All over-head activities (such as those of professional drapery hangers, ceiling plaster-ers, baseball pitchers) and acromial loading (hose-carrying firemen, those carrying heavy purses by straps over the shoulder, mail delivery persons) pursued over a long period can induce changes (bony spurring, bursal destruction) with impingement signs and symptoms.

PROBLEM SPOTS IN THE SHOULDER REGION
(Anterior view)

Acromion
Coracoacromial ligament
Clavicle
Coracoid process
Subacromial bursa
Muscle tendon under bursa
Articular capsule
Long tendon of biceps brachii
Scapula
Humerus

ANTERIOR VIEW

BURSA_E
LIGAMENT*∗

MOVERS OF SHOULDER JOINT

DELTOID꞉ PECTORALIS MAJOR꞉
LATISSIMUS DORSI꞉ TERES MAJOR꞉
CORACOBRACHIALIS꞉ BICEPS BRACHII꞉
TRICEPS BRACHII (LONG HEAD)꞉

CN: (1) Begin with both posterior views; note that the biceps and triceps are not shown on the lateral view.
(2) When coloring the muscles below, note the actions of different parts of the deltoid (A) and pectoralis major (B).

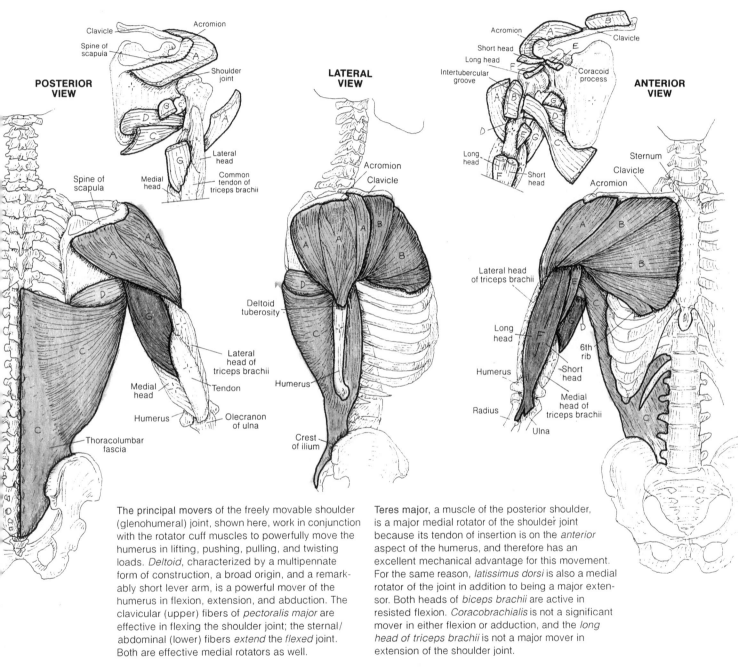

The principal movers of the freely movable shoulder (glenohumeral) joint, shown here, work in conjunction with the rotator cuff muscles to powerfully move the humerus in lifting, pushing, pulling, and twisting loads. *Deltoid*, characterized by a multipennate form of construction, a broad origin, and a remarkably short lever arm, is a powerful mover of the humerus in flexion, extension, and abduction. The clavicular (upper) fibers of *pectoralis major* are effective in flexing the shoulder joint; the sternal/abdominal (lower) fibers *extend* the *flexed* joint. Both are effective medial rotators as well.

Teres major, a muscle of the posterior shoulder, is a major medial rotator of the shoulder joint because its tendon of insertion is on the *anterior* aspect of the humerus, and therefore has an excellent mechanical advantage for this movement. For the same reason, *latissimus dorsi* is also a medial rotator of the joint in addition to being a major extensor. Both heads of *biceps brachii* are active in resisted flexion. *Coracobrachialis* is not a significant mover in either flexion or adduction, and the *long head of triceps brachii* is not a major mover in extension of the shoulder joint.

MOVEMENTS OF THE HUMERUS AT THE SHOULDER JOINT

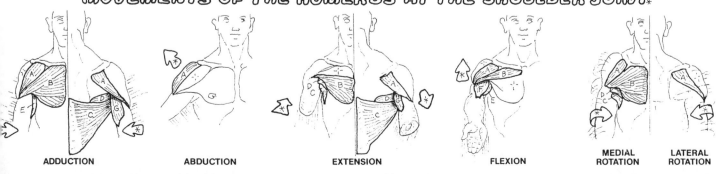

ADDUCTION   **ABDUCTION**   **EXTENSION**   **FLEXION**   **MEDIAL ROTATION**   **LATERAL ROTATION**

MOVERS OF ELBOW & RADIOULNAR JOINTS

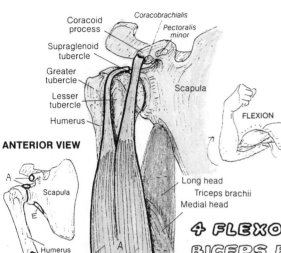

ANTERIOR VIEW

Coracoid process
Coracobrachialis
Pectoralis minor
Supraglenoid tubercle
Greater tubercle
Lesser tubercle
Humerus
Scapula

FLEXION

Long head
Triceps brachii
Medial head

Scapula
Humerus
Long head
Short head

Radial tuberosity

Ulna
Radius

Medial epicondyle

Bicipital aponeurosis (cut)

Radius
Ulna
Interosseous ligament

Styloid process

CN: Use the same colors for biceps brachii (A) and triceps brachii (E) as you did for those muscles on Plate 56. (1) Color the four flexors and their attachment sites on the drawings to their left. Do the same for the extensors on the right. (2) Color the supinators and pronators below. the arrows demonstrating their actions, and their attachment sites at upper left.

4 FLEXORS.
BICEPS BRACHII A
BRACHIALIS B
BRACHIORADIALIS C
PRONATOR TERES D

The principal flexors of the elbow joint are *brachialis* and *biceps brachii*, of which the former has the best mechanical advantage. Yet it's the bulge of a contracted biceps that gets all the visual attention! The tendon of biceps inserts at the tuberosity of the radius, making the muscle a supinator of the forearm as well. With the limb supinated, the biceps works to fulfill flexion of the elbow and supination of the elbow. Take away the supinating function (flexing the pronated elbow), and the appearance of biceps is disappointing (in most of us!). Note the additional attachment of the bicipital aponeurosis into the deep fascia of the common flexor group (not shown) in the forearm. *Brachioradialis* is active in flexion of the elbow and in rapid extension where it counters the centrifugal force produced by that movement. *Pronator teres* assists in elbow flexion as well as pronation.

POSTERIOR VIEW

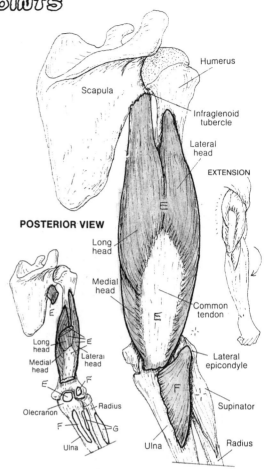

Scapula
Humerus
Infraglenoid tubercle
Lateral head

EXTENSION

Long head
Medial head
Common tendon
Lateral epicondyle
Supinator
Radius

Long head
Medial head
Lateral head
Olecranon
Radius
Ulna

2 EXTENSORS.
TRICEPS BRACHII E
ANCONEUS F

The principal extensor of the elbow joint is the three-headed *triceps brachii* with its massive tendon of insertion. The smaller *anconeus* assists in this function. Triceps is a powerful antagonist to the elbow flexors.

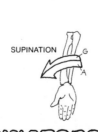

SUPINATION

2 SUPINATORS.
BICEPS BRACHII A
SUPINATOR G

Biceps brachii is the more powerful supinator of the elbow, but *supinator* is important in maintaining supination. Supinator arises from the lateral aspect of the elbow, passing obliquely downward and forward to a rather broad insertion on the upper lateral and anterior surface of the radius. A bundle of fibers from the upper lateral ulna passes behind the radius to join the lateral fibers of supinator.

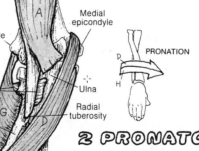

Lateral epicondyle
Medial epicondyle
Ulna
Radial tuberosity
Radius

Interosseous ligament

PRONATION

2 PRONATORS.
PRONATOR TERES D
PRONATOR QUADRATUS H

Pronator quadratus is the principal pronator of the elbow joint, superior in its mechanical advantage to *pronator teres*. Pronating the forearm (palm down) involves medial rotation of the radius. Since only the radius can rotate in the forearm, the pronators clearly cross the radius on the anterior side of the forearm, and their origin is ulnar.

ANTERIOR VIEW

MOVERS OF WRIST & HAND JOINTS

FLEXORS *[1]

DEEP LAYER

FLEX. DIGITORUM PROFUNDUS_A
FLEX. POLLICIS LONGUS_B

INTERMEDIATE LAYER

FLEX. DIGITORUM SUPERFICIALIS_C

SUPERFICIAL LAYER

FLEX. CARPI ULNARIS_D
PALMARIS LONGUS_E
FLEX. CARPI RADIALIS_F

ANTERIOR VIEW

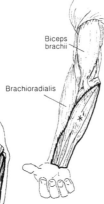

Biceps brachii

Brachioradialis

CN: A more detailed view of the tendons of these muscles (with the same subscripts) can be seen among the intrinsic muscles of the hand on the next plate. (1) Begin with the flexors; note the deeper muscles have been omitted from the superficial view. Color gray the entire flexor mass in the smaller illustration. (2) Continue with the extensors, coloring gray the entire extensor mass in the smaller illustration.

Humerus
Radius
Interosseous membrane
Supinator
Medial epicondyle
Ulna
Pronator teres

C
A
B

Pronator quadratus
Ulna
Carpals
Metacarpal
Palmar aponeurosis (cut)
F E D
A

DEEP **INTERMEDIATE** **SUPERFICIAL**

The flexors of the wrist (carpus) and fingers (digits) take up most of the anterior compartment of the forearm, arising as a group from the medial epicondyle, the upper radius and ulna, and the intervening interosseous membrane. The deep layer of muscles in the anterior forearm (*flexor pollicis longus* or FPL in the radial half, *flexor digitorum profundus* or FDP in the ulnar half) lie in contact with the radius and ulna. The superficial layer of muscles (wrist flexors: the *"carpi" muscles and palmaris longus*) is seen just under the skin and thin superficial fascia. The intermediate layer (*flexor digitorum superficialis*, FDP) lies between the superficial and deep groups. In the anterior (palmar) fingers, note how the tendons of FDS, which insert on the sides of the middle phalanges, split at the level of the proximal phalanges, permitting the deeper (posterior) tendons of FDP to pass on through to the bases of the distal phalanges.

EXTENSORS *[2]

DEEP LAYER

EXT. INDICIS_G
EXT. POLLICIS LONGUS_H
EXT. POLLICIS BREVIS_I

SUPERFICIAL LAYER

EXT. CARPI ULNARIS_J
EXT. DIGITI MINIMI_K
EXT. DIGITORUM_L
EXT. CARPI RADIALIS LONGUS_M
EXT. CARPI RADIALIS BREVIS_N

ABDUCTOR POLLICIS LONGUS_O

The extensors of the wrist and fingers arise from the lateral epicondyle and upper parts of the bones and interosseous membrane of the forearm, forming an extensor compartment on the posterior side of the forearm. The wrist extensors insert on the distal carpal bones or metacarpals, while the finger extensors form an expansion of tendon over the middle and distal phalanges to which the small intrinsic muscles of the hand insert. The wrist extensor muscles are critical to hand function: grasp a finger of one hand with your fingers and an extended wrist of the other; now try it with wrist fully flexed. Note that the power of the hand exists only with an extended wrist.

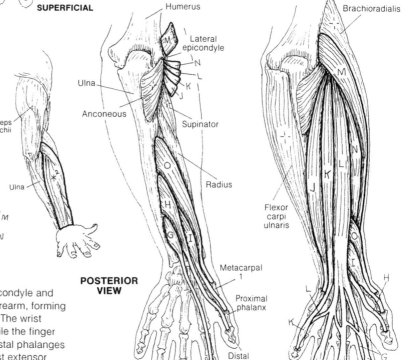

Humerus
Lateral epicondyle
Ulna
Anconeous
Supinator
Triceps brachii
Radius
Ulna
M N L K J
O
H G I
Metacarpal 1
Proximal phalanx
Distal phalanx
Index finger

POSTERIOR VIEW

DEEP

Brachioradialis
M
N L
J K
Flexor carpi ulnaris
O
I
L
H
K G
L

SUPERFICIAL

MOVERS OF HAND JOINTS (INTRINSICS)

CN: The extrinsic muscles that move the wrist and finger joints were covered on Plate 58, and their tendons are shown in dark line and labeled here for identification and study, *but not for coloring.* If possible, use different colors on this plate. (1) Color the muscles of the two anterior views, as well as the flexor retinaculum (gray). (2) Color the posterior view. (3) In the illustration of finger abduction (at the bottom) note that the little finger is not moved by the dorsal interosseous (U).

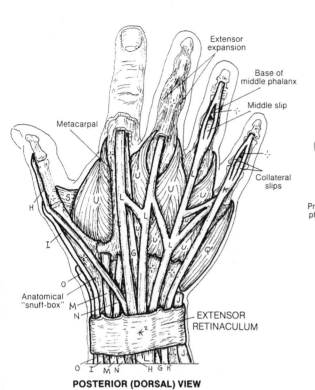

Extensor expansion
Base of middle phalanx
Middle slip
Metacarpal
Collateral slips
Anatomical "snuff-box"
EXTENSOR RETINACULUM

POSTERIOR (DORSAL) VIEW

Fibrous sheath
Proximal phalanx
Palmar lig.
Pisiform bone
FLEXOR RETINACULUM
(Transverse carpal lig.)
Palmar aponeurosis (cut)
Tendon of brachioradialis

ANTERIOR (PALMAR) VIEW

Proximal phalanx
Metacarpal bone
Carpal tunnel
Radius
Median nerve
Ulna

DEEP PALMAR VIEW

THENAR EMINENCE *'
OPPONENS POLLICIS ᴘ
ABDUCTOR POLLICIS BREVIS ǫ
FLEXOR POLLICIS BREVIS ʀ

Note the palpable bulge of muscle (thenar eminence) just proximal to the thumb on your own hand. Integrated with the action of the other thumb movers, these three muscles make possible complex movements of the thumb. The thenar muscles arise/insert in the same general area as one another; however, their different orientation orders different functions.

HYPOTHENAR EMINENCE *
OPPONENS DIGITI MINIMI ᴘ'
ABDUCTOR DIGITI MINIMI ǫ'
FLEXOR DIGITI MINIMI BREVIS ʀ'

These muscles move the 5th digit; they are complementary to the thenar muscles in attachment and function. The function of opposition is basic to some of the complex grasping functions of the hand.

DEEP MUSCLES
ADDUCTOR POLLICIS ѕ
PALMAR INTEROSSEUS ᴛ
DORSAL INTEROSSEUS ᴜ
LUMBRICAL ᵥ

Adductor pollicis, in concert with the first dorsal interosseous muscle, provides great strength in grasping an object between thumb and index finger... try it. The *interossei* and *lumbrical muscles* insert into expanded finger extensor tendons (extensor expansion; see posterior view) forming a complex mechanism for flexing the metacarpophalangeal joints and extending the interphalangeal joints. By their phalangeal insertions, the interossei abduct/adduct certain digits.

ACTIONS OF INTRINSIC MUSCLES
ON THE THUMB *

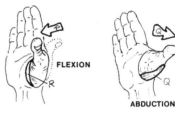
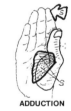

FLEXION   **ABDUCTION**   **ADDUCTION**

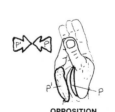

OPPOSITION   **CIRCUMDUCTION**

ON THE FINGERS

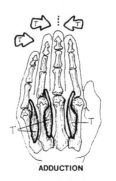
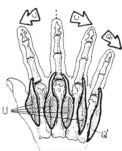

ADDUCTION   **ABDUCTION**

REVIEW OF MUSCLES

CN: Begin with the three muscles labeled with an "A" color. Color each wherever it appears before going on to the next. Write the name of the muscle in color on the appropriate line. If your point isn't sharp enough, write the name in black pencil and circle the identifying letter next to it in color. Try to visualize the function of these superficial muscles; some of them have more than the one function assigned to them by these categories.

(See appendix A in the back of the book for answers)

MUSCLES ACTING PRIMARILY ON THE SCAPULA

A _____
A^1 _____
A^2 _____

MUSCLES MOVING THE SHOULDER JOINT

B _____
B^1 _____
B^2 _____
B^3 _____
B^4 _____
B^5 _____
B^6 _____

MUSCLES MOVING ELBOW & RADIOULNAR JOINTS

C _____
C^1 _____
C^2 _____
C^3 _____
C^4 _____
C^5 _____

MUSCLES MOVING WRIST & HAND JOINTS

D _____
D^1 _____
D^2 _____
D^3 _____
D^4 _____
D^5 _____
D^6 _____
D^7 _____

FOREARM MUSCLES MOVING THE THUMB

E _____
E^1 _____
E^2 _____

THENAR MUSCLES MOVING THE THUMB

F _____
F^1 _____
F^2 _____

HYPOTHENAR MUSCLES MOVING THE 5TH DIGIT

G _____
G^1 _____
G^2 _____

OTHER MUSCLES ACTING ON THE THUMB & FINGERS

H _____
H^1 _____
H^2 _____

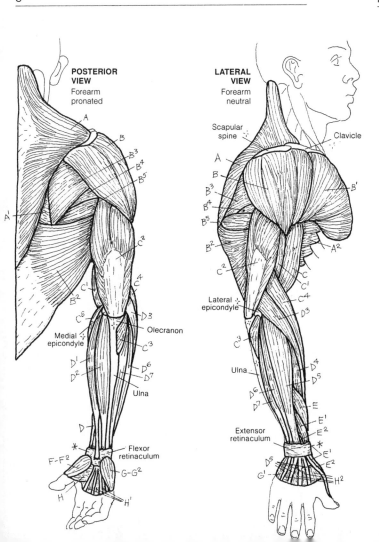

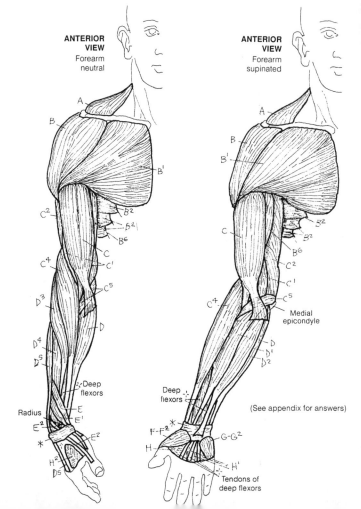

MUSCLES OF THE GLUTEAL REGION

CN: In the posterior and lateral views (superficial dissections), the upper fibers of the illiotibial tract (x¹) have been cut away, exposing gluteus medius. (1) Color each muscle in all views, including the directional arrows, before going on to the next. The origin of piriformis (E) cannot be seen in these views, but see Plate 52. A better view of the origin of obturator internus (F) also can be seen on Plate 52.

3 GLUTEAL MUSCLES
GLUTEUS MAXIMUS A
GLUTEUS MEDIUS B
GLUTEUS MINIMUS C
TENSOR FASCIAE LATAE D

The gluteal muscles are arranged in three layers: the most superficial is *gluteus maximus*. The large sciatic nerve runs deep to it, as every student nurse has learned well. Its thickness varies. Gluteus maximus extends the hip joint during running and walking up-hill, but does not act in relaxed walking. The intermediately placed, more lateral *gluteus medius* is a major abductor of the hip joint and an important stabilizer (leveler) of the pelvis when the opposite lower limb is lifted off the ground.

6 DEEP, LATERAL ROTATORS
PIRIFORMIS E
OBTURATOR INTERNUS F
OBTURATOR EXTERNUS G
QUADRATUS FEMORIS H
GEMELLUS SUPERIOR I
GEMELLUS INFERIOR J

The deepest layer of gluteal muscles is the *gluteus minimus* and the *lateral rotators* of the hip joint. They cover up/fill the greater and lesser sciatic notches. These muscles generally insert at the posterior aspect of the greater trochanter of the femur. The gluteal muscles (less gluteus maximus) correspond to some degree with the rotator cuff of the shoulder joint: lateral rotators posteriorly, abductor (gluteus medius) superiorly, medial rotators (gluteus medius and minimus, tensor fasciae latae) anteriorly.

ILIOTIBIAL TRACT *

The iliotibial tract, a thickening of the deep fascia (fascia lata) of the thigh, runs from ilium to tibia and helps stabilize the knee joint laterally. The muscle *tensor fasciae latae*, a frequently visible and palpable flexor and medial rotator of the hip joint, inserts into this fibrous band, tensing it. Despite its major flexor function, this anterolaterally-placed muscle is considered a part of the more posterior gluteal group; it shares its insertion into the iliotibial tract with gluteus maximus, and it is supplied by the superior gluteal nerve and artery.

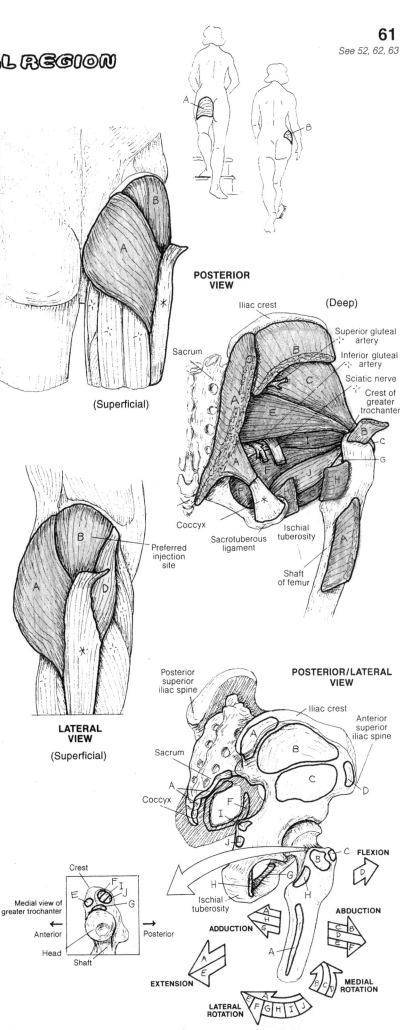

POSTERIOR VIEW

(Superficial)

(Deep)

Iliac crest

Superior gluteal artery

Inferior gluteal artery

Sciatic nerve

Crest of greater trochanter

Sacrum

Coccyx

Sacrotuberous ligament

Ischial tuberosity

Shaft of femur

LATERAL VIEW

(Superficial)

Preferred injection site

POSTERIOR/LATERAL VIEW

Posterior superior iliac spine

Iliac crest

Anterior superior iliac spine

Sacrum

Coccyx

Crest

Medial view of greater trochanter

Anterior

Posterior

Head

Shaft

FLEXION

ABDUCTION

ADDUCTION

MEDIAL ROTATION

EXTENSION

LATERAL ROTATION

Ischial tuberosity

MUSCLES OF THE POSTERIOR THIGH

HAMSTRINGS*
SEMIMEMBRANOSUS_A
SEMITENDINOSUS_B
BICEPS FEMORIS_C

Tight hamstrings limit flexion of hip
when knee joint is extended.

CN: (1) Color each hamstring muscle in the deep view before going on to the superficial. Then color the diagrams of flexion and extension. (2) Color gray the outline of the muscles in the drawings at upper right.

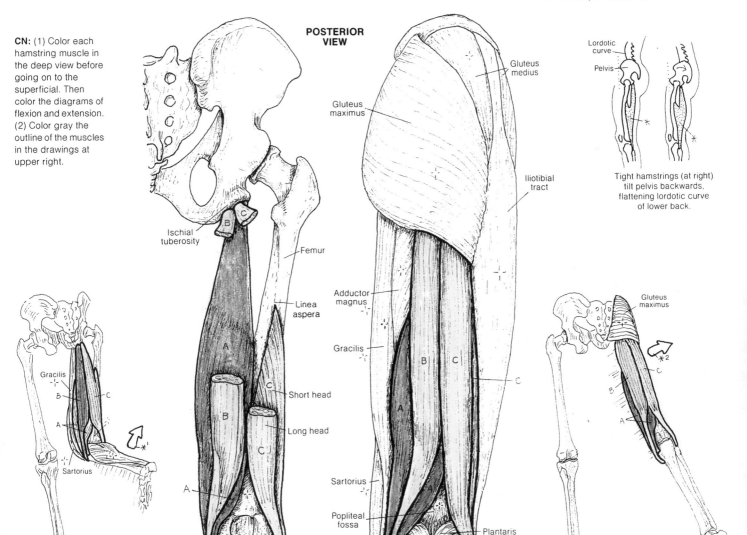

POSTERIOR VIEW

Gluteus medius

Gluteus maximus

Iliotibial tract

Ischial tuberosity

Femur

Linea aspera

Adductor magnus

Gracilis

Short head

Long head

Sartorius

Popliteal fossa

Plantaris

Tibia

Head of fibula

Gastrocnemius

Gracilis

Sartorius

FLEXORS OF THE KNEE JOINT*¹

DEEP

SUPERFICIAL

Gluteus maximus

EXTENSORS OF THE HIP JOINT*²

Lordotic curve

Pelvis

Tight hamstrings (at right) tilt pelvis backwards, flattening lordotic curve of lower back.

The hamstring muscles are equally effective at both extension of the hip joint and flexion of the knee joint. Unlike the hip extensor gluteus maximus, the hamstrings are active during normal walking. In relaxed standing, both gluteus maximus and the hamstrings are inactive. In knee flexion, the hamstrings act in concert with sartorius, gracilis, and gastrocnemius (Plates 63 and 66). Long tendons of the hamstrings can be palpated just above the partially flexed knee on either side of the midline.

Reduced hamstring stretch ("tight hamstrings") limits hip flexion with the knee extended; flexion of the knee permits increased hip flexion. Try this on yourself. Tight hamstrings, by their ischial origin, pull the posterior pelvis down, lengthening the erector spinae muscles and flattening the lumbar lordosis, potentially contributing to limitation of lumbar movement and back pain. Tight hamstrings often cause posterior thigh pain on straight leg raise testing (subject is supine, lower limbs horizontal; one heel is lifted, progressively flexing the hip joint with knee extended). This pain from muscle stretch may be confused with sciatic nerve/nerve root pain, which normally shoots into the leg and foot.

Gluteus maximus

Powerful extensors of the hip joints.

MUSCLES OF THE MEDIAL THIGH

CN: Color one muscle at a time in the five main views before going to the next one. Note that the attachment sites on the *posterior* surface of the femur are represented by dotted lines.

PECTINEUS A
ADDUCTOR BREVIS B
ADDUCTOR LONGUS C
ADDUCTOR MAGNUS D
GRACILIS E
OBTURATOR EXTERNUS F

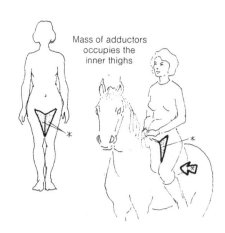

Mass of adductors occupies the inner thighs

ANTERIOR VIEW

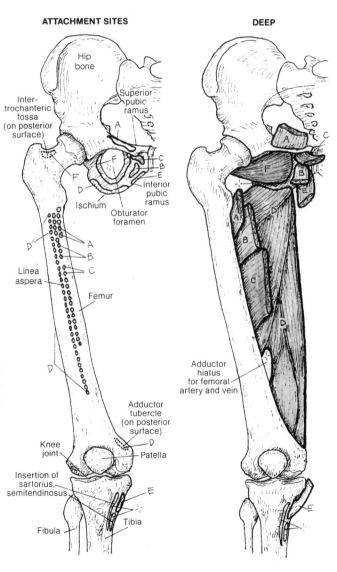

ATTACHMENT SITES

Hip bone

Inter-trochanteric fossa (on posterior surface)

Superior pubic ramus

Ischium

Inferior pubic ramus

Obturator foramen

Linea aspera

Femur

Adductor tubercle (on posterior surface)

Knee joint

Insertion of sartorius, semitendinosus

Fibula

Patella

Tibia

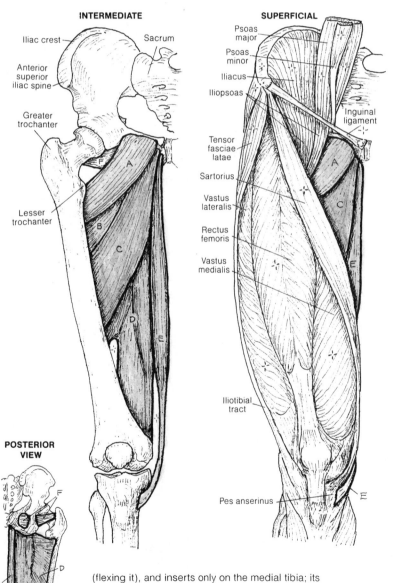

DEEP

Adductor hiatus for femoral artery and vein

POSTERIOR VIEW

Linea aspera

Adductor hiatus

INTERMEDIATE

Iliac crest

Sacrum

Anterior superior iliac spine

Greater trochanter

Lesser trochanter

SUPERFICIAL

Psoas major

Psoas minor

Iliacus

Iliopsoas

Inguinal ligament

Tensor fasciae latae

Sartorius

Vastus lateralis

Rectus femoris

Vastus medialis

Iliotibial tract

Pes anserinus

The **medial thigh muscles** consist of the hip joint *adductors* (A through E) and *obturator externus*, a lateral rotator of that joint. The latter was colored on Plate 61 as one of the deep gluteal muscles, as its tendon passes into that region. However, it is compartmentalized by fasciae in the medial thigh, covers the external surface of the obturator foramen in the deep upper medial thigh, and receives the same innervation as the adductors. The *gracilis* is the longest of the adductor group, crosses the medial knee

(flexing it), and inserts only on the medial tibia; its tendon joins the tendons of sartorius and semitendinosus to form an insertion shaped like a goose's foot (hence called the pes anserinus). The *adductor magnus* is the most massive of the group (see posterior view). In its lower half, adductor magnus fibers give way to the passage of the femoral vessels (adductor hiatus). All the adductors, except gracilis, insert on the vertical rough line (linea aspera) on the posterior surface of the femur.

MUSCLES OF THE ANTERIOR THIGH

SARTORIUS_A

QUADRICEPS FEMORIS-:-
RECTUS FEMORIS_B
VASTUS LATERALIS_C
VASTUS INTERMEDIUS_D
VASTUS MEDIALIS_E

ILIOPSOAS_F

PATELLAR LIGAMENT_G*

CN: The patellar ligament (G*) is colored gray but the patella is left uncolored. (1) Begin with the deep view of the thigh and then complete the superficial view. (2) On the far left, color the visualized portions of the quadriceps that are antagonists to the hamstring group. (3) Complete the action diagrams along the right margin.

FLEXORS OF THE HIP JOINT *

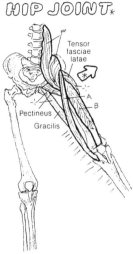

Tensor fasciae latae
Pectineus
Gracilis
F

ANTERIOR VIEW

DEEP

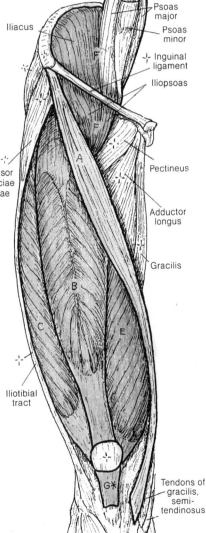

Anterior superior iliac spine
Anterior inferior iliac spine
Symphysis pubis
Hip joint
Tendon of quadriceps
Patella
Knee joint
Tibial tuberosity

SUPERFICIAL

Iliacus
Psoas major
Psoas minor
-:- Inguinal ligament
Iliopsoas
Pectineus
Adductor longus
Gracilis
Tensor fasciae latae
Iliotibial tract
Tendons of gracilis, semi-tendinosus

LATERAL VIEW

HAMSTRING MUSCLES
QUADRICEPS FEMORIS
Femur
Patella
G*
Head of fibula
Tibial tuberosity

FLEXOR OF THE KNEE JOINT -:-

EXTENSORS OF THE KNEE JOINT -:-

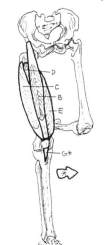

The sartorius ("tailor's" muscle; so-called because of the role of this muscle in enabling a crossed-legs sitting posture) is a flexor and lateral rotator of the hip joint and a flexor of the knee joint, as you can infer from its illustrated attachments. *The quadriceps femoris* muscle arises from four heads. The *vastus medialis* and *lateralis* arise from the linea aspera on the posterior aspect of the femur; the *vastus intermedius* arises from the anterior femoral shaft. All four converge onto the superior aspect (base) of the patella to form the patellar tendon. Some tendon fibers continue over the patellar surface to join the ligament below. At the inferior aspect (apex) of the patella, the tendinous fibers continue to the tibial tuberosity.

The tendon between the patella and the tibial tuberosity is called the *patellar ligament*. *Rectus femoris*, a strong hip joint flexor, is the only member of quadriceps to cross that joint. Quadriceps femoris is the only knee extensor. The significance of its role becomes crystal clear to those having experienced a knee injury; the muscles tend to atrophy and weaken rapidly with disuse, and "quad" exercises are essential to maintain structural stability of the joint. The *iliopsoas* is the most powerful flexor of the hip, having a broad thick muscle belly and attaching at the lesser trochanter at the proximal end of the femoral shaft. Recall Plate 50 for its posterior abdominal origin.

MUSCLES OF THE ANTERIOR & LATERAL LEG

CN: Take care with the narrow attachment sites of the anterior leg. Although the muscles A, B, and C arise from the interosseous ligament as well as the tibia and the fibula, the ligament has been left out of the attachments illustration for purposes of simplification. Attachment sites on the plantar surface of the foot are shown at upper right.

The muscles of the leg are arranged into anterior-lateral, lateral, and posterior compartments. The bony ridge (anterior margin) of the tibia creates two oblique surfaces, the anterolateral of which relates to the anterior leg muscles; the anteromedial surface is bony (ouch!) and devoid of muscle. The lateral compartment fibular muscles largely arise from the fibula and the interosseous ligament between the tibia and fibula.

ANTERIOR LEG
TIBIALIS ANTERIOR_A
EXTENSOR DIGITORUM LONGUS_B
EXTENSOR HALLUCIS LONGUS_C
FIBULARIS TERTIUS_D

LATERAL LEG
FIBULARIS LONGUS_E
FIBULARIS BREVIS_F

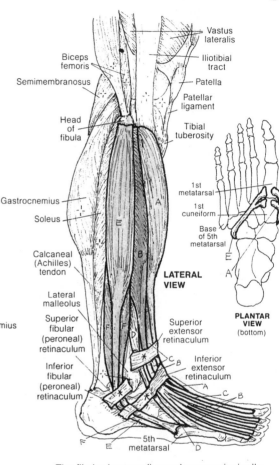

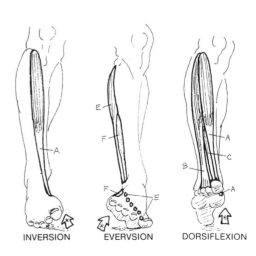

The fibular (peroneal) muscles are principally evertors of the foot, and are especially active during plantar flexion, as in walking on the toes or pushing off with the great toe. Fibularis tertius arises in the fibular compartment but is actually part of extensor digitorum.

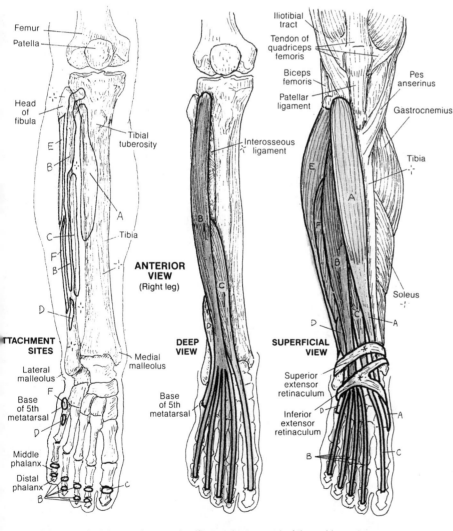

All of the anterior leg muscles are dorsiflexors (extensors) of the ankle; *extensors hallucis and digitorum longus* are toe extensors; *tibialis anterior* is an invertor of the subtalar joints as well, and *fibularis tertius* (the 5th tendon of extensor digitorum) is an evertor. Due to rotation of the lower limb during embryonic development, these extensors are anterior to the bones in the anatomical position (unlike the upper limb wrist extensors). Tibialis anterior is particularly helpful in lifting the foot up during the swing phase of walking to avoid striking the toes.

MUSCLES OF THE POSTERIOR LEG

TIBIALIS POSTERIOR g
FLEXOR DIGITORUM LONGUS h
FLEXOR HALLUCIS LONGUS i
POPLITEUS j
PLANTARIS k
SOLEUS l
GASTROCNEMIUS m

CN: The muscles to be colored on this plate are labeled G–M; any other letter label found here (A–F from Pl. 65; N–Y from Pl. 67) is for identification only, and those muscles should be left uncolored. You may repeat colors used for muscles on Plate 57 on this and/or the next plate. (1) Color one muscle at a time in each of the posterior views. Note that the plantaris (K), the soleus (L), and the gastrocnemius (M) all insert into the same tendon (tendocalcaneus), which receives the color M. (2) Color the upper and lower medial views.

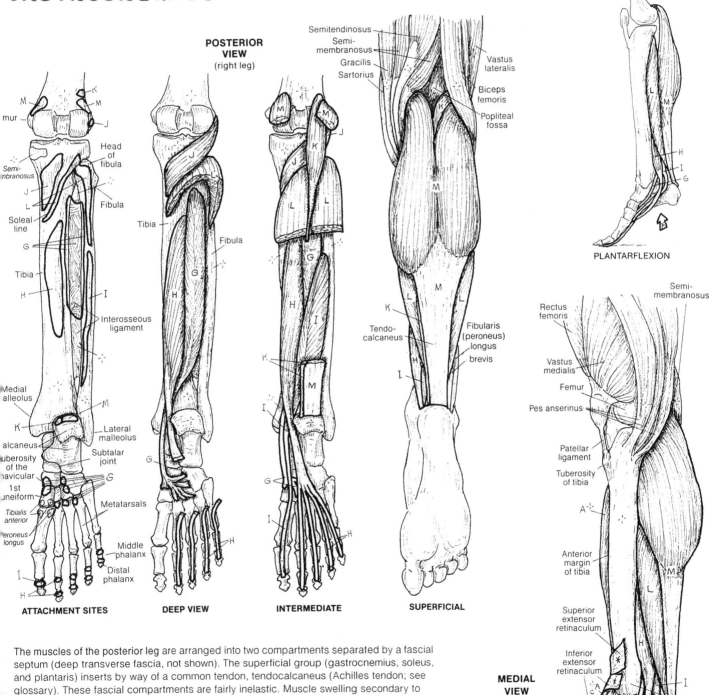

POSTERIOR VIEW (right leg)

ATTACHMENT SITES

DEEP VIEW

INTERMEDIATE

SUPERFICIAL

PLANTARFLEXION

MEDIAL VIEW

The muscles of the posterior leg are arranged into two compartments separated by a fascial septum (deep transverse fascia, not shown). The superficial group (gastrocnemius, soleus, and plantaris) inserts by way of a common tendon, tendocalcaneus (Achilles tendon; see glossary). These fascial compartments are fairly inelastic. Muscle swelling secondary to vascular insufficiency can result in serious muscle compression/death (compartment syndrome) in the absence of fascial (surgical) decompression.

The major calf muscle is gastrocnemius, which flexes the knee and, with its two fellows, plantarflexes the ankle joint. In knee flexion it is aided by *popliteus*, which also rotates the tibia medially. The other deep flexors plantarflex the ankle joint (both toe and great toe *flexors* and *tibialis posterior*), flex the toes (the flexors), and invert the foot (tibialis posterior).

MUSCLES OF THE FOOT (INTRINSICS)

CN: Feel free to use the colors used for the letter labels on plates 65 and 66. Those letters are presented here for identification, and the muscles they refer to are not meant to be colored. Also note that plantar surface attachment sites for those extrinsic foot muscles have been omitted in the illustration of the fourth layer but can be found on the two preceding plates. (1) Begin with the fourth layer and complete each illustration before going on to the next.

The dorsal intrinsic muscles of the foot (those that arise and insert within the dorsum of the foot) are limited to two *small extensors* of the toes, shown at right, most of the extensor function being derived from extrinsic extensors.

The intrinsic muscles of the plantar region of the foot are shown here in four layers. The *plantar interossei*, wedged between the metatarsal bones, constitute the deepest (4th) layer. They adduct toes 3–5, flex the metatarsophalangeal (MP) joints of these toes, and contribute to extension of the interphalangeal (IP) joints of these toes through the mechanism of the extensor expansion. The *dorsal interossei* abduct toes 3–5 and facilitate the other actions of the plantar interossei.

The third layer of muscles acts on the great toe (hallux) and 5th digit (digiti minimi). The second layer includes the *quadratus plantae*, inserting into the lateral border of the common tendon (H) of flexor digitorum longus (FDL). It assists that muscle in flexion of the toes. The *lumbricals* arise from the individual tendons of FDL and insert into the medial aspect of the extensor expansion (dorsal aspect). They flex the MP joints and extend the IP joints of toes 2–5 via the extensor expansion.

The superficial (first) layer consists of the *abductors* of the 1st and 5th digits and the *flexor digitorum brevis*. The plantar muscles are covered by the thickened deep fascia of the sole, the plantar aponeurosis, extending from calcaneus to the fibrous sheath of the flexor tendons.

DORSAL SURFACE
(Right foot)

EXTENSOR DIGITORUM BREVIS_N

EXTENSOR HALLUCIS BREVIS_O

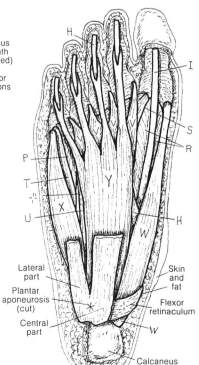

Extensor expansion
Base of 5th metatarsal
Lateral malleolus
Superior extensor retinaculum
Medial malleolus
Inferior extensor retinaculum

3 PLANTAR INTEROSSEI_P

4 DORSAL INTEROSSEI_Q

FLEX. HALLUCIS BREVIS_R

ADDUCTOR HALLUCIS_S

FLEX. DIGITI MINIMI BREV._T

QUADRATUS PLANTAE_U

4 LUMBRICALS_V

ABDUCTOR HALLUCIS_W

ABDUCTOR DIGITI MINIMI_X

FLEX. DIGITOR. BREVIS_Y

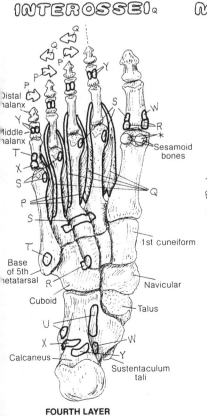

Distal phalanx
Middle phalanx
Sesamoid bones
1st cuneiform
Base of 5th metatarsal
Navicular
Cuboid
Talus
Calcaneus
Sustentaculum tali

FOURTH LAYER

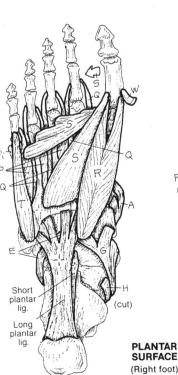

Short plantar lig.
Long plantar lig.

THIRD LAYER

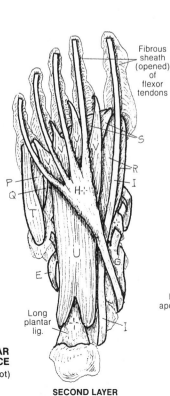

Fibrous sheath (opened) of flexor tendons
Long plantar lig.

PLANTAR SURFACE
(Right foot)

SECOND LAYER

Lateral part
Plantar aponeurosis (cut)
Central part
Skin and fat
Flexor retinaculum
Calcaneus

FIRST LAYER (superficial)

REVIEW OF MUSCLES

CN: Follow the same procedure as you did with Plate 60. Be sure to look for the muscle in each of the 4 drawings before committing its name to writing. Try to visualize each muscle's function as you color and name it. Several muscles (B, B^3–B^7) cross both hip and knee joints and are effective at both joints.

(See appendix A in the back of the book for answers)

MUSCLES ACTING PRIMARILY ON THE HIP JOINT

A _____

A^1 _____

A^2 _____

A^3 _____

A^4 _____

A^5 _____

A^6 _____

A^7 _____

MUSCLES ACTING PRIMARILY ON THE KNEE JOINT

B _____

B^1 _____

B^2 _____

B^3 _____

B^4 _____

B^5 _____

B^6 _____

B^7 _____

MUSCLES ACTING PRIMARILY ON THE ANKLE JOINTS

C _____

C^1 _____

C^2 _____

C^3 _____

C^4 _____

C^5 _____

C^6 _____

C^7 _____

C^8 _____

MUSCLES ACTING PRIMARILY ON BELOW THE ANKLE JOINTS

D _____

D^1 _____

MUSCLES ACTING ON DIGITS OF THE FOOT

E _____

E^1 _____

E^2 _____

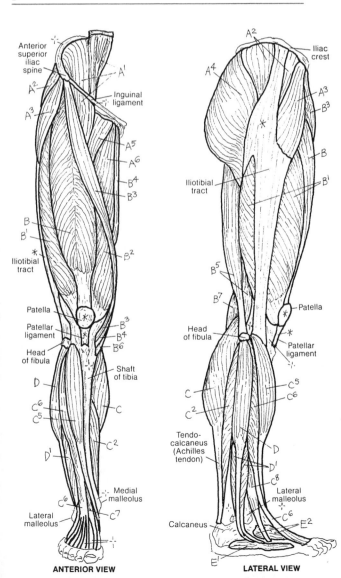

ANTERIOR VIEW

LATERAL VIEW

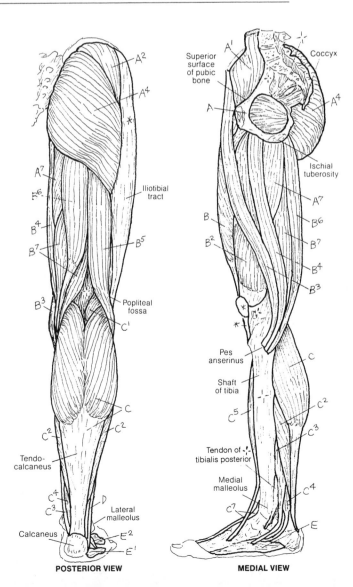

POSTERIOR VIEW

MEDIAL VIEW

FUNCTIONAL OVERVIEW

FLEXOR_A
EXTENSOR_B
ABDUCTOR_C
ADDUCTOR_D
ROTATOR_E
SCAPULAR
    STABILIZER_F
EVERTOR_G
INVERTOR_H

CN: Use light colors throughout (especially for A and B). Deeper muscles are not included in the large illustrations. (1) Color all of the muscle groups in the anterior view before going on to the posterior view at right. Only the muscles on one side of the figure have been labeled. As you color the muscle, also color its opposite. (2) Color the small diagram below.

Upon coloring these functional groups, note the spatial relationship of adductors to abductors and evertors to invertors. Take particular note of the extensors and flexors. Recall that extension of weightbearing joints is an anti-gravity function, and extensor muscles of these joints tend to keep the standing body vertically straight. Note the line of gravity and its relationship to the vertebral, hip, knee, and ankle joints. The center of gravity of an average human being standing with perfect posture is just anterior to the motion segment of S1–S2. Flexion of the neck and torso moves the center of gravity forward, loading the posterior cervical, thoracic, and lumbar paraspinal (extensor) muscles. The actors moving the vertebral, hip, knee, and ankle joints make possible erect standing and walking/running posture.

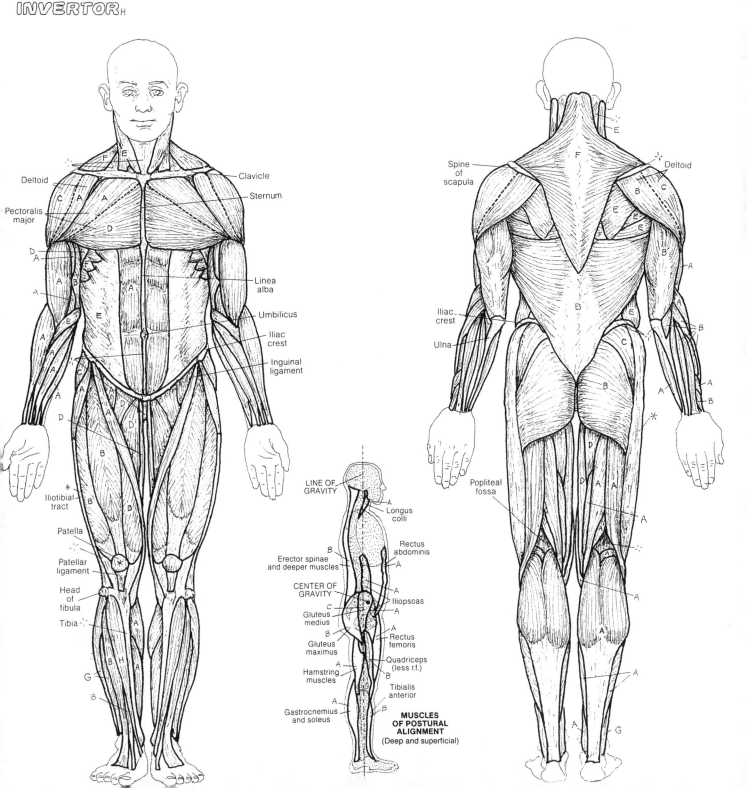

MUSCLES
OF POSTURAL
ALIGNMENT
(Deep and superficial)

ORGANIZATION

CN: Use very light colors for A and C. The numbers in parentheses following the titles under Spinal Nerves refer to the number of nerves in each of the regions listed. (1) In the central illustration, the spinal cord has been brought out of the vertebral column to show its regions in relation to the vertebrae. Spinal nerves, depicting regional limits, are shown with arrowheads pointing to the same spinal nerves emerging from the vertebral column. Avoid coloring the filum terminale—it is not a spinal nerve. (2) At upper right, color the cranial nerves. (3) At lower right, color over the lines representing the spinal nerves and their branches on the left side of the figure. Color the autonomic ganglia on the right side of the spinal cord.

CENTRAL NERVOUS SYSTEM (CNS)

BRAIN
CEREBRUM A
BRAINSTEM B
CEREBELLUM C
SPINAL CORD D
REGIONS D-
CERV. G THOR. H LUM. I SAC. J CO. K

The nervous system consists of neurons arranged into a highly integrated central part (central nervous system, or CNS) and bundles of neuronal processes (nerves) and islands of neurons (ganglia) largely outside the CNS making up the peripheral part (peripheral nervous system, or PNS). These neurons are supported by neuroglial cells and a rich blood supply. Neurons of the CNS are interconnected to form centers (nuclei; gray matter) and axon bundles (tracts; white matter). The brain is the center of sensory awareness and movement, emotions, rational thought and behavior, foresight and planning, memory, speech, and language and interpretation of language.

The spinal cord, an extension of the brain and part of the CNS, begins at the foramen magnum of the skull, traffics in ascending/descending impulses, and is a center for spinal reflexes, source of motor commands for muscles below the head, and receiver of sensory input below the head.

PERIPHERAL NERVOUS SYSTEM (PNS)

CRANIAL NERVES (12 PAIR) E
SPINAL NERVES & BRANCHES F
CERVICAL (8) G'
THORACIC (12) H'
LUMBAR (5) I'
SACRAL (5) J'
COCCYGEAL (1) K'

AUTONOMIC NERVOUS SYS.
SYMPATHETIC DIV. L
PARASYMPATHETIC DIV. M

The PNS consists largely of bundles of sensory and motor axons (nerves) radiating from the brain (*cranial nerves*) and spinal cord (*spinal nerves*) segmentally and bilaterally and reaching to all parts of the body (visceral and somatic) through a classic pattern of distribution. *Branches* of spinal nerves are often called peripheral nerves. Nerves conduct all sensations from the body to the brain and spinal cord; they conduct motor commands to all the skeletal muscles of the body. The *autonomic nervous system* (ANS) is a subset of ganglia and nerves in the PNS dedicated to visceral movement and glandular secretion and to the conduction of visceral sensations to the spinal cord and brain.

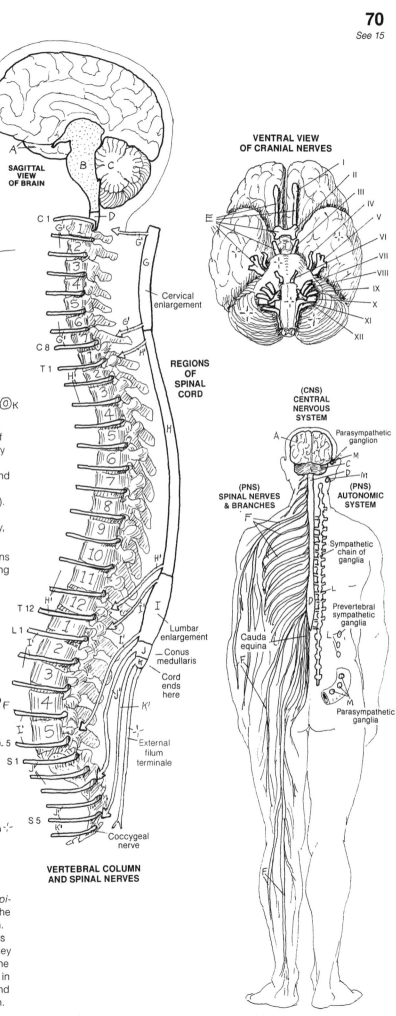

SAGITTAL VIEW OF BRAIN

Cervical enlargement

REGIONS OF SPINAL CORD

Lumbar enlargement

Conus medullaris

Cord ends here

External filum terminale

Coccygeal nerve

VERTEBRAL COLUMN AND SPINAL NERVES

VENTRAL VIEW OF CRANIAL NERVES

(CNS) CENTRAL NERVOUS SYSTEM

Parasympathetic ganglion

(PNS) SPINAL NERVES & BRANCHES

(PNS) AUTONOMIC SYSTEM

Sympathetic chain of ganglia

Prevertebral sympathetic ganglia

Cauda equina

Parasympathetic ganglia

FUNCTIONAL CLASSIFICATION OF NEURONS

CN: Use light colors throughout the plate.
Do not color the summary diagram at the top of
the page until completing the rest of the plate.

Neurons generally function in one of three modes: They conduct impulses
from receptors in the body to the central nervous system or CNS (sensory
or afferent neurons); they conduct motor command impulses from the
CNS to muscles of the body (motor or efferent neurons); or they form a
network of interconnecting neurons in the CNS between motor and sensory
neurons (interneurons). If the sensory or motor neurons relate to musculoskeletal structures or the skin and fascia, the prefix "somatic" may be applied
(somatic afferent/somatic efferent). If these neurons are related to organs with
hollow cavities (viscera), the prefix "visceral" may be applied (visceral
afferent/visceral efferent).

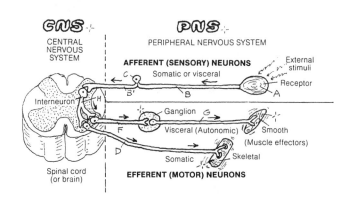

PNS:
SENSORY NEURON
RECEPTOR A
AXON
(PERIPHERAL PROCESS) B
CELL BODY C
AXON
(CENTRAL PROCESS) B'

Sensory neurons conduct impulses from sensory *receptors* to synapses
in the CNS. The receptors may be sensitive to touch, pressure, pain, joint
position, muscle tension, chemical concentration, light, or other mechanical
stimulus, basically providing information on the external or internal environment and related changes. Sensory neurons are unipolar neurons, with certain
exceptions (bipolar neurons), and are characterized by *peripheral processes*
("axons"), *cell bodies*, and *central processes* ("axons").

PNS:
SOMATIC MOTOR N.
DENDRITE D
CELL BODY C'
AXON
MOTOR END PLATE E

Motor neurons conduct impulses from *cell bodies* located in the CNS through
axons that leave the CNS and subsequently divide into branches, each of
which becomes incorporated into the cell membrane of a muscle cell (*motor
end plate*). Here the neuron releases its neurotransmitter, which induces the
muscle cell to shorten.

PNS:
AUTONOMIC MOTOR N.
PREGANGLIONIC NEURON F
POSTGANGLIONIC NEURON G

Autonomic motor neurons function as paired units connected at a ganglion
by a synapse. The first or *preganglionic neuron* arises in the CNS, and its axon
embarks for a ganglion located some distance from the CNS. There it synapses with the cell body or dendrite of a *postganglionic neuron* whose axon
proceeds to the effector organ: smooth muscle, cardiac muscle, or glands.

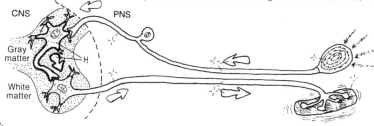

CNS:
INTERNEURON
(ASSOCIATION N.) H

Interneurons are found mostly in the CNS. They make up the bulk of the neurons of the brain and spinal cord. They come in a variety of shapes and sizes.
Many of them are directly related to incoming (sensory) impulses and others to
outgoing motor commands. Others serve to integrate sensory or ascending
input with higher centers to effect an appropriate motor output.

SYNAPSES & NEUROTRANSMITTERS

CN: Use light colors for A, B, and C. (1) In the upper drawing, each of the synapses shown has two parts. Color only the ones labeled with subscripts (A, B. C). Color the nerve impulse title (D, at the top) and the related directional arrows. (2) Color the numbered steps in the lower drawing. Note the change of color in the presynaptic membrane between exocytosis (H) and endocytosis (K).

BASIC TYPES OF SYNAPSES

AXO_A AXONIC_A
AXO_A SOMATIC_B
AXO_A DENDRITIC_C

Connections between and among neurons are called synapses. The great majority are non-contact connections in which chemical neurotransmitters carry the impulse from one neuron to another. Electrical synapses (where electrically charged atoms or ions pass from one neuron to another by way of protein channels; not shown) also exist in the brain and embryonic nervous tissue but are far less common. Most synapses are *axodendritic*; that is, the axon of one neuron synapses with the dendrite or dendritic spine of another neuron. The neuron in front of the synapse is said to be presynaptic. The second neuron is said to be postsynaptic. Another common synapse is *axosomatic*, where the axon of one neuron and the cell body (soma) of another neuron communicate by way of neurotransmitters. Other, more infrequently seen synapses are illustrated here as well. For example, note the complex of synapses (a glomerulus) between three axons and a dendritic spine, all surrounded by a neuroglial sheath.

Synapses permit the conduction of electrochemical impulses among myriad neurons almost instantly. Synapses vary from simple reflex arcs (see Plate 85) to polysynaptic pathways in the brain and spinal cord that involve millions of synapses. A single motor neuron of the spinal cord may have as many as 10,000 synapses on its body and dendrites! Multiple synapses greatly increase the available options of nervous activity. The ability to integrate, coordinate, associate, and modify sensory input and memory to achieve a desired motor command is directly related to the number of synapses in the pathway.

TYPICAL SYNAPSE

PRESYNAPTIC AXON_A
PRESYNAPTIC MEMBRANE_E
SYNAPTIC VESICLE_F
NEUROTRANSMITTER_G
FRAGMENT_G'
EXOCYTOSIS_H
SYNAPTIC CLEFT_I
POSTSYNAPTIC MEMBRANE_J
RECEPTOR_G'
ENDOCYTOSIS_K

Here we show a typical axodendritic synapse. (1) The *presynaptic axon* transmits the electrochemical impulse toward the synapse. As the impulse reaches the axon terminal, calcium ion (Ca^{++}) channels/gates are opened in the cell membrane, and extracellular Ca^{++} pours into the axon terminal. (2) *Synaptic vesicles*, loaded with *neurotransmitter* (e.g., acetylcholine, norepinephrine), influenced by the incoming Ca^{++}, migrate toward the *presynaptic membrane* and fuse with it (3). Following fusion, neurotransmitter is spilled from the vesicles into the tiny synaptic cleft (*exocytosis*). Neurotransmitter molecules bind to receptor proteins on the *postsynaptic membrane* of the dendrite; ion channels are opened, and the altered membrane potential (impulse) is propagated along the dendrite (4). Inactivated neurotransmitter *fragments* are taken up by the presynaptic membrane (5; *endocytosis*), enclosed in a synaptic vesicle, and resynthesized (6).

The electrical activity of the postsynaptic membrane may be facilitated or inhibited by the neurotransmitter. If sufficiently excited by multiple facilitory synapses, the postsynaptic neuron will depolarize and transmit an impulse to the next neuron or effector (muscle cell, gland cell). Sufficiently depressed by multiple inhibitory synapses, the neuron will not depolarize and will not transmit an impulse.

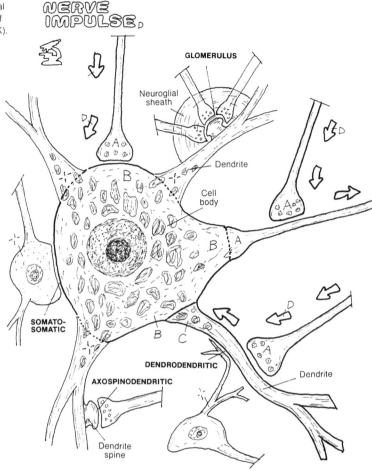

NERVE IMPULSE_D

GLOMERULUS

Neuroglial sheath

Dendrite

Cell body

SOMATO-SOMATIC

DENDRODENDRITIC

AXOSPINODENDRITIC

Dendrite spine

Dendrite

TYPES OF SYNAPSES

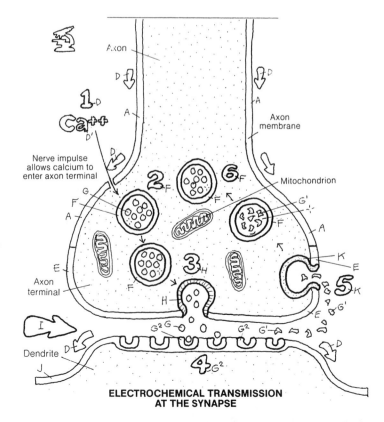

Axon

Nerve impulse allows calcium to enter axon terminal

Axon membrane

Mitochondrion

Axon terminal

Dendrite

ELECTROCHEMICAL TRANSMISSION AT THE SYNAPSE

CEREBRAL HEMISPHERES

CN: Use light colors for B, E, I, and J. (1) Color the coronal section; most of the frontal lobe and part of the temporal lobe have been removed. Color the cerebral cortex (A) gray. In the two large hemispheres the stippled areas of specialized function are parts of lobes, but receive their own colors. Color the arrows identifying the major fissures and sulcus. (3) Color gray the diagram illustrating how the convolutions provide increased surface area in a smaller space.

CEREBRAL CORTEX (GRAY MATTER)A*

FRONTAL LOBEB
PRINCIPAL SPEECH AREAc
PRIMARY MOTOR AREA (PRECENTRAL GYRUS)D
PARIETAL LOBEE
PRIMARY SENSORY AREA (POSTCENTRAL GYRUS)F
TEMPORAL LOBEG
AUDITORY AREAH
OCCIPITAL LOBEI
VISUAL AREAJ

MAJOR FISSURES/SULCUS
LONGITUDINAL FISSUREK
CENTRAL SULCUSL
LATERAL FISSUREM

SUBCORTICAL WHITE MATTERN

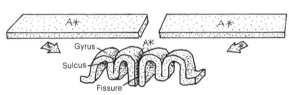

CORTICAL CONVOLUTIONS: INCREASED SURFACE AREA

The paired cerebral hemispheres (cerebrum), derivatives of the embryonic telencephalon (see Plate 169), consist of four major elements: (1) an outer cerebral cortex of gray matter, the topography of which reveals fissures (deep grooves), gyri (hills), and sulci (furrows); (2) underlying white matter consisting of numerous tracts destined for or leaving the cortex and oriented along three general directions (Plate 74); (3) discrete masses of gray matter at the base of the cerebrum (basal nuclei) that subserve motor areas of the cortex (Plate 74); (4) paired cavities called lateral ventricles (Plate 80). The cerebral cortex is the most highly evolved area of the brain. About 2–4 mm (roughly 1/6 inch) thick, the cortex is divided into lobes distinctly bordered by sulci; the lobes are generally related to the cranial bones that cover them: frontal, parietal, temporal, occipital. The exception is the limbic lobe (part of which is shown); it incorporates parts of other (frontal, temporal, parietal) lobes.

Cortical mapping experiments (based on electrical stimulation and clinical/pathologic data) have been the principal methods by which functions of the cortex have been discovered. All parts of the cortex are concerned with storage of experience (memory), exchange of impulses with other cortical areas (association), and the two-way transmission of impulses with subcortical areas (afferent/efferent projections).

The frontal lobe is concerned with intellectual functions such as reasoning and abstract thinking, aggression, sexual behavior, olfaction (smell), articulation of meaningful sound (speech), and voluntary move-

ment (precentral gyrus). The central sulcus separates the frontal lobe from the parietal lobe. The parietal lobe is concerned with body sensory awareness, including taste (postcentral gyrus), the use of symbols for communication (language), abstract reasoning (e.g., mathematics), and body imaging. The temporal lobe is partly limbic and here is concerned with the formation of emotions (love, anger, aggression, compulsion, sexual behavior); the non-limbic portion of the temporal lobe is concerned with interpretation of language and awareness and discrimination of sound (hearing; auditory area); it constitutes a major memory processing area. The occipital lobe is concerned with receiving, interpreting, and discriminating visual stimuli from the optic tract and associating those visual impulses with other cortical areas (e.g., memory).

The limbic lobe or system is the oldest part of the cortex, in evolutionary terms. It is the center of emotional behavior. The limbic neurons occupy parts of the inferior and medial cortices of each hemisphere, and some subcortical areas as well. Certain limbic areas are closely related topographically to the olfactory tracts.

The cerebral hemispheres appear structurally as mirror images of one another; functionally they are not. The speech area develops fully only on one side, usually the left. In general, the left hemisphere tends to deal with certain higher functions (mathematical, analytical, verbal) while the right concentrates on visual, spatial, and musical orientations. The matter of cerebral "dominance" (left hemisphere, left speech center, righthandedness, or vice versa) is quite controversial.

TRACTS/NUCLEI OF CEREBRAL HEMISPHERES

CN: Use very light colors for F and G. (1) Color gray the various sections of cerebral cortex without coloring the cortical surfaces.

CEREBRAL CORTEX A*
SUBCORTICAL AREAS -:-
BASAL NUCLEI -:-
CAUDATE NUCLEUS B
LENTICULAR NUCLEUS C
LATERAL VENTRICLE D
WHITE MATTER TRACTS -:-
COMMISSURES E
CORPUS CALLOSUM E'
PROJECTION TRACTS F
CORONA RADIATA F'
INTERNAL CAPSULE F²
ASSOCIATION TRACTS G

Below the cerebral cortex, the cerebral hemispheres embody centrally placed cavities, masses of gray matter at the base of the cerebrum, and bundles of white matter. These structures can be colored in the coronal section at upper right cerebrum (1).

The basal nuclei are discrete, bilateral islands of gray matter on either side of and above the diencephalon (Plate75). They consist primarily of the tail-shaped *caudate nucleus* and the lens-shaped *lenticular nucleus* (2). These structures can be colored in illustrations 1, 2, 3, and 4. The lenticular nucleus is further divided into the medial globus pallidus and the more lateral putamen (1). Each of these nuclei is part of the extra-pyramidal system (Plate 79). They have extensive connections among themselves, with the cerebral cortex, and with nuclei of the diencephalon. They are concerned with the maintenance of muscle tone and the programming of subconscious, sequential postural adjustments. They monitor and mediate descending motor commands from the cerebral cortex.

The subcortical white matter of the hemispheres is arranged into bundles or bands (tracts) of largely myelinated axons essentially arranged in three axes (5). They conduct impulses among various areas of the cortex. The largest commissure is the *corpus-callosum*, forming a roof over the subcortical nuclei (1). It is bent caudally at both anterior and posterior ends (genu and splenium) (4, 6). Association tracts connect anterior and posterior cerebral cortices (5). They exist as both short and long tracts (6).

The most spectacular tract is the fan-shaped array of fibers called the *corona radiata* (5, 7). This projection system radiates caudally from all areas of the cortex. It narrows into a curved band (internal capsule; 1, 2, 3, 4) as it descends between the caudate nucleus and the thalamus medially, and the putamen laterally. The term "internal capsule" refers to the inner wall of the figurative encapsulation of the basal nuclei (1, note external capsule). The axons of the projection tract continue through the diencephalon into the brain stem and spinal cord; many make connections en route (Plates 78, 79).

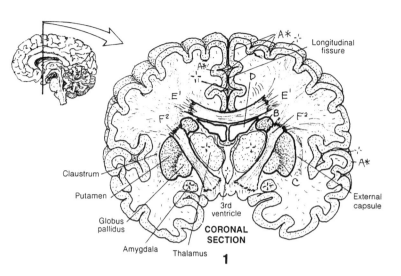

CORONAL SECTION
1

Longitudinal fissure — Claustrum — Putamen — Globus pallidus — Amygdala — Thalamus — 3rd ventricle — External capsule

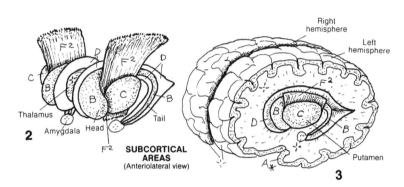

SUBCORTICAL AREAS
(Anteriolateral view)
2

Thalamus — Amygdala — Head — Tail — Putamen — Right hemisphere — Left hemisphere
3

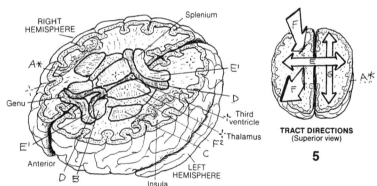

HORIZONTAL SECTION
4

RIGHT HEMISPHERE — Splenium — Genu — Third ventricle — Thalamus — Anterior — Insula — LEFT HEMISPHERE

TRACT DIRECTIONS
(Superior view)
5

MEDIAL VIEW
(Right hemisphere section)
7

Putamen — Anterior

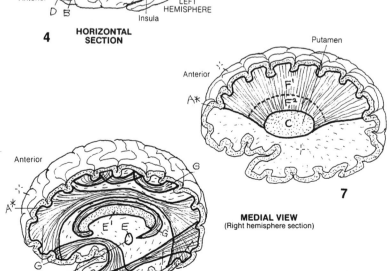

6

Anterior — Anterior commissure

DIENCEPHALON

CN: Use light colors for A and B, and a very bright color for C. (1) Color each structure wherever it appears before going on to the next title. (2) Although not colored, the neighboring relations of the diencephalic structures are important and have been identified by name. These should be given special attention.

DIENCEPHALON ÷
THALAMUS A
HYPOTHALAMUS B
EPITHALAMUS
(PINEAL GLAND) C
THIRD VENTRICLE D

The diencephalon, the smaller of two derivatives of the early forebrain, fits between but is not part of the surrounding cerebral hemispheres (see drawings 2 and 3). It consists largely of paired masses of nuclei and related tracts of white matter arranged around the thin, purse-like third (III) ventricle (2 and 3). The nature of this cavity can be seen in Plate 80.

On each side of the third ventricle, note the *thalamus*, *subthalamus*, and *hypothalamus* (2 and 3). The *epithalamus* or *pineal gland* is a midline structure seeming to hang off the posterior thalamus. The relationship of these nuclei to the basal nuclei and internal capsule should be carefully studied while coloring to ensure orientation (recall Plate 74).

The thalamus (1–4) consists of several groups of cell bodies and processes that, in part, process all incoming impulses from sensory pathways (except olfactory). It has broad connections with the motor, general sensory, visual, auditory, and association cortices. Not surprisingly, the corticothalamic (cortex to thalamus) fibers contribute significantly to the *corona radiata*. Still other thalamic nuclei connect to the *hypothalamus* and other brainstem nuclei. Thalamic activity (1) integrates sensory experiences resulting in appropriate motor responses, (2) integrates specific sensory input with emotional (motor) responses (e.g., a baby crying in response to hunger), and (3) regulates and maintains the conscious state (awareness), subject to facilitating/inhibiting influences from the cortex. *Subthalamic nuclei* (3) are concerned with motor activity and have connections with the basal ganglia.

The hypothalamus (1, 3, and 4) consists of nuclear masses and associated tracts on either side of the lower third ventricle. The hypothalamus maintains neuronal connections with the frontal and temporal cortices, thalamus, neurohypophysis, and brainstem. Its neurosecretions (hormones) are also directed to the *adenohypophysis* via the hypophyseal portal system. In addition, the hypothalamus is concerned with emotional behavior, regulation of the autonomic (visceral) nervous system and related integration of visceral (autonomic) reflexes with emotional reactions, and activation of the drive to eat (hunger) and the subsequent feeling of satisfaction (satiety) following fulfillment of that drive. Finally, it mediates descending impulses related to both reflexive and skilled movement—all of this in an area the size of four peas!

The epithalamus (pineal gland) (1) consists primarily of the pineal body and related nuclei and tracts that have connections with the thalamus, hypothalamus, basal nuclei, and the medial temporal cortex. It produces melatonin (a pigment-enhancing hormone), the synthesis of which is related to diurnal cycles or rhythms (body activity in day or sunlight as opposed to dark or nocturnal periods). It may influence the onset of puberty through inhibition of testicular/ovarian function. Remarkably, the pineal is the only unpaired structure in the brain.

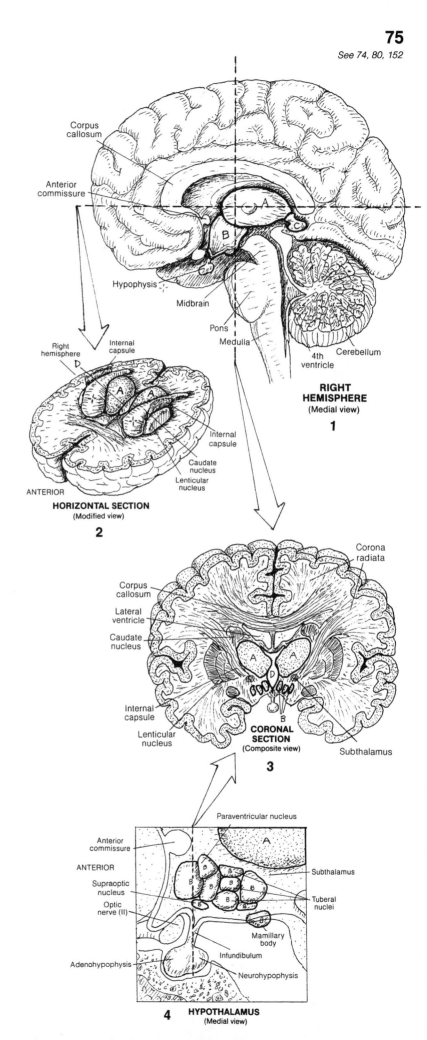

RIGHT HEMISPHERE
(Medial view)
1

Corpus callosum
Anterior commissure
Hypophysis
Midbrain
Pons
Medulla
4th ventricle
Cerebellum

Right hemisphere
Internal capsule
Internal capsule
Caudate nucleus
Lenticular nucleus
ANTERIOR
HORIZONTAL SECTION
(Modified view)
2

Corona radiata
Corpus callosum
Lateral ventricle
Caudate nucleus
Internal capsule
Lenticular nucleus
Subthalamus
CORONAL SECTION
(Composite view)
3

Paraventricular nucleus
Anterior commissure
ANTERIOR
Supraoptic nucleus
Optic nerve (II)
Adenohypophysis
Subthalamus
Tuberal nuclei
Mamillary body
Infundibulum
Neurohypophysis
4 HYPOTHALAMUS
(Medial view)

BRAIN STEM/CEREBELLUM

CN: Use darker colors for C, E, and M and the lightest for K. (1) As you color each structure in as many views as it is shown, take particular note of the orientation of the view. (2) Note that the fourth ventricle is located in both parts of the hindbrain and receives the same color in both parts. The diencephalon has been presented on the previous plate and is shown here only for orientation.

BRAIN STEM /

DIENCEPHALON A
MIDBRAIN B
CEREBRAL AQUEDUCT C
SUPERIOR COLLICULUS B'
INFERIOR COLLICULUS B²
CEREBRAL PEDUNCLE F
SUP. CEREBELLAR PEDUNCLE D

HINDBRAIN /
4TH VENTRICLE E
PONS F
MID. CEREBELL. PED. G
MEDULLA OBLONGATA H
INF. CEREBELL. PED. I

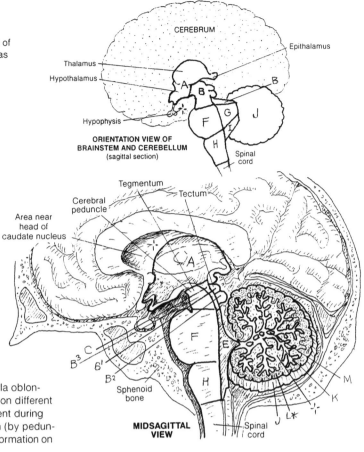

ORIENTATION VIEW OF
BRAINSTEM AND CEREBELLUM
(sagittal section)

MIDSAGITTAL
VIEW

The brain stem includes the diencephalon, midbrain, pons, and medulla oblongata. Throughout the brain stem, the brain cavity (Plates 80, 82) takes on different shapes, a reflection of the kind of differential growth the brain underwent during development (Plate 169). The cerebellum is attached to the brain stem (by peduncles) but is not considered a part of the brain stem. See Plate 75 for information on the diencephalon.

In the midbrain, the *cerebral peduncles* are composed of long descending tracts that originate in the cerebral cortex, descend through the internal capsule (recall Plate 74), and continue caudally to the pons and medulla (for cranial nerves) and the spinal cord (for spinal nerves). Immediately posterior to these tracts in the midbrain is the tegmentum, an area of neurons associated with the reticular formation and cranial nerves III and IV, and multiple tracts. The superior cerebellar peduncles transmit fibers to the *cerebellum* from the spinal cord, and fibers to the thalamus and medulla from the cerebellum. The superior colliculi are centers for visual reflexes; the inferior colliculi make possible auditory reflexes (e.g., rapid, involuntary movements in response to visual and auditory stimuli).

The pons is characterized by its massive anterior bulge consisting of stalks of white matter that bridge the 4th ventricle (pons = bridge) to reach the cerebellum as the middle cerebellar peduncles. These fibers largely arise from neurons in the pons—neurons that convey impulses from both motor and sensory areas of the cerebral cortex. Cranial nerve nuclei V, VI, VII, and VIII are located here. Both ascending and descending tracts pass through here, including the neurons of the reticular formation.

The medulla contains life-sustaining control centers of respiration, heart rate, and vasomotor function. Nuclei for cranial nerves VIII, IX, X, XI, and XII exist here. The inferior cerebellar peduncles carry fibers to the cerebellum from the spinal cord and brain stem vestibular and reticular systems, as well as fibers from the cerebellum to the vestibular system.

CEREBELLUM J
ARBOR VITAE K
CEREBELLAR CORTEX L*
DEEP CEREB. NUCLEUS M

The cerebellum consists of two hemispheres, with a cortex of gray matter on its surface (*cerebellar cortex*), central masses of motor-related (*deep cerebellar*) nuclei, and bands of white matter forming a treelike appearance (*arbor vitae* = tree of life) when the cerebellum is cut in section. The cerebellum is attached to the brain stem by the three cerebellar peduncles. The cerebellum is concerned with equilibrium and position sense, fine movement, control of muscle tone, and overall coordination of muscular activity in response to proprioceptive input and descending traffic from higher centers.

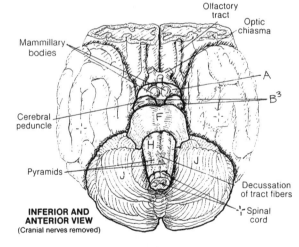

INFERIOR AND
ANTERIOR VIEW
(Cranial nerves removed)

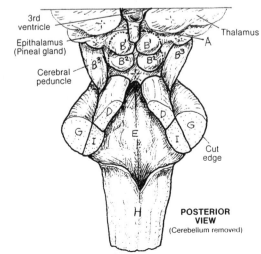

POSTERIOR
VIEW
(Cerebellum removed)

SPINAL CORD

CN: Use bright colors for A–C (except where indicated by asterisk (*) or no-color symbol (-¦-). Use gray for D*, medium dark colors for E–I, and light colors for K–M. (1) In the upper drawing, color B* gray over the nerve roots within the dura mater and outside the spinal cord. (2) Color the cord sections taken at various levels. (3) Color the meninges of the spinal cord. What is not shown (due to space limitations) is the presence of the arachnoid and subarachnoid space (and cerebrospinal fluid) around the nerve roots. (4) Do not color the structures within the subarachnoid space or the central canal in the drawing at the bottom of the plate.

SPINAL CORD A
MENINGES -¦-
PIA MATER A'
INTERNAL FILUM TERMINALE A²
SUBARACHNOID SPACE B*
ARACHNOID B'
DURA MATER c
EXTERNAL FILUM TERMINALE c'
EPIDURAL SPACE c²-¦-

The spinal cord is the lower extension of the central nervous system. It takes off from the medulla oblongata at the foramen magnum of the skull and ends as the *conus medullaris* at the vertebral level of L1 or L2. It bulges slightly in the lower cervical and lumbar segments where it gives off the roots of spinal nerves destined for the upper and lower limbs, respectively. The cord is ensheathed by *three coverings (meninges)*: the thin, vascular *pia mater* closely applied to the spinal cord, the translucent *arachnoid* separated from the pia by the subarachnoid space, and the tough, fibrous *dura mater* that is a prolongation of the dura surrounding the brain.

The pia forms triangular sheets that project away from the cord between the pairs of nerve roots. These sheets extend to the dura and are called *denticulate ligaments*. The pia extends below the conus medullaris as the thin cord-like *filum terminale internum*, ending at the dural sac at the vertebral level of S2. The subarachnoid space is filled with *cerebrospinal fluid (CSF)*; the space ends inferiorly as the dural sac; its CSF-containing cavity is the lumbar cistern. Outside the dura is the *epidural space*, containing fat and veins.

GRAY MATTER D*
POSTERIOR HORN E
ANTERIOR HORN F
LATERAL HORN (T1-L2) G
INTERMEDIATE ZONE H
GRAY COMMISSURE I

WHITE MATTER J-¦-
POSTERIOR FUNICULUS K
LATERAL FUNICULUS L
ANTERIOR FUNICULUS M

The spinal cord consists of a central mass of gray matter arranged into the form of an H and a peripheral array of white matter *(funiculi)* consisting of descending and ascending tracts. The amount of white matter decreases as the cord progresses distally, seen especially well in the sacro-coccygeal region. The gray *posterior horns* (actually columns when seen in three dimensions) receive the central processes of sensory neurons (recall Plate 71) and direct incoming impulses to the adjacent white matter for conduction to other cord levels or higher centers. The *anterior horns* include lower motor neurons that represent the "final common pathway" for motor commands to muscle. *Lateral horns* exist only in the thoracic and upper lumbar cord and include autonomic motor neurons supplying smooth muscle (in vessels and viscera) and glands. It is in the gray matter that spinal reflexes occur in conjunction with facilitory and inhibitory influences from higher centers.

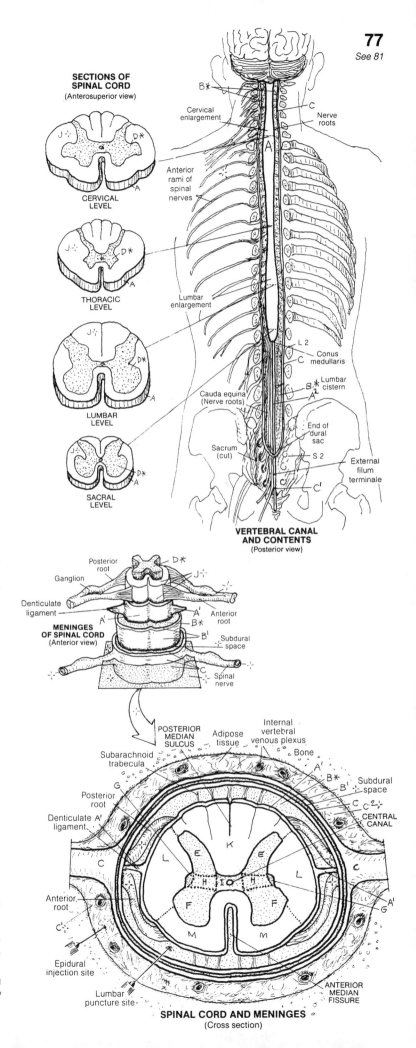

ASCENDING TRACTS

CN: Use bright colors for A–C and a light color for F. (1) Color the pain/temperature pathway, which is shown on one side only for visual simplicity. Note that the sensory cortex and the thalamus are to be colored gray. (2) In the muscle stretch/position sense pathways, note there are two different cerebellar peduncles, each receiving a different color.

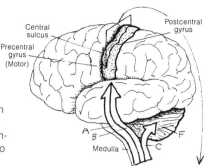

Ascending pathways consist of linearly arranged neurons, the axons of which travel in a common bundle (tract) conducting impulses toward the thalamus, cerebral cortex, or cerebellum. In the examples shown here, each of the pathways begins with a sensory neuron. These sensory pathways permit body surface sensations and muscle/tendon stretch information (below the head) to reach brain stem and cerebellar centers for response and cortical centers for awareness.

PAIN/TEMPERATURE A

SENSORY NEURON A¹
LAT. SPINOTHALAMIC TRACT A²
THALAMUS *¹
THALAMOCORTICAL TRACT A³
SENSORY CORTEX *²

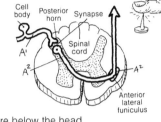

Pain and temperature receptors on the body surface and elsewhere below the head generate impulses that travel to the spinal cord by axons of *sensory neurons* (1st-order neuron). The central process ("axon") of each sensory neuron enters the posterior horn and synapses with the 2nd-order neuron whose axon crosses (decussates) to the contralateral side, enters the lateral funiculus, and ascends as part of the *lateral spinothalamic tract*. This neuron ascends to the thalamus, where it synapses with relay (3rd-order) neurons, the axons of which traverse the internal capsule and corona radiata (*thalamocortical tract*) to reach the postcentral gyrus of the cerebral cortex ("sensory cortex").

TOUCH/PRESSURE B

SENSORY NEURON B¹
N. CUNEATUS & GRACILIS B²
INT. ARCUATE FIBERS B³
MED. LEMNISCUS B⁴
THALAMUS *¹
THALAMOCORTICAL TRACT B⁵
SENSORY CORTEX *²

Touch and pressure receptors below the head generate electrochemical impulses that travel to the spinal cord through *sensory neurons* that enter the posterior horn and join/ ascend the posterior funiculus (posterior columns) to the medulla. Here they synapse with 2nd-order neurons in the *nuclei cuneatus* and *gracilis*. The axons of these neurons sweep to the opposite side (as *internal arcuate fibers*) to form an ascending bundle (*medial lemniscus*) in the brain stem that terminates in the thalamus. There these axons synapse with 3rd-order relay neurons whose axons reach the postcentral gyrus of the cerebral cortex via the *thalamocortical tract.*

MUSCLE STRETCH / POSITION SENSE c

SENSORY NEURON C¹
POST. SPINOCEREBELLAR TRACT C²
INF. CEREBELLAR PED. D
ANT. SPINOCEREBELLAR TR. C³
SUP. CEREBELLAR PED. E
CEREBELLAR CORTEX F

Impulses from muscle spindles and other proprioceptors (receptors responsive to muscle stretch/loads) are conducted by *sensory neurons* to the spinal cord. Single receptor input is conducted by 2nd-order neurons that ascend the ipsilateral lateral funiculus (*posterior spinocerebellar tract*) and enter the cerebellum via the *inferior cerebellar peduncle*. More global proprioceptive input ascends the contralateral anterior *spinocerebellar tract* and enters the cerebellum via the *superior cerebellar peduncle*. By these and similar pathways that function in the absence of awareness, the cerebellum maintains an ongoing assessment of body position, muscle tension, muscle overuse, and movement. In turn, it mediates descending impulses from cortical and subcortical centers destined for motor neurons.

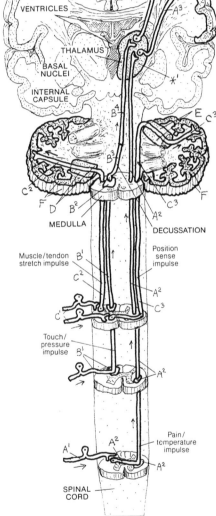

CEREBRAL CORTEX, CEREBELLUM, AND SPINAL CORD

(Schematic)

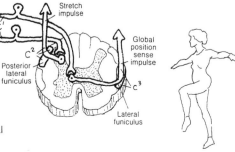

DESCENDING TRACTS

CN: Use light colors for H, I, and K. (1) Color the pyramidal tract in the sagittal view. (2) Color the pyramidal tract in the schematic coronal section at upper right, including the percentage figures. (3) Color the extra-pyramidal system.

PYRAMIDAL TRACT A / RELATED AREAS

MOTOR CORTEX *

CORTICOSPINAL TRACT A'

LAT. A² / ANT. CORTICOSPINAL TRACT A³

MEDULLARY PYRAMID A⁴

LATERAL FUNICULUS B

ANTERIOR FUNICULUS C

FINAL COMMON PATHWAY

LOWER MOTOR NEURON D

EFFECTOR E

The principal neural pathway for voluntary movement is the *corticospinal tract*. Its neuronal cell bodies are in the pre-central gyrus of each frontal lobe (motor cortex). The axons of these neurons descend—without synapse—through the corona radiata, internal capsule, cerebral peduncles, pons, and medulla into the spinal cord. Pathways are often named according to their origin and their termination, and in that order; hence, cortico- (referring to cortex) spinal (referring to the spinal cord). The corticospinal tracts form bulges, called pyramids, on the anterior surface of the medulla, hence the name "*pyramidal tract*." Eighty percent of these tracts cross (decussate) to the contralateral side in the medulla (*decussation of the pyramids*); 20% do not. In the spinal cord, many corticospinal fibers terminate on inter-neurons (recall Plate 71) at the base of the posterior horn (not shown); the majority end by synapsing with anterior horn motor neurons. Corticospinal input to the lower motor (anterior horn) neurons, however, is only one input for desired skeletal muscle function.

EXTRAPYRAMIDAL SYSTEM

PONTINE RETICULOSPINAL TRACT F

VESTIBULOSPINAL TRACT G

INTERNEURON H

Each lower motor neuron receives axons from multiple descending tracts, many of which conduct impulses related to body position, memory, and a host of other commands necessary for any given movement at any given time. These collective inputs from the cerebral cortex, basal nuclei, cerebellum, and elsewhere arrive at the appropriate lower motor neurons by a number of descending pathways, none of which pass through the medullary pyramids (hence, extrapyramidal system or tracts). Two major extrapyramidal tracts are shown here: the *reticulospinal tract* from the brain stem reticular nuclei and the *vestibulospinal tract* from the vestibular nuclei in the brain stem. Other tracts include the rubrospinal and tectospinal tracts (not shown, but see glos-sary). The synaptic connections of these axons with each lower motor neuron (often by way of *interneurons*) are in the thousands. Depending on the neurotransmitter produced by the presynaptic neuron, the synapse may facilitate or inhibit production of an excitatory impulse from the lower motor neuron. Discharge of the lower motor neuron, or lack of it, is dependent on the sum of the facilitory and inhibitory impulses impinging on it at any moment. Once generated, the electrochemical impulse moving down the axon of the lower motor neuron reaches the effector without further mediation. Thus, the anterior horn motor neuron is truly the final common pathway for the ultimate expression of all nervous activity: muscular contraction.

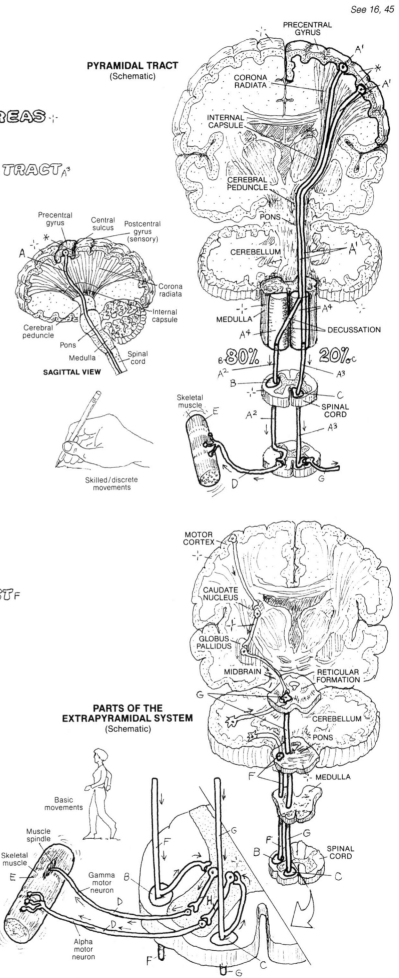

PYRAMIDAL TRACT
(Schematic)

PRECENTRAL GYRUS

CORONA RADIATA

INTERNAL CAPSULE

CEREBRAL PEDUNCLE

PONS

CEREBELLUM

MEDULLA

DECUSSATION

B 80% 20% C

SPINAL CORD

Skeletal muscle

Skilled/discrete movements

Precentral gyrus

Central sulcus

Postcentral gyrus (sensory)

Corona radiata

Internal capsule

Cerebral peduncle

Pons

Medulla

Spinal cord

SAGITTAL VIEW

MOTOR CORTEX

CAUDATE NUCLEUS

GLOBUS PALLIDUS

MIDBRAIN

RETICULAR FORMATION

CEREBELLUM

PONS

MEDULLA

SPINAL CORD

PARTS OF THE EXTRAPYRAMIDAL SYSTEM
(Schematic)

Basic movements

Muscle spindle

Skeletal muscle

Gamma motor neuron

Alpha motor neuron

VENTRICLES OF THE BRAIN

CN: Use a light color for A. (1) Color the drawings of ventricular development; note that the neural cavity in the "8 weeks" drawing corresponds to the color pattern in the lower illustrations. (2) Color the lateral and superior views of the fully developed ventricles. (3) Color the coronal and modified sagittal sections revealing the relationship of the ventricles to surrounding structure.

VENTRICLE DEVELOPMENT:
NEURAL CAVITY OF THE:
FOREBRAIN A
TELENCEPHALON B
DIENCEPHALON C
MESENCEPHALON D
HINDBRAIN E
METENCEPHALON F
MYELENCEPHALON G
SPINAL CORD H

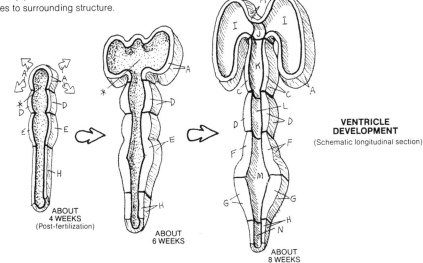

ABOUT 4 WEEKS (Post-fertilization)

ABOUT 6 WEEKS

ABOUT 8 WEEKS

VENTRICLE DEVELOPMENT
(Schematic longitudinal section)

The central nervous system develops from a hollow neural tube near the dorsal surface of the embryo (Plate 169). The neural cavity undergoes extraordinary revision in association with development of the brain regions. The shape of the cavity in each brain region reflects the local changes and mechanical pressures imposed by the developing brain. The ventricles may be identified by name, roman numerals, or arabic numerals.

The cavity of the developing forebrain expands into the lateral (first and second) ventricles with the outpocketings of the growing cerebral hemispheres. Each projection of the lateral ventricles reflects the direction of growth of that part of the cerebrum. The lateral ventricles retain connections to the neural cavity of the diencephalon by means of the bilateral tubular *interventricular foramina*.

The neural tube of the diencephalon is compressed by the developing bilateral thalami into a thin *third ventricle*. The front of the third ventricle is drawn out anteriorly and caudally into an infundibular recess in the region of the hypothalamus. The tubular infundibulum projects into the hypophysis (Plate 152). Posteriorly, the third ventricle projects a small recess into the pineal gland.

The neural cavity of the mesencephalon or midbrain undergoes relatively little distortion during development, retaining its tubular shape as the *cerebral aqueduct*.

The neural cavity of the metencephalon or hindbrain undergoes lateral and posterior expansion as a consequence of the developing cerebellum, forming the diamond-shaped 4th ventricle. The thin roof of these two lateral recesses develops breaks (called apertures) that permit the passage of cerebrospinal fluid (CSF) from the 4th ventricle into the subarachnoid space. Another, more median, aperture, near the beginning of the spinal canal, permits passage of CSF into the subarachnoid space.

The 4th ventricle narrows caudally to become the *central canal* of the spinal cord. This canal is narrow and is often occluded.

In the medial walls of the lateral ventricles and the roof of the 3rd and 4th ventricles, the pia mater comes in contact with the single layer of neuroglia-derived cells that line the ventricles (ependymal cells) to form a delicate, highly vascular tissue called the *choroid plexus*, which secretes CSF into the ventricles.

DERIVATIVES:
LATERAL VENTRICLE (1&2) I
INTERVENTRICULAR FORAMEN J
3RD VENTRICLE K
CEREBRAL AQUEDUCT L
4TH VENTRICLE M
CENTRAL CANAL N

CHOROID PLEXUS O

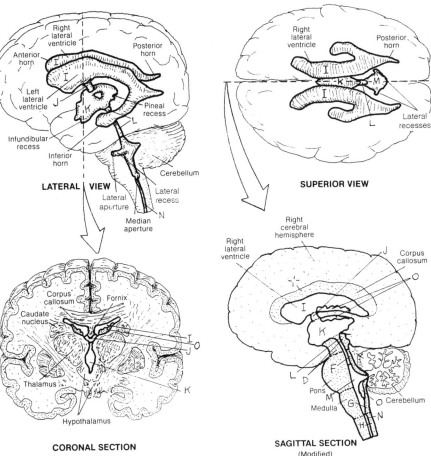

LATERAL VIEW

Right lateral ventricle, Anterior horn, Left lateral ventricle, Infundibular recess, Inferior horn, Posterior horn, Pineal recess, Cerebellum, Lateral recess, Lateral aperture, Median aperture

SUPERIOR VIEW

Right lateral ventricle, Posterior horn, Lateral recesses

CORONAL SECTION

Corpus callosum, Fornix, Caudate nucleus, Thalamus, Hypothalamus

SAGITTAL SECTION
(Modified)

Right cerebral hemisphere, Right lateral ventricle, Corpus callosum, Pons, Medulla, Cerebellum

MENINGES

CN: Use very light colors for A–D. (1) Begin with the dura mater of the brain (A) and spinal cord (A^2). (2) Color the meninges of the brain in the enlargement of the coronal section (meninges are not present in the smaller drawing). In the same enlargement, color the superior sagittal sinus gray (3) Color the infoldings of the dura mater (B–D). These coverings have been thickened and separated from each other for coloring purposes. Color over the darkened sinuses located within the falx cerebri (C). Much of the left half of the overlying inner dura layer (B) has been cut away to reveal inner structures. Visualize how the right and left cerebral hemispheres would fit within these coverings.

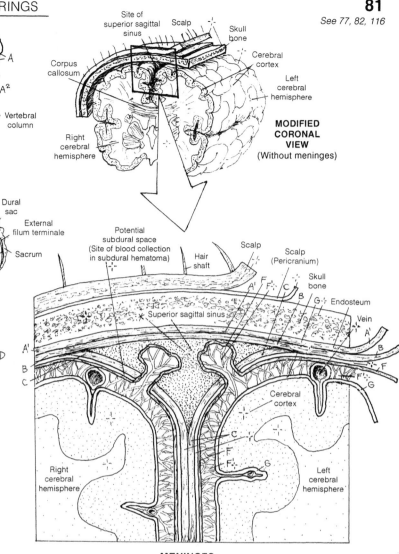

MODIFIED CORONAL VIEW
(Without meninges)

DURA MATER OF BRAIN AND SPINAL CORD

CRANIAL MENINGES

DURA MATER A
OUTER (PERIOSTEAL) LAYER A'
INNER (MENINGEAL) LAYER B
FALX CEREBRI C
TENTORUMI CEREBELLI D
FALX CEREBELLI E (N.S.)
ARACHNOID F
SUBARACHNOID SPACE F
PIA MATER G

SPINAL DURA MATER A^2

The brain and spinal cord are enveloped in fibrous coverings called *meninges*. The meninges of the spinal cord, which were presented in Plate 77, are the inferior extent of the cranial membranes presented here.

The outermost covering of the brain and spinal cord is the *dura mater*. It has two layers: the *outer periosteal layer* lining the internal surface of the cranium and vertebral canal (endosteum), and the inner or meningeal layer, split off from the endosteum, enclosing the entire brain (cranial dura mater) and spinal cord (spinal dura mater). Three partitions form from the cranial dura. The falx cerebri (1) is formed from the joining of two layers of dura. Superiorly, the two layers arise from the cranial roof and enclose the superior sagittal sinus. Inferiorly, the falx is formed from the dura on the floor of the anterior cranial fossae and posteriorly from the two sides of the "tent-like" tentorium cerebelli. The falx descends between the two cerebral hemispheres in the longitudinal cerebral fissure, its free edge ending just above the corpus callosum.

The tentorium cerebelli (2) supports the occipital lobes and separates them from the cerebellum set deeply in the posterior cranial fossae. The free edges of the tentorium create a notch (incisura) for the midbrain, and run anteriorly to the dorsum sellae (posterior wall of the sella turcica). Notice the dural roof of the sella (diaphragma sellae), perforated to transmit the infundibulum; see Plate 152. Extending vertically downward from the midline of the tentorium is the falx cerebelli (3; not shown), separating the cerebellar hemispheres. It is continuous with the dura lining the posterior cranial fossa.

The filmy, vulnerable arachnoid lies deep to and flush with the inner dura. The arachnoid is separated from the deeper pia mater by the subarachnoid space, filled with cerebrospinal fluid (CSF). This space becomes voluminous at various locations (cisterns; Plate 82). The pia is a vascular layer of loose fibrous connective tissue, supporting the vessels reaching the brain (and spinal cord). It is inseparable from the surface of the brain and cord.

MENINGES OF THE BRAIN
(Modified coronal section)

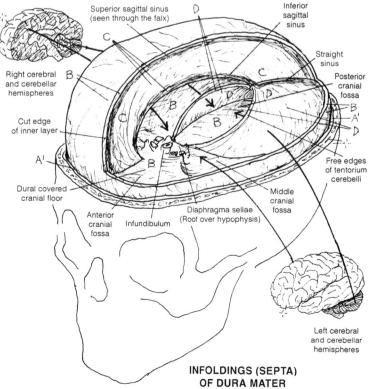

INFOLDINGS (SEPTA) OF DURA MATER
(Brain and skull cap removed)

CIRCULATION OF CEREBROSPINAL FLUID (CSF)

CN: Use the same colors as were used on the previous plate for the three meninges. Use blue for L and light colors for E through H, J, and K. (1) Color the large illustration and the coronal section simultaneously, paying close attention to the arrows of directional flow. Note that both layers of dura (A) are given one color. (2) The four cisterns, part of the subarachnoid space, all receive one color (J¹), including the lumbar cistern at lower right. (3) Color the median and lateral apertures of the IV ventricle.

Scalp
Bone
Right cerebral cortex
Left cerebral cortex
Falx cerebri
CORONAL SECTION

Right lateral ventricle
Left lateral ventricle
VENTRICLES

SCHEME OF CSF CIRCULATION
(Modified sagittal view)

Right cerebral hemisphere
Right lateral ventricle
Arachnoid villus
Periosteal dura
Meningeal dura
Scalp
Bone
SUPERIOR CISTERN
INTERPEDUNCULAR CISTERN
Cerebellum
Pons
Medulla
PONTINE CISTERN
CEREBELLO-MEDULLARY CISTERN
Spinal cord

MENINGES
DURA MATER A
ARACHNOID B
PIA MATER C

CSF CIRCULATION
CHOROID PLEXUS D
LATERAL VENTRICLE E
INTERVENTRICULAR FORAMEN F
3RD VENTRICLE E¹
CEREBRAL AQUEDUCT G
4TH VENTRICLE E²
CENTRAL CANAL H
MEDIAN I LATERAL APERTURE I¹
SUBARACHNOID SPACE J
CISTERN J¹
ARACHNOID VILLUS B¹
SUPERIOR SAGITTAL SINUS K

LUMBAR CISTERN
(Cauda equina removed)
L 2
End of cord
Internal filum terminale
LUMBAR CISTERN
S 2
Sacrum
External filum terminale

Cerebrospinal fluid (CSF) is a clear, largely acellular fluid secreted by the *choroid plexus* (70%) and vessels near the ventricular walls (30%) into the *lateral, third,* and *fourth ventricles.* About 150 ml of CSF circulates through the ventricles and around the *subarachnoid spaces* (including *cisterns*). CSF flow through the *central canal* is minimal to nonexistent. Although the fluid is an exudate of plasma from the capillaries (in the pia mater enfolded with ependymal cells lining the ventricles), it has significantly less density and protein than plasma. CSF drains into the subarachnoid space via *median* and *lateral apertures* located in the roof of the fourth ventricle. Cisterns are dilated subarachnoid spaces formed at flexures of the brain. The most notable of the cisterns is the lumbar cistern, in which float the lumbar and lower nerve roots (cauda equina). This

cistern is a frequent site of puncture (at a level of about the 4th lumbar vertebra) for withdrawal and diagnostic testing of CSF. Anesthetic agents and radiopaque dyes also can be introduced at this site. Cerebrospinal fluid is resorbed by cauliflower-shaped outpocketings of arachnoid called *villi.* These villi project into the *superior sagittal sinus,* one of the large veins draining the brain.

CSF suspends the brain and spinal cord within the *dura mater* in virtually a no-load condition, preserving their structural integrity. That is, the brain and spinal cord do not experience a gravitational force from the base of the skull or the sacrum. In forceful movements of the head and torso, up to a point, the CNS is protected from striking the skull or vertebral column by its fluid-filled container functioning as a cushion.

CRANIAL NERVES

CN: Use light colors throughout. (1) Beginning with the first cranial nerve, color the title on the left; the large Roman numeral, the cranial nerve (cut), and the related function arrow at lower left; and the Roman numeral and accompanying illustration at upper right. The illustrations generally depict target organs/areas. (2) Note carefully the direction of the function arrows at lower left (sensory/afferent is incoming; motor/efferent is outgoing). (3) The accessory nerve (XI) has two roots: a spinal root and a cranial root that travels with the vagus nerve (X).

CRANIAL NERVES:

OLFACTORY (I)ᵢ
OPTIC (II)ᵢᵢ
OCULOMOTOR (III)ᵢᵢᵢ
TROCHLEAR (IV)ᵢᵥ
TRIGEMINAL (V)ᵥ
ABDUCENS (VI)ᵥᵢ
FACIAL (VII)ᵥᵢᵢ
VESTIBULOCOCHLEAR (VIII)ᵥᵢᵢᵢ
GLOSSOPHARYNGEAL (IX)ᵢₓ
VAGUS (X)ₓ
ACCESSORY (XI)ₓᵢ
HYPOGLOSSAL (XII)ₓᵢᵢ

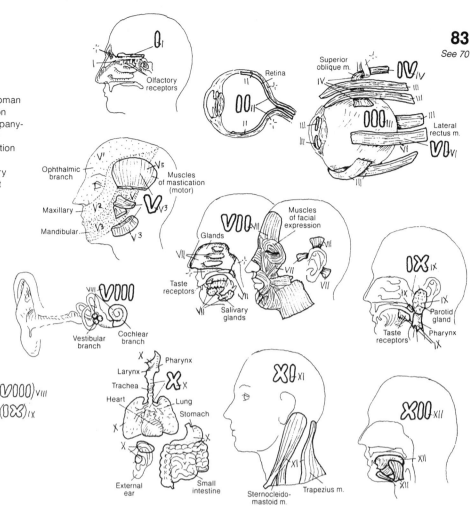

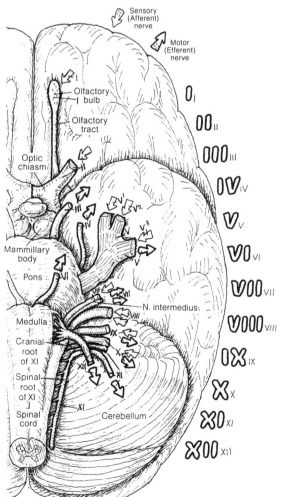

ANTERIOR-INFERIOR SURFACE
(Left brain, brainstem, and cerebellum)

Cranial nerves I and II are derived from the forebrain; all others arise from the brain stem. *V* = *visceral*, referring to smooth muscle, glands, and organs with hollow cavities; *S* = *somatic*, referring to the skin, eye, skeletal, facial, and skeletal muscles; *A* = *afferent or sensory*; *E* = *efferent or motor*. All motor nerves depicted include proprioceptive fibers (sensory for muscle, tendon, and joint movement).

I   VA: smell-sensitive (olfactory) receptors in roof/walls of nasal cavity.

II SA: light-sensitive (visual) receptors in the retina of the eye.

III SE: to extrinsic eye muscles (exc. lat. rectus and sup. oblique); VE: parasympathetic to ciliary and pupillary sphincter (eye) muscles via ciliary ganglion in the orbit.

IV SE: to superior oblique muscle of the eye.

V   SA: from face via three divisions indicated; VE: to muscles of mastication, tensor tympani, tensor veli palatini, mylohyoid, and digastric muscles.

VI SE: to lateral rectus muscle of the eye.

VII VA: from taste receptors ant. tongue; SA: from ext. ear; VE parasympathetic to glands of nasal/oral cavity, lacrimal gland (via pterygopalatine ganglion in fossa of same name), submandibular/sublingual salivary glands (via submandibular ganglion in region of same name); VE: to facial muscles, stapedius (mid. ear), stylohyoid, post. digastric muscles.

VIII SA: cochlear part is sound-sensitive; vestibular part is sensitive to head balance and movement (equilibrium).

IX VA: from taste receptors post. one-third tongue; SA: from ext. ear and ext. auditory canal; VA: from mucous membranes of posterior mouth, pharynx, auditory tube, and middle ear; from pressure and chemical receptors in carotid body and common carotid artery; VE: to sup. constrictor m. of the pharynx, stylopharyngeus; VE: parasymp. to parotid gland (via otic ganglion in infratemporal fossa).

X   VA: from taste receptors at base of tongue and epiglottis; SA: from ext. ear and ext. aud. canal; VA: from pharynx, larynx, thoracic and abdominal viscera; VE: to muscles of palate, pharynx, and larynx; VE: parasymp. to muscles of thoracic and abdominal viscera (via intramural ganglia).

XI Cranial root: joins vagus (VA to laryngeal muscles); spinal root (C1–C5): innervates trapezius and sternocleidomastoid muscles.

XII SE: to extrinsic and intrinsic muscles of tongue.

SPINAL NERVES & NERVE ROOTS

CN: Use very light colors for D through G. (1) Begin with the upper illustration. Color all three pairs of spinal nerves as they emerge from the intervertebral foramina (M). (2) Color the cross-sectional view in the center. (3) Color the spinal nerve axons and the arrows representing direction of impulse flow.

SPINAL NERVE ROOT
POSTERIOR ROOT A
SENSORY AXON B
CELL BODY C
POSTERIOR ROOT GANGLION D
ANTERIOR ROOT E
MOTOR AXON F
CELL BODY G

SPINAL NERVE H RAMUS H'

Spinal nerves are collections of axons of sensory and motor neurons located in or adjacent to the spinal cord. They are the spinal equivalent of cranial nerves. Spinal nerves arise from nerve roots that come directly off the spinal cord. The spinal nerves and their roots are arranged segmentally (from cervical to coccygeal) and bilaterally along the length of the spinal cord. The central relations of these spinal nerves/roots can be recalled in Plates 78 and 79. The spinal nerves branch soon after they are formed into *anterior and posterior rami*.

Axons of sensory neurons that form the major part of the *posterior root* are called central processes (see drawing of spinal nerve axons). The cell bodies of these neurons form the posterior root ganglia and are located in or near the intervertebral foramina, except for the sacral and coccygeal nerves, whose ganglia are in the vertebral canal. The peripheral processes of the sensory neurons join with the *axons of motor neurons* to form the spinal nerves.

The cell bodies of the motor neurons are multipolar and exist in the anterior horns of the spinal cord. Their axons emerge from the cord to form the anterior roots of the spinal nerves.

The nerve roots join to form the spinal nerves in the region of the *intervertebral foramina*. The nerve roots are progressively longer from cervical to coccygeal regions because the spinal cord does not fill the vertebral canal; it ends at the level of the 1st lumbar vertebra. Thus, some spinal nerve roots are quite long, remaining within the vertebral canal before reaching the lumbar, sacral, and coccygeal intervertebral foramina. The collection of these long nerve roots forms the "cauda equina" (recall Plate 77).

NERVE ROOT RELATIONS
VERTEBRA
BODY I
LAMINA J
ARTICULAR PROCESS K
VERTEBRAL CANAL L
LATERAL RECESS L'
INTERVERTEBRAL FORAMEN M

Spinal nerves and their roots have fairly tight quarters. The relations of these nerves and roots can best be appreciated in the cross-sectional view. Nerve roots are vulnerable to irritation (radiculitis) from encroaching, hypertrophic bone in the lateral recesses (degenerative joint disease), from bulging intervertebral discs (degenerative disc disease), or from cysts, meningeal tumors, and so on. With compression of axons or blood vessels supplying the axons, functional deficits can result (radiculopathy: sensory loss, motor loss, and/or tendon reflex change).

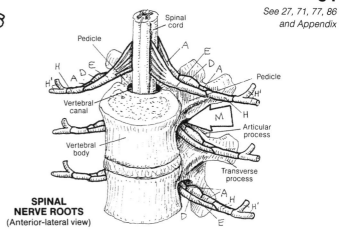

**SPINAL
NERVE ROOTS**
(Anterior-lateral view)

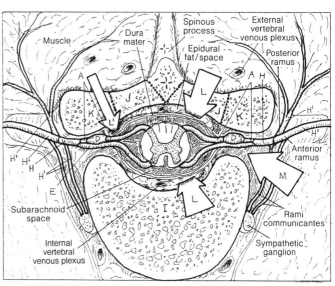

CROSS SECTION THROUGH T9
(Seen from above)

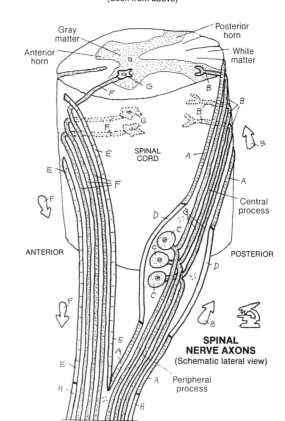

**SPINAL
NERVE AXONS**
(Schematic lateral view)

SPINAL REFLEXES

CN: Use light colors for A and C, and use the same colors you used on Pl. 84 for structures D–F. (1) Color the upper two illustrations simultaneously, in numerical sequence 1–6, including the arrows. The small arrows at the end of the muscle segments indicate contraction or stretch. (2) Color the lower two illustrations similarly. Note that the motor neuron synapsing with the inhibitory interneuron, and the related effector, are not colored.

MONOSYNAPTIC REFLEX

STRETCH RECEPTOR (N-T ORGAN) A
STRETCH RECEPTOR (MUSCLE SPINDLE) A'
SENSORY NEURON A²
SPINAL CORD B
MOTOR NEURON C
END PLATE C'
EFFECTOR MUSCLE D

SPINAL NERVES / ROOTS

SPINAL NERVE E
BRANCH E'
POSTERIOR ROOT F
GANGLION F'
ANTERIOR ROOT G

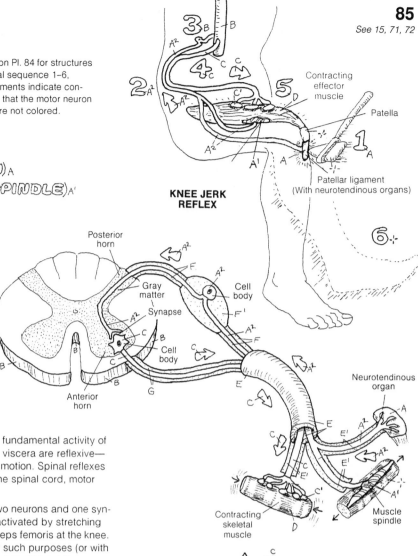

KNEE JERK REFLEX

A reflex is an involuntary muscle response to a stimulus. It is a fundamental activity of the nervous system; most body movements and movement of viscera are reflexive— e.g., heart rate, respiratory rate, peristalsis of gastrointestinal motion. Spinal reflexes involve sensory receptors, sensory neurons, interneurons of the spinal cord, motor neurons, and effectors.

The simplest spinal reflex is a monosynaptic reflex involving two neurons and one synapse (myotatic [stretch] or deep tendon reflex). The reflex is activated by stretching the tendon of a specific muscle, such as the tendon of quadriceps femoris at the knee. This can be done with the sharp tap of a small mallet used for such purposes (or with the 5th-digit side of a hand). The *receptors* responsive to such a stretch are the neurotendinous organs in the patellar ligament and the muscle spindles in the belly of the quadriceps muscle. Muscle spindles are encapsulated, specialized muscle fibers within muscle bellies that have nerve endings sensitive to muscle stretch. Impulses generated in these receptors (1) are conducted by *sensory neurons* (2) to the *spinal cord* (3); these synapse in the gray matter with the anterior horn *motor neurons* (4). The motor neuron conducts impulses to the *end plates* of the *effector* muscle (5). The muscle contracts sufficiently, in the case of the knee reflex ("jerk"), to extend the knee joint momentarily (6).

POLYSYNAPTIC REFLEX

PAIN RECEPTOR A³
SENSORY NEURON A²
INTERNEURON H
FACILITATORY (+) H'
INHIBITORY (−) H²
(+) MOTOR NEURON C / EFFECTOR D
(−) MOTOR NEURON C' / EFFECTOR D'

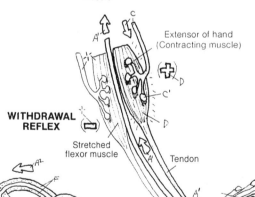

WITHDRAWAL REFLEX

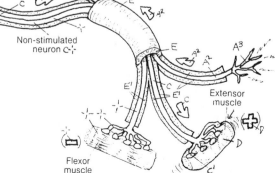

Polysynaptic reflexes range from simple withdrawal reflexes to complex reflexes involving several segments of the spinal cord and brain. The complexity of a polysynaptic reflex relies on the number of interneurons in the reflex and the number of synaptic contacts. In this case, temperature and pain receptors respond to a sharp increase in heat; sensory neurons conduct the impulse to the spinal cord. An interneuron receives the impulse. Branches of the interneuron excite two interneurons, one facilitatory and one inhibitory. The excitatory interneuron facilitates the firing of the motor neuron that induces the extensor muscle to contract, lifting the fingers from the flame. Simultaneously, the inhibitory neuron depresses the firing of the 2nd motor neuron (C3), and the antagonist flexor muscle is stretched without contracting, permitting the fingers to be withdrawn from the flame.

DISTRIBUTION OF SPINAL NERVES & THORACIC SPINAL NERVE

CN: (1) Begin with the upper illustration, which is an introduction to the major nerve plexuses (detailed on the following three plates) formed by spinal nerves. Note that each nerve (shown emerging from the left half of the spinal cord) receives the color of the plexus to which it contributes; exceptions are nerves T1 and L4, which make two contributions but receive the color of their main grouping. Thoracic nerves (C) give rise to intercostal nerves (O), represented above in the company of each rib and treated in more detail below. (2) Color the example of nerve coverings, taken from a cutaneous nerve (F[1]) in the cross-sectional view to the right. (3) Color the larger view and review the introduction to these structures on Plate 84.

Thirty-one pairs of spinal nerves supply the body structure with sensory and motor innervation, except for areas covered by cranial nerves. From above to below, there are 8 cervical spinal nerves (C1–C8), 12 thoracic (T1–T12), 5 lumbar (L1–L5), 5 sacral (S1–S5), and one coccygeal (Co1). There is one more nerve than vertebrae in the cervical spine; C1 passes above the C1 vertebra, C8 passes below the C7 vertebra. Thus, spinal nerves after C6 pass below the vertebra of the same number; above C7 they pass above the vertebra of the same number.

Spinal nerves arise from roots; once formed, they split into rami (see Plate 84 and the cross section below right). The anterior rami of all spinal nerves (except thoracic) form interconnecting networks or plexuses outside the vertebral column. The posterior rami do not contribute to plexuses. Peripheral nerves are branchings from the plexuses and are directed to geographically related parts of the body. The nerves of the cervical plexus (C1–C4) can be colored in Plate 87, the nerves of the brachial plexus (C5–T1) in Plate 88, and the nerves of the lumbar plexus (L1–L4) and sacral plexus (S1–S4) in Plate 89. The coccygeal plexus (S4, S5, Co1) is not shown.

The anterior rami of thoracic spinal nerves form intercostal nerves, not plexuses (see Plate 50), although T1 contributes a branch to the brachial plexus. An idealized cross section through the thorax reveals the ring-like distribution of a "typical" thoracic spinal nerve (see below right). The anterior ramus of one thoracic nerve supplies a segment of the cutaneous, subcutaneous, and musculoskeletal areas of the torso, and the smaller posterior ramus (along with posterior rami of cervical and lumbar spinal nerves) supplies its posterior wall and that of the neck. Note the formation of the cutaneous branches to appreciate the innervation of the skin around the body (see also Plate 90).

A cross section through any nerve reveals coverings similar to those of muscle (Plate 44). These fibrous envelopes ensure physical security for the individual axons (endoneurium), fascicles of neurons (perineurium), and the entire nerve (epineurium continuous with deep fascia). These coverings also physically secure the vessels (vasa vasorum) and nerves (vasa nervosum) supplying the axons.

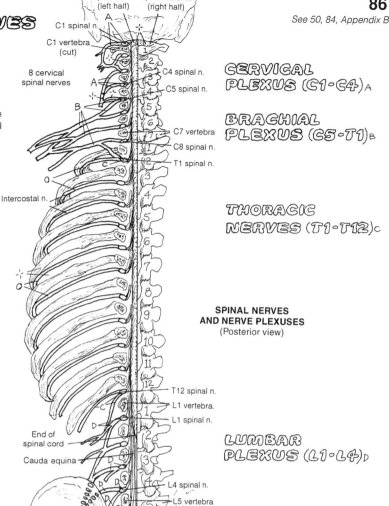

SPINAL NERVES AND NERVE PLEXUSES
(Posterior view)

CERVICAL PLEXUS (C1-C4)A

BRACHIAL PLEXUS (C5-T1)B

THORACIC NERVES (T1-T12)c

LUMBAR PLEXUS (L1-L4)D

SACRAL PLEXUS (L4,L5-S4)E

Spinal cord (left half)
Spinal column (right half)
C1 spinal n.
C1 vertebra (cut)
8 cervical spinal nerves
C4 spinal n.
C5 spinal n.
C7 vertebra
C8 spinal n.
T1 spinal n.
Intercostal n.
T12 spinal n.
L1 vertebra.
L1 spinal n.
End of spinal cord
Cauda equina
L4 spinal n.
L5 vertebra
S1 spinal n.
Hip bone
Sacrum
Coccyx

NERVE COVERINGS.
EPINEURIUMF
PERINEURIUMG
ENDONEURIUMH
AXONI

POSTERIOR ROOTJ
ANTERIOR ROOTK

THORACIC SPINAL NERVEc
POSTERIOR RAMUSL
LATERAL (MUSCULAR) BRANCHM
MEDIAL (CUTANEOUS) BRANCHN
ANTERIOR RAMUSO
(INTERCOSTAL NERVE)O
LAT. CUTANEOUS BR.F[1]
ANT. CUTANEOUS BR.P

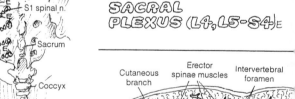

NERVE SECTION

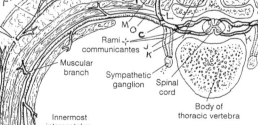

Cutaneous branch
Erector spinae muscles
Intervertebral foramen
Rami communicantes
Muscular branch
Sympathetic ganglion
Spinal cord
Body of thoracic vertebra
Skin
Innermost intercostal m.
Internal intercostal m.
External intercostal m.
Superficial fascia
Cutaneous branch
Muscular branch
Transversus thoracic m.
Sternum

PATTERN OF A TYPICAL THORACIC SPINAL NERVE
(Cross section through mid-thorax, viscera removed)

CERVICAL PLEXUS & NERVES TO THE NECK

CN: Use dissimilar colors for A–E. Label C has been omitted to avoid confusion with C1–C5 (spinal nerves). (1) It will be helpful to follow the text as you color the large schematic. Color each C and its respective numeral, as well as the directional arrows. Where two roots form a nerve, that nerve (and its title) receives both colors. The phrenic nerve (F), formed by three nerves, receives its own color. (2) The sternocleidomastoid muscle, which lies above the spinal nerves, has been removed from the schematic but does appear in the cutaneous nerves illustration. A darkly outlined rectangle provides a frame of reference for the material covered in the schematic. (3) Color the four nerves of the cervical plexus, and C5 (bottom illustration).

CERVICAL PLEXUS

C1 SPINAL NERVE$_A$ & BRANCH$_{A'}$
ANSA CERVICALIS: SUP. ROOT$_{A^2}$

C2 SPINAL N.$_B$ & BR.$_{B'}$

C3 SPINAL N.$_D$ & BR.$_{D'}$
ANSA CERVICALIS: INF. ROOT$_{B^2+D^2}$
GREAT AURICULAR N.$_{B^3+D^3}$
LESSER OCCIPITAL N.$_{B^4+D^4}$
TRANSVERSE CERVICAL N.$_{B^5+D^5}$

C4 SPINAL N.$_E$ & BR.$_{E'}$
SUPRACLAVICULAR NS.$_{D^2+E^2}$

PHRENIC N.$_F$

ACCESSORY N. (XI CRANIAL)$_G$

C5 SPINAL N.$_H$ & BR.$_{H'}$
ROOT OF BRACHIAL PLEXUS$_{H^2}$

The distribution of spinal nerves from just below the neck to the lower abdomen is segmental and bilateral (Plate 86). In the neck and limbs, the distribution of spinal nerves is more irregular and occurs by way of interconnecting branches (*plexus*, a network) solely from the anterior rami.

The cervical plexus arises in the deep lateral neck, formed from the anterior rami of cervical spinal nerves 1 through 4 (C1–C4). C1 sends a loop to C2; from this loop the deepest cervical muscles are innervated. Other C1 fibers pass along a length of the *hypoglossal (XII)* cranial nerve to supply the geniohyoid and thyrohyoid muscles. A branch of C1 (superior root of the *ansa cervicalis*) turns inferiorly from the hypoglossal nerve to descend to the level of C4, where it is joined by a confluence of fibers from the C2 and C3 spinal nerves (inferior root of the ansa). The fibers of this loop supply the infrahyoid muscles.

Fibers from C2 and C3 give origin to three important cutaneous nerves emerging from the anterior border of the mid-posterior triangle: the *great auricular* nerve destined for the external ear, the *lesser occipital* nerve to the posterolateral scalp, and the *transverse* (cutaneous) nerve of the neck supplying the skin and fascia over the anterior triangle (recall Plate 48).

Note that fibers from C3 and C4 send medial, intermediate, and lateral supraclavicular nerves to the skin in a broad area centered over the clavicle (anterolateral neck, shoulder, and anterior upper chest). They also project fibers to the trapezius muscle. Recall from Plate 83 that the primary nerve supply for this muscle is the spinal root of the *accessory (XI)* cranial nerve. Branches from C3 and C4 join with a branch from C5 to form the *phrenic* nerve, which innervates the thoracic diaphragm (recall Plate 50; see Plate 135). C5 is a major contributor to the upper trunk of the brachial plexus.

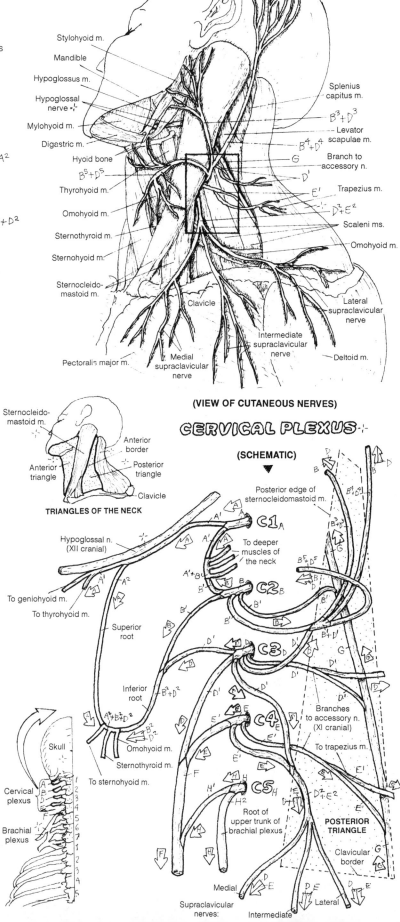

(VIEW OF CUTANEOUS NERVES)

TRIANGLES OF THE NECK

CERVICAL PLEXUS

(SCHEMATIC)

BRACHIAL PLEXUS & NERVES TO THE UPPER LIMB

CN: Use light colors for A–D. (1) In the upper illustration, color the letters and numbers identifying the five roots of the brachial plexus. Note but do not color the small branches of the plexus as you color the plexus itself. Note in the lower illustration that the entire plexus is colored gray. (2) As you color each of the major nerves arising from the plexus, color it in the lower illustration as well. As you color each nerve, try to visualize it on your own limb.

BRACHIAL PLEXUS & MAJOR BRANCHES

ROOTS C5, C6ₐ
   UPPER TRUNKʙ
ROOT C7ₐ'
   MIDDLE TRUNKʙ'
ROOTS C8, T1ₐ²
   LOWER TRUNKʙ²
_____

ANTERIOR DIVISIONc
  LATERAL CORD (C5-C7)ᴅ
    MUSCULOCUTANEOUS N.ᴇ
    BR. TO MEDIAN N.ꜰ
  MEDIAL CORD (C8-T1)ᴅ'
    BR. TO MEDIAN N.ꜰ
  MEDIAN N.ꜰ'
  ULNAR N.ɢ
_____

POSTERIOR DIVISION (C5-T1)c'
  POSTERIOR CORDᴅ²
    AXILLARY N. (C5-C6)ʜ
    RADIAL N. (C5-T1)ɪ

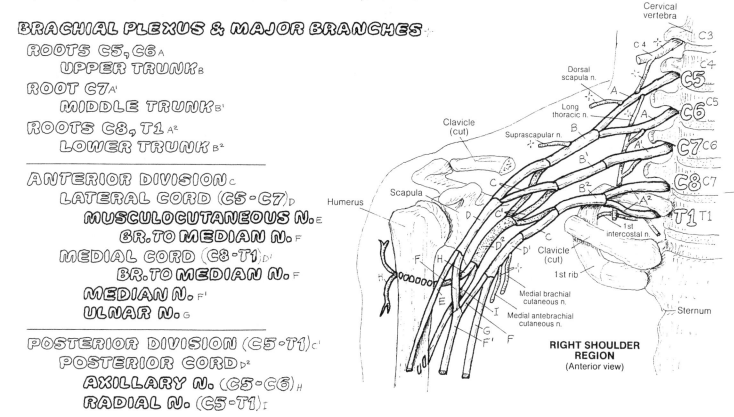

RIGHT SHOULDER REGION
(Anterior view)

The major nerves to the structures of the upper limb arise from the brachial plexus, formed from the anterior rami of spinal nerves C5–T1 (plus or minus one level). These rami form the *roots of the plexus*. In the pattern illustrated, further branching and joining of fibers in the neck, supraclavicular area, and axilla result in the formation of the five major nerves of the upper limb.

The brachial plexus is subject to injury (plexopathy) from excessive stretching or traction (e.g., rapid, forceful pulling of the upper limb) and compression (e.g., long-term placement of body weight on axillary or armpit cushions of crutches). In such injuries, there is great variation in degree of deficit, signs, and symptoms.

The musculocutaneous nerve (C5–C7) supplies the anterior arm muscles and is cutaneous in the forearm. Packaged in muscle, it is rarely traumatized. C5 and/or C6 nerve root compression can weaken these muscles. The *median nerve* (C5–C8, T1; "carpenter's nerve") supplies the anterior forearm muscles and the thenar muscles. It can be compressed at the carpal tunnel (recall Pl. 35), resulting in some degree of sensory deficit to fingers 1–3 and weakness in thumb movement (carpal tunnel

syndrome). Similar complaints can be associated with a C6 nerve root compression.

The ulnar nerve (C8–T1; "musician's nerve") supplies certain muscles of the forearm and most intrinsic muscles of the hand. It is subject to trauma as it rounds the elbow in the cubital tunnel, possibly resulting in ulnar-side finger pain, hand weakness, or abnormal little finger position. Similar complaints can be associated with a C8 nerve root compression. The *axillary nerve* (C5–C6) wraps around the neck of the humerus to supply deltoid and teres minor. It is vulnerable in fractures of the humeral neck, possibly resulting in a weak or paralyzed deltoid muscle. The *radial nerve* (C5–C8, T1) supplies the triceps, brachioradialis, and posterior forearm (extensor) muscles moving the wrist and hand. It is subject to damage as it rounds the mid-shaft of the humerus; significant nerve loss here results in "wrist drop" and loss of ability to work the hand (try moving your fingers with your wrist flexed hard). A C7 radiculopathy is characterized by a weak triceps, loss of the triceps jerk (reflex), and numbness of the middle finger. See the appendix for a listing of upper limb muscles and their nerve supply.

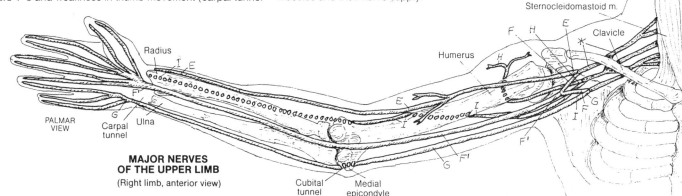

MAJOR NERVES OF THE UPPER LIMB
(Right limb, anterior view)

LUMBAR PLEXUS & NERVES TO THE LOWER LIMB

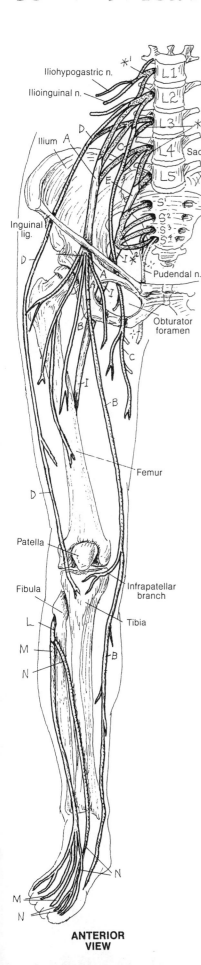

ANTERIOR VIEW

CN: Use a bright color for J. (1) Begin with the anterior view. Color the lumbar and sacral plexuses gray; note that they have been dotted for easy identification. Note the longest branch of the femoral nerve: saphenous nerve. (2) Color the posterior view, which includes almost entirely the sciatic nerve and its branches. The heel of the foot has been lifted to view the plantar nerves.

LUMBAR PLEXUS (L1-L4) *¹
FEMORAL N. A
SAPHENOUS N. B
OBTURATOR N. C
LAT. FEMORAL CUTAN. N. D

LUMBOSACRAL TRUNK (L4-L5) E

SACRAL PLEXUS (L4-S4) *²
POST. FEMORAL CUTAN. N. F
SUPERIOR GLUTEAL N. G
INFERIOR GLUTEAL N. H

SCIATIC N. (L4-S3) I
TIBIAL N. (L4-S3) J
MED. K LAT. PLANTAR N. K'
COMMON FIBULAR N. (L4-S2) L
SUPERFICIAL FIBULAR N. M
DEEP FIBULAR N. N

The lumbar plexus, formed from the anterior rami of L1–L4 spinal nerves, is located against the muscles of the posterior abdominal wall. The *femoral nerve* (L2–L4) passes through the psoas major muscle in its descent, emerging lateral to the muscle in the pelvis. As the nerve passes under the inguinal ligament, it lies on the muscle's anterior surface. The femoral nerve breaks up into a leash of nerves in the proximal thigh, supplying the four heads of the quadriceps femoris muscle and the sartorius muscle. Medially, the cutaneous *saphenous nerve* descends to the medial knee and beyond to the ankle. In mid-thigh, it passes through the adductor canal into the posterior femoral compartment, with the femoral artery and vein (recall Plate 63). The *obturator nerve* (L2–L4) passes along the lateral pelvic wall on the obturator internus muscle. It penetrates the obturator foramen to enter the medial thigh, supplying the adductor muscles. Both femoral and obturator nerves are subject to trauma or compression within the pelvis.

The lumbosacral trunk (L4, L5) joins with the sacral spinal nerves to form the *sacral plexus* (L4–S4). From this plexus, the *superior gluteal nerve* (L4, L5, S1) passes through the greater sciatic foramen, above the piriformis muscle, to supply gluteus medius (and sometimes minimus). The *inferior gluteal nerve* (L5, S1, S2) comes into the gluteal region above piriformis to supply gluteus maximus.

The sciatic nerve joins the posterior femoral cutaneous nerve and the inferior gluteal nerve to pass through the greater sciatic foramen under the piriformis muscle, deep to gluteus maximus (but not innervating it). It descends between the ischial tuberosity and the greater trochanter of the femur. Within the posterior femoral compartment, above the knee, the sciatic nerve splits into the tibial and common *fibular (peroneal) nerves.* The *tibial nerve* supplies the posterior leg muscles and the plantar muscles of the foot. The common fibular nerve supplies the lateral leg muscles (superficial fibular nerve) and the muscles of the anterolateral leg compartment (deep fibular nerve).

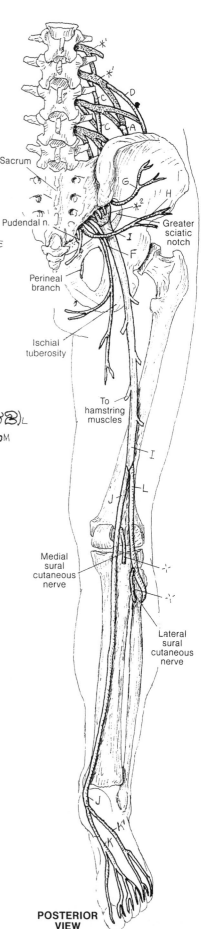

POSTERIOR VIEW

DERMATOMES

CN: (1) Begin with the diagram at left, depicting sensory innervation of an area of skin (dermatome) and the degree of overlap among contiguous spinal nerve cutaneous branches and the dermatomes they supply. Color gray the three spinal nerves and the rectangular borders of the related dermatomes. Note the overlap. (2) Use very light colors for the five groups of dermatomes. Use one color for all dermatomes with the letter V, another color for the dermatomes marked with a C, and so on with T, L, and S. Suggestion: carefully outline the collection of C dermatomes with the color used for C, then color in the enclosed area, focusing on the skin areas serviced by the related spinal nerve; repeat with T, L, and S dermatomes.

SPINAL NERVE* DERMATOME*'

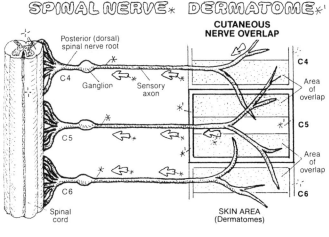

DERMATOMES OF

TRIGEMINAL NERVE v
V¹–V³ v
CERVICAL NERVES a
C2–C8 a
THORACIC NERVES b
T1–T12 b
LUMBAR NERVES c
L1–L5 c
SACRAL NERVES d
S1–S5 d

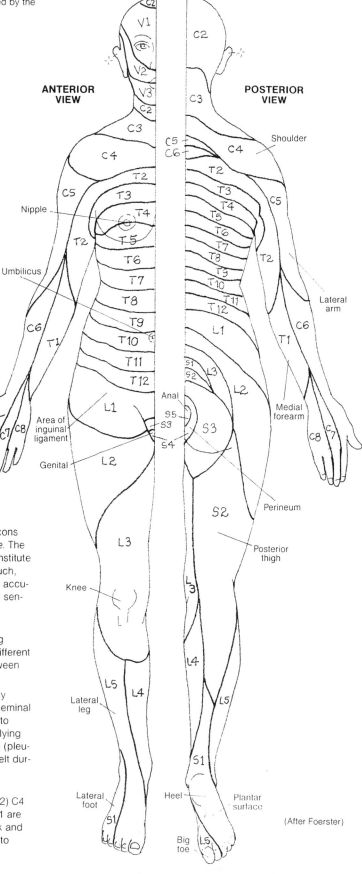

ANTERIOR VIEW

POSTERIOR VIEW

(After Foerster)

A **dermatome** is an area of skin (cutaneous area) supplied by the sensory axons of a single *spinal nerve* or a single division of the *trigeminal (V cranial) nerve*. The body surface is globally covered by sensory receptors. The dermatomes constitute a map of cutaneous innervation. Testing of general sensations (hot, cold, touch, pressure) and pain can help determine deficits in specific dermatomes. The accuracy of dermatomal representation has been corroborated in cases of spinal sensory root/nerve deficit (radiculopathy), trigeminal nerve irritation (trigeminal neuralgia), and spinal cord deficits (myelopathy).

In the case of spinal nerves and the trigeminal nerve, there is overlap among cutaneous branches of neighboring sensory axons. Thus, two branches of different spinal nerves or divisions of the trigeminal nerve cover the border zone between pairs of contiguous dermatomes.

In the case of pain, it is important to understand that dermatomes reflect only cutaneous pain and pain referred to the skin (e.g., visceral pain, spinal or trigeminal sensory nerve root pain). Commonly, pain of visceral origin may be referred to cutaneous areas served by the same spinal sensory nerve(s) as those supplying the visceral structure. For example, the pain of an inflamed lining of the lung (pleurisy), which is innervated by C3–C5 spinal nerves (phrenic nerve), may be felt during deep inspiration in the cervical dermatomes C3–C5 (usually along the supraclavicular nerve distribution).

Finally, note that (1) C1 has no dermatome because it has no sensory root; (2) C4 and T2 dermatomes overlap the chest wall because the spinal nerves C5–T1 are largely committed to the upper limb; and (3) the same is true in the low back and perineum with respect to spinal nerves L4–S2, which are largely committed to the lower limb.

SENSORY RECEPTORS

CN: Use your lightest colors for A and E. (1) Begin with the overview of a sensory pathway. (2) Color the general exteroceptors. Note that each receptor is connected to a sensory neuron (B) of a different color. (3) Color the proprioceptors in the lower illustration. Color over the entire muscle spindle, but not the surrounding muscle fibers.

Sensory receptors provide information to the brain about the internal and external environment of the body. Most receptors are transducers: they convert mechanical, chemical, electrical, or light stimuli to electrochemical impulses that can be conducted by the nervous system. Once generated, informational or sensory impulses travel to the CNS via sensory neurons, ultimately reaching the thalamus. Here impulses are relayed to the sensory cortex (conscious interpretation) or to motor centers for appropriate (reflexive) response.

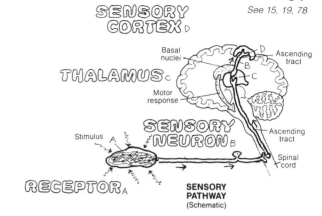

SENSORY CORTEX D

THALAMUS C

Basal nuclei

Ascending tract

Motor response

Ascending tract

SENSORY NEURON B

Stimulus A

Spinal cord

RECEPTOR A

SENSORY PATHWAY (Schematic)

EXTEROCEPTORS -

SPECIAL N.S. -

GENERAL (CUTANEOUS) -

FREE NERVE ENDINGS / AXON D

MERKEL (TACTILE) CELL / AXON D'

ENCAPSULATED ENDINGS E -

MEISSNER (TACTILE) CORPUSCLE / AXON E'

RUFFINI (DEFORMATION) ENDINGS / AXON E²

Exteroceptors are located near the body surface. Special exteroceptors (not shown) include photoreceptors of the retina (light stimuli; Plate 95), taste receptors (chemical stimuli; Plate 100), and auditory receptors (sound stimuli; Plate 98). General exteroceptors are cutaneous sensory endings. They are either encapsulated or free. Free nerve endings, either single or in networks, are found in the epidermis and virtually all of the connective tissues of the body. Free endings may serve as thermoreceptors (heat/cold), mechanoreceptors (light touch), or pain receptors (nociceptors). Free endings may be specialized, as with the Merkel cell endings (see Plate 18) and the spiral endings around hair follicles sensitive to hair movement.

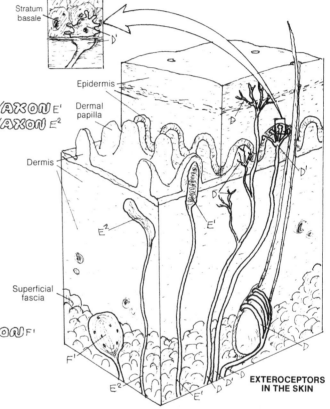

Stratum basale

Epidermis

Dermal papilla

Dermis

Superficial fascia

EXTEROCEPTORS IN THE SKIN

PROPRIOCEPTORS F -

PACINIAN (PRESSURE) CORPUSCLE / AXON F'

MUSCLE SPINDLE / MIXED AXONS F²

NEUROTENDINOUS ORGAN / AXON F³

Proprioceptors are found in deeper tissues (e.g., superficial fascia, deep fascia, tendons, ligaments, muscles, joint capsules) of the musculoskeletal system. They are sensitive to stretch, movement, pressure, and changes in position. The *Pacinian corpuscles* are large lamellar bodies acting as mechanoreceptors: distortion of their onion skin–like lamellae induces generation of an electrochemical impulse. *Muscle spindles*, sensitive to stretch, consist of two types of special muscle fibers (nuclear bag and nuclear chain) entwined with spiral or flower-spray sensory endings. Stretch of these spindles (and the skeletal muscle in which they are located) induces discharge in the sensory fibers. These impulses reach the cerebellum. Reflexive motor commands tighten the special muscle fibers and increase resistance of the skeletal muscle to stretch. By these spindles, the CNS controls muscle tone and muscle contraction. *Neurotendinous organs* (Golgi) are nerve endings enclosed in capsules located at muscle/tendon junctions or in tendons. They are induced to generate electrochemical impulses in response to tendon deformation or stretch.

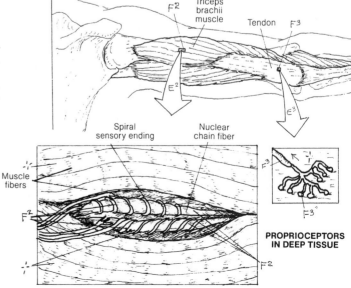

Triceps brachii muscle

Tendon F³

Spiral sensory ending

Nuclear chain fiber

Muscle fibers

PROPRIOCEPTORS IN DEEP TISSUE

INTEROCEPTORS N.S. -

Interoceptors (not shown) are free or encapsulated nerve endings, often in association with special epithelial cells, located in the walls of vessels and viscera. These receptors include chemoreceptors, baroreceptors (pressure), and nociceptors, They are not usually sensitive to the stimuli to which exteroceptors react.

SYMPATHETIC DIVISION (1)

CN: This plate is part one of a two-part presentation of the sympathetic division, and many structures with the same titles and subscripts on this and the next plate should receive the same color. (1) Begin with the schematic of the spinal cord segments containing the cell bodies of preganglionic neurons. These neurons (not shown) leave the spinal cord to enter or pass through the sympathetic chain. (2) Color the sympathetic chain and relations at upper right. (3) Color the pathways of the preganglionic and postganglionic neurons below. (4) Color the inset illustration.

SPINAL CORD SEGMENTS T1-L2 A
PREGANGLIONIC CELL BODY B
PREGANGLIONIC AXON B'
WHITE COMM. RAMUS C+
SPLANCHNIC NERVE D
PREVERTEBRAL GANGLION E
SYMPATHETIC CHAIN F
POSTGANGLIONIC CELL BODY G
POSTGANGLIONIC AXON G'
GRAY COMM. RAMUS H*
SPINAL NERVE I

The autonomic nervous system (ANS; also visceral nervous system or VNS) is a part of the peripheral nervous system (PNS), responsible for the innervation of smooth muscle and glands in viscera and skin and of specialized cardiac muscle. It is a motor system uniquely characterized by two-neuron linkages and motor ganglia (pre- and post-ganglionic neurons). Sensory impulses from viscera are conducted by typical sensory neurons not generally described with the ANS but considered part of the VNS. The sympathetic (thoracolumbar) division of the ANS is concerned with degrees of "fight or flight" responses to stimuli: pupillary dilatation, increased heart and respiratory rates, increased blood flow to brain and skeletal muscles, and other related reactions.

The cell bodies of preganglionic neurons are restricted to the lateral horns of the *spinal cord segments T1 through L2*. The axons of these neurons leave the cord via the anterior roots, join with *spinal nerves* for a very short distance, and turn medially to enter the *sympathetic chain of ganglia* via the *white communicating rami* (white because the axons are myelinated and "white"). The chain is located bilaterally alongside the vertebral column (see inset illustration). Once in the chain, the preganglionic axons can take one or more of four courses: (1) synapse with the *postganglionic neuron* at the same level it entered the chain; (2) ascend and synapse at a higher level of the chain; (3) descend and synapse at a lower level of the chain; (4) pass straight through the chain, forming a nerve that runs from the chain to the front of the vertebral column (splanchnic nerve), and synapse with a *postganglionic neuron* there (*prevertebral ganglia*).

The postganglionic neuron within the chain leaves via the *gray communicating ramus* to join the *spinal nerve*. There are gray rami bilaterally at every segment of the spinal cord; white rami exist only from T1 to L2. Gray rami are so called because the resident axons are unmyelinated and collectively have a duller color than those of the white rami. Postganglionic axons from prevertebral ganglia travel in a plexus configuration to the viscus they supply. Plate 92 puts this division into a more meaningful perspective.

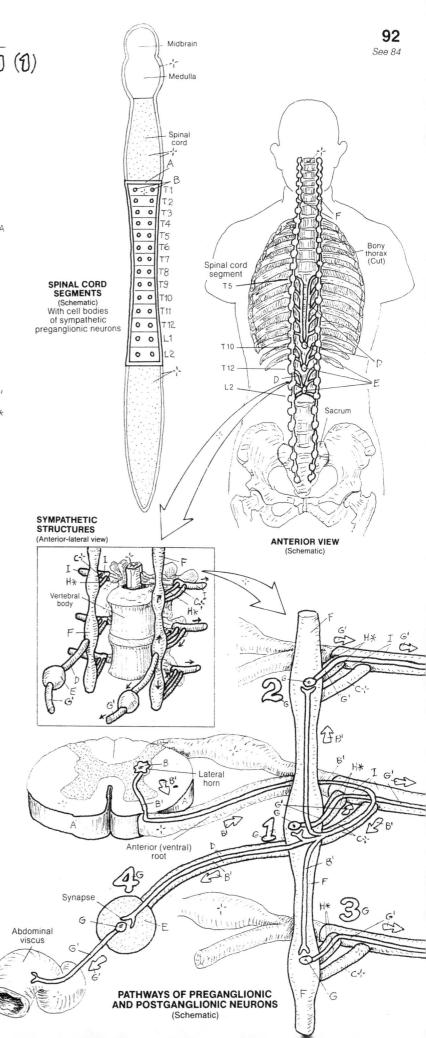

SPINAL CORD SEGMENTS (Schematic) With cell bodies of sympathetic preganglionic neurons

Midbrain
Medulla
Spinal cord
T1 T2 T3 T4 T5 T6 T7 T8 T9 T10 T11 T12 L1 L2

Spinal cord segment T5
Bony thorax (Cut)
T10
T12
L2
Sacrum

ANTERIOR VIEW (Schematic)

SYMPATHETIC STRUCTURES (Anterior-lateral view)
Vertebral body

Lateral horn
Anterior (ventral) root

Synapse
Abdominal viscus

PATHWAYS OF PREGANGLIONIC AND POSTGANGLIONIC NEURONS (Schematic)

SYMPATHETIC DIVISION (2)

CN: Use the same colors as you used on the preceding plate for preganglionic neurons (B), splanchnic nerves (D), and postganglionic neurons (G), all of which have been given the subscripts they had on Plate 92. First orient yourself to this diagram. Note the spinal cord in the center with sympathetic chains of ganglia on either side. Not all connections of both chains are shown. Here, the pathways on the left are to the skin. Pathways on the right are to viscera in the

head and body cavities. Start with the preganglionic neurons on the left (B) and color the chain and related parts (G, G^3) on the left. Then read the related text. Color the preganglionic neurons (B) on the right and the splanchnic nerves (D) to the abdominal viscera. Color the postganglionics (G, G^1, G^2) to the head and thorax, and then the postganglionics (G^4, G^5) from the prevertebral ganglia to the abdominal and pelvic/perineal organs.

PREGANGLIONIC NEURONS ᴮ
SPLANCHNIC N. ᴰ
POSTGANGLIONIC NEURONS ᴳ
TO HEAD & NECK ᴳ'
TO THORACIC VISCERA ᴳ²
TO SKIN ᴳ³
SWEAT GLANDS ᴳ³
ARRECTOR PILI ᴳ³
BLOOD VESSELS ᴳ³
TO ABDOMINAL VISCERA ᴳ⁴
TO PELVIC/ PERINEAL VISC. ᴳ⁵

Sympathetic innervation of skin (and viscera as well) begins with the *preganglionic neurons* in the thoracolumbar part of the spinal cord. The axons leave the cord via the anterior rami of spinal nerves, enter and leave the spinal nerves to join the white communicating rami. These rami bring the axons into the sympathetic chain. Axons from the upper thoracic cord ascend the chain up to the highest ganglion (superior cervical ganglion at the level of the first cervical vertebra). Axons from the lower thoracic and upper lumbar cord enter the chain and descend as far as the lowest ganglion (ganglion impar at the level of the coccyx). At every level of the chain (roughly coincident with spinal cord segments), the preganglionic axons synapse with *postganglionic neurons*. The postganglionic axons leave the chain via the gray communicating rami, enter the spinal nerves from C1 through Co1, and reach the skin via cutaneous branches of these nerves. These axons induce secretory activity in sweat glands, contraction of arrector pili muscles, and vasoconstriction in skin arterial vessels.

Postganglionics to the head (vessels and glands) leave the superior cervical ganglia and entwine about arteries enroute to the head (in the absence of spinal nerves) to reach their target organs. *Postganglionics to the heart and lungs* leave the upper ganglia of the chain, reaching these organs via cardiac nerves and the pulmonary plexus. These neurons act on heart muscle and the cardiac conduction system to increase heart rate; they induce relaxation of bronchial musculature, facilitating easier breathing.

Preganglionics to abdominal and pelvic viscera leave the cord at levels T5–L2, enter the white communicating rami, and pass through the sympathetic chain without synapsing. They form three pairs of *splanchnic nerves* between the chain and the prevertebral ganglia on the aorta. These axons synapse with the postganglionic neurons in the prevertebral ganglia. The axons of these neurons reach for smooth muscle, inducing contraction of sphincters and decreasing intestinal motility, relaxing bladder muscle and constricting the urinary sphincter. These axons stimulate the adrenal medulla to secrete mostly epinephrine and some norepinephrine, stimulate secretion of glands and muscle contraction in the male genital ducts (ejaculation), and stimulate uterine contractions.

SYMPATHETIC DIVISION
(Schema showing the following)

1 Bilateral chains of ganglia with white rami connections to spinal cord.
  Each chain has both connections (**2**) and (**3**), and they are mirror-images of one another.
2 Left chain with postganglionic connections to skin.
3 Right chain with splanchnic nerves/postganglionic connections to viscera.

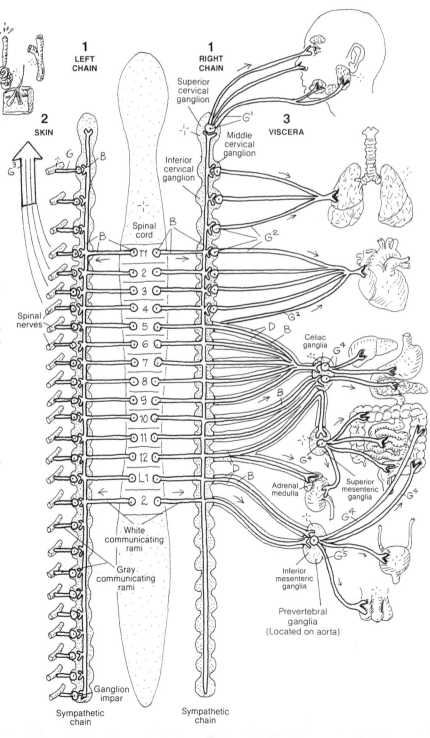

PARASYMPATHETIC DIVISION

CN: Continue using the same colors you used on Plates 91 and 92 for subscripts B, D, and G. Use a bright color for E. This drawing shows the parasympathetic scheme on one side of the body only (nerve distribution is identical for both sides). (1) Start with the preganglionic neurons in the head and work through the postganglionic neurons, noting the structures innervated. Note particularly the extensive pattern associated with the vagus nerve. (2) Continue with the sacral preganglionics and postganglionics, noting the target organs. (3) Color the diagram describing ganglia location in the two ANS divisions.

PREGANGLIONIC NEURONS.

- III CRANIAL N.ₑ B¹
- VII CRANIAL N.ₑ B²
- IX CRANIAL N.ₑ B³
- X CRANIAL B⁴
- PELVIC SPLANCHNIC N. D

GANGLIA E

- CILIARY, E¹
- PTERYGOPALATINE, E²
- SUBMANDIBULAR, E³
- OTIC, E⁴
- INTRAMURAL E⁵

POSTGANGLIONIC NEURONS G

- EYE, G¹
- NASAL/ORAL CAVITIES, G²
- SALIVARY GLANDS, G³
- THORACIC/ABDOMINAL VISC., G⁴
- PELVIC/PERINEAL VISCERA G⁵

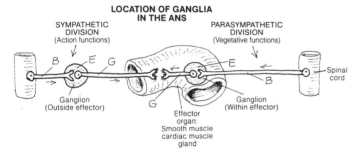

**LOCATION OF GANGLIA
IN THE ANS**

SYMPATHETIC DIVISION (Action functions) — PARASYMPATHETIC DIVISION (Vegetative functions)

The parasympathetic division of the ANS is concerned with vegetative functions—e.g., it encourages secretory activity on the body's mucous and serous membranes, promotes digestion by increased peristalsis and glandular secretion, and induces contraction of the urinary bladder.

The parasympathetic preganglionic neuronal cell bodies in the head are located in the brain stem associated with certain cranial nerves. The preganglionic axons leave the brain stem with their cranial nerve and synapse at one of the cranial *ganglia*. The *postganglionic neurons* tend to be short, terminating in salivary glands and other glands of the nasal and oral cavities. The preganglionic fibers associated with the *vagus (X cranial) nerve* are unusually long, descending the neck, the esophagus, and through the esophageal hiatus to the gastrointestinal tract. The axons of these neurons extend as far as the descending colon. The ganglia are in the muscular walls of the organ they supply (*intramural ganglia*); the postganglionic axons are very short, terminating in smooth muscle and glands.

The cell bodies of the sacral preganglionic neurons are located in the lateral horns of sacral segments 2, 3, and 4 of the spinal cord. Their axons leave the cord via the anterior rami but form their own nerves, called the *pelvic splanchnic nerves* (nervi erigentes). These nerves project to the pelvis, mix with sympathetic postganglionics in the pelvic plexus, and depart for their target organs. They synapse with the postganglionic neurons in intramural ganglia in the walls of the organ supplied. These fibers stimulate contraction of rectal and bladder musculature and induce vasodilatation of vessels to the penis and clitoris (erection).

The parasympathetic and sympathetic divisions of the autonomic nervous system are not antagonistic. Their respective activities are coordinated and synchronized to achieve dynamic stability of body function during a broad range of life functions, such as eating, running, fear, or relaxation.

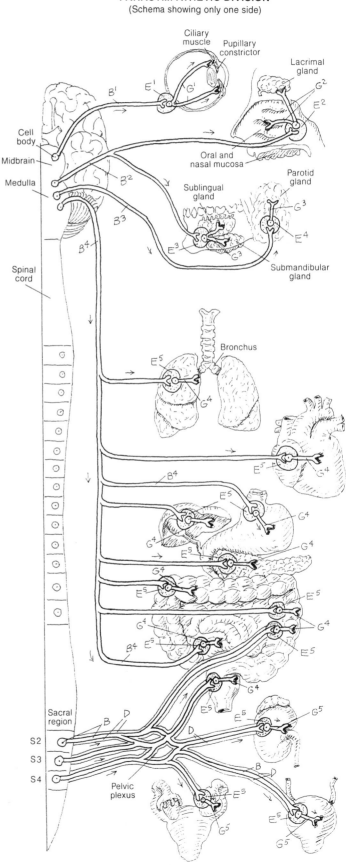

PARASYMPATHETIC DIVISION
(Schema showing only one side)

VISUAL SYSTEM (1)

CN: Use orange for E, yellow for G, red for M, blue for N, and very light colors for C, H, I, J, and K. (1) Color the sagittal section of the eyeball and the uppermost illustrations simultaneously. Arteries (M) and veins (N) are too narrow to be colored on the surface of the retina in the sagittal section. (2) When coloring the retinal layers, color gray the arrows (in dark outlines) representing the nerve impulse.

EYE LAYERS

SCLERA A /
CORNEA A'
CHOROID B
CILIARY BODY C
PROCESS C'
IRIS D
RETINA E
OPTIC DISC F
MACULA LUTEA G
FOVEA CENTRALIS R

FLUIDS

VITREOUS BODY (HUMOR) H
AQUEOUS HUMOR I

OTHER STRUCTURES

LENS J
SUSPENSORY LIG. K
PUPIL L
OPTIC NERVE F'
RETINAL ARTERY M / BRANCH M'
RETINAL VEIN N / BRANCH N'

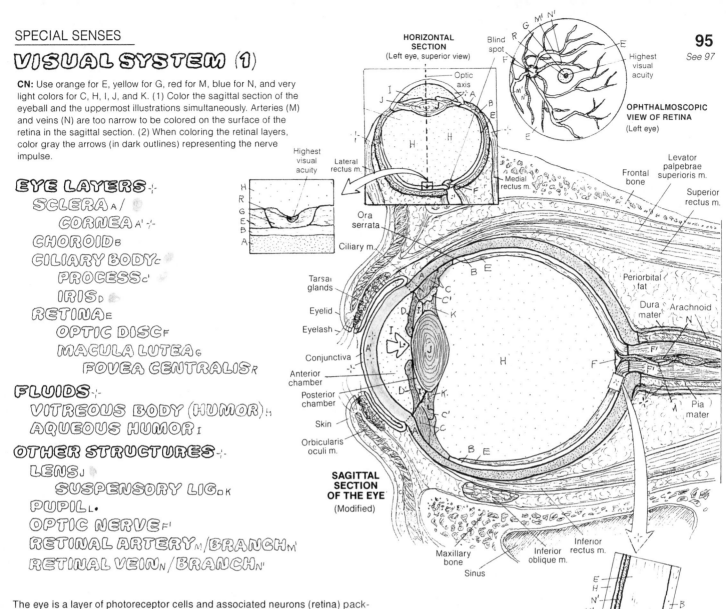

HORIZONTAL SECTION (Left eye, superior view)

OPHTHALMOSCOPIC VIEW OF RETINA (Left eye)

Highest visual acuity

SAGITTAL SECTION OF THE EYE (Modified)

LAYERS OF RETINA E

AXON F² / NERVE FIBER LAYER F³
GANGLION CELL O / LAYER O'
BIPOLAR CELL P / LAYER P'
ROD CELL Q   LAYER Q'
CONE CELL R'   LAYER R²
PIGMENTED EPITHELIAL LAYER R

The eye is a layer of photoreceptor cells and associated neurons (retina) packaged within a white, fibrous, rubberlike protective globe (*sclera*) that is transparent in front (*cornea*). The cornea, composed of five layers of epithelia and fibrous tissue, is the chief refractive medium of the eye, focusing light rays onto the retina. The *lens* (tightly packed, encapsulated non-elastic lens fibers derived from epithelial cells) also refracts light, and up to middle age, it can vary its shape (and refractive index). The *aqueous humor* (extracellular fluid) filling the anterior and posterior chambers of the eye, and the more gelatinous (99% water) *vitreous humor* taking up 80% of the globe's volume, function as refractive media. The inner surface of the posterior two-thirds of the sclera is lined with a vascular, highly pigmented layer (*choroid*) that absorbs and prevents scattering of light. The choroid thickens anteriorly as the pigmented, fibromuscular *ciliary body* that surrounds the lens. The ciliary body projects outpocketings (*processes*) to which *suspensory ligaments* from the lens attach. On the anterior aspect of the ciliary body, a thin, pigmented, epithelial and fibromuscular layer (*iris*) circumscribes the hole (*pupil*) in front of the lens.

The retina lines the posterior half of the interior of the globe, and a bit more, ending anteriorly at the ora serrata. At the retinal end of the optic axis, there is a yellow pigmented area (*macula lutea*) within which is a depressed area called the *fovea centralis*. Under lighted conditions, this is the center of greatest visual acuity (clarity of form and color), reflecting a dense accumulation of color-sensitive cells (cones). About 3 mm to the nose side of the macula, axons of the nerve fiber layer stream out through the optic disc to become the *optic nerve*. The *optic disc* is devoid of light-sensitive cells and is therefore a blind spot. The *pigmented layer* of the retina (refreshing pigment to the adjacent rods/cones) is closest to the choroid. The photoreceptor layer consists of *cone cells* (sensitive to form and color) and color-insensitive *rod cells* possessing great sensitivity to light. *Bipolar cells* receive and mediate input from rod and cone cells and conduct resultant impulses to the *ganglion cell layer*. Among these two more-peripheral layers are interwoven numerous horizontal cells (not shown for visual clarity) that influence neuronal activity. The axons of the ganglion cells, the final common pathway of retinal activity, form the fibers of the optic nerve.

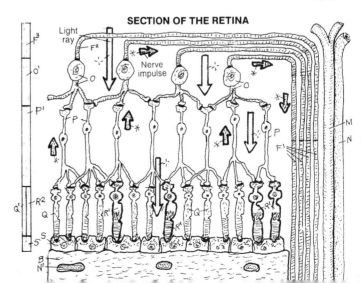

SECTION OF THE RETINA

VISUAL SYSTEM (2)

CN: Use the same colors as were used on the previous plate (with different subscripts) for structures J, K, L, M, N¹, and O. Use light colors for A, G, H, and I. Note that various structures in the central illustration also appear in the illustration below it.

ACCESSORY STRUCTURES:

LACRIMAL APPARATUS
LACRIMAL GLAND A
TEAR A'
DUCT B
LACRIMAL PUNCTA C
CANAL D
LACRIMAL SAC E
NASOLACRIMAL DUCT F
INFERIOR MEATUS OF NASAL CAVITY G
TARSAL PLATE/GLAND H
CONJUNCTIVA I

Fluid (tears) interfacing the conjunctivae of the eyelid (palpebra) and the cornea facilitate easy movement of the lids over the cornea without inducing irritation. Tears also function as a vehicle for moving epithelial debris and microorganisms from the corneal surface and undersurface of the eyelids into the nasal cavity via the lacrimal apparatus. Thus, there is an anatomic basis for blowing your nose after a good cry. The absence of tears can cause remarkable pain and even blindness. The principal gland for tears is the *lacrimal gland*, located in the anterior, superior and lateral (temporal) aspect of the orbit. Other glands and sources of tears include unicellular (goblet) glands of the conjunctiva and *tarsal glands* of the lids. Episodic blinking (rapid cycle of lid approximation and retraction) maintains a film of tears on the conjunctiva and resists "dry eye." Routine closing of the lids occurs with muscle relaxation; energetic closure requires the orbicularis oculi muscle. Retraction of the eyelids is accomplished by smooth muscle fibers (tarsal muscle of Müller; sympathetic innervation) and the levator palpebrae muscle in the upper lid.

SECRETION/DRAINAGE OF AQUEOUS HUMOR:

FLOW OF AQUEOUS HUMOR J
SCLERA K
CORNEA K'
CILIARY BODY L
PROCESS L'
POSTERIOR CHAMBER J'
IRIS M
ANTERIOR CHAMBER J²
CANAL OF SCHLEMM N
VEIN N'
VITREOUS BODY O
INTRAOCULAR PRESSURE (IOP) P

Aqueous humor is a fluid in the anterior and posterior chambers of the eye, secreted by cells of the *ciliary processes* (see lowest drawing). Fluid and electrolytes also enter by diffusion from the *ciliary body*. Aqueous humor is a clear, plasma-like fluid (but constituted differently). It is filtered into the *canal of Schlemm* (scleral venous sinus), a modified vein filled with fibrous trabeculae, located at the sclero-corneal junction. Fluid in the canal drains into nearby *veins*. Obstruction to drainage is one of several causes of increased *intraocular pressure*, in which the increasing pressure in the anterior/posterior chambers presses on the lens, which presses on the *vitreous* (99% water) body. As water cannot be compressed, pressure is applied to the contiguous retina. Unrelenting pressure compresses vessels to the axons and neurons of the retina, damages neurons, and can result in blindness (glaucoma).

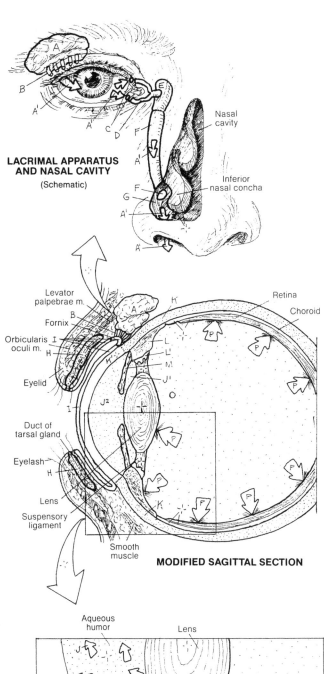

LACRIMAL APPARATUS AND NASAL CAVITY
(Schematic)

Nasal cavity

Inferior nasal concha

Levator palpebrae m.

Fornix

Orbicularis oculi m.

Eyelid

Duct of tarsal gland

Eyelash

Lens

Suspensory ligament

Smooth muscle

Retina

Choroid

MODIFIED SAGITTAL SECTION

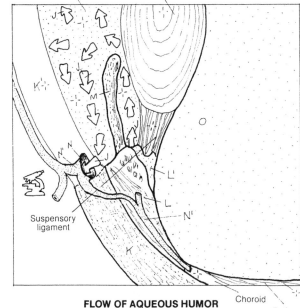

Aqueous humor

Lens

Suspensory ligament

Choroid

Retina

FLOW OF AQUEOUS HUMOR

VISUAL SYSTEM (3)

CN: Use light colors for A–F, H, and I. Use contrasting colors for J and K. (1) After coloring each eye muscle, color its functional arrow in the upper diagram. (2) In the drawing on ciliary action, color only the contracted ciliary muscles. (3) Carefully color the diagram below, noting that only the first titles (visual field) receive J and K colors. The rest of the titles are left uncolored, but use the two colors on the structures to which they refer.

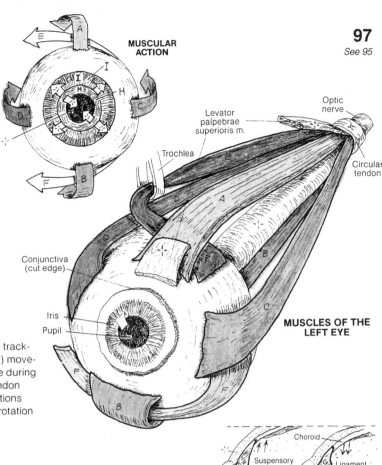

MUSCULAR ACTION

Levator palpebrae superioris m.

Optic nerve

Trochlea

Circular tendon

Conjunctiva (cut edge)

Iris

Pupil

MUSCLES OF THE LEFT EYE

EXTRAOCULAR MUSCLES:
SUPERIOR RECTUS (ELEV.)ₐ
INFERIOR RECTUS (DEPR.)ᵦ
LATERAL RECTUS (ABD.)ᵧ
MEDIAL RECTUS (ADD.)ᵨ
SUPERIOR OBLIQUE (ROT. R.)ₑ
INFERIOR OBLIQUE (ROT. L.)ꜰ

The extraocular (extrinsic) muscles of the eye provide for a remarkable tracking capacity of the eye. CNS mechanisms permit conjugate (binocular) movement of both eyes. Slowed, incomplete, or absent movement of one eye during tracking movements suggests cranial nerve dysfunction or muscle/tendon incarceration, as might occur in an orbital plate fracture. The true functions of these muscles is more complex than shown, one reason being eye rotation and torsion requiring multiple muscle action. Deviation from co-equal alignment of the eyes is called strabismus.

INTRINSIC MUSCLES:
CILIARYᴳ
SPHINCTER PUPILLAEᴴ
DILATOR PUPILLAEᴵ

The intrinsic muscles are located in the ciliary body (ciliary muscle) and the iris (pupillary dilator and sphincter). Contraction of the *ciliary muscles* (1) wrinkles the ciliary body tissue and puts slack in the processes, giving laxity to the suspensory ligaments (2) and permitting the lens to round up on its own accord (tension in lens fibers) (3). These muscles function (by parasympathetic innervation) during near vision in which greater refractivity is desired. The *dilator pupillae* consists of myoepithelial cells that pull the iris toward the ciliary body, dilating the pupil (sympathetic innervation). The *sphincter pupillae* circumscribes the inner iris; its contraction constricts the iris, narrowing the pupil (parasympathetic innervation). See the uppermost drawing.

Choroid

Suspensory ligament

Ligament relaxed

Iris

Lens more convex

CILIARY MUSCLE ACTION

Muscle at rest

Muscle contracts

VISUAL PATHWAYS:
VISUAL FIELD J / VISUAL FIELD κ
LIGHT WAVE (J',K')
RETINA (J²,K²)
OPTIC NERVE (J³,K³) CHIASMA (J⁴,K⁴) TRACT (J⁵,K⁵)
LATERAL GENICULATE BODY (J⁶,K⁶)
SUPERIOR COLLICULUS (J⁷,K⁷)
OPTIC RADIATION (J⁸,K⁸)
VISUAL CORTEX (J⁹,K⁹)

As you color the lower diagram, note that the axons (K^2) from the *retinas* on the temporal side of the optic axis do not cross at the *chiasma*. Note further that an expanding tumor of the hypophysis is likely to impair visual acuity in the temporal visual fields only ("tunnel vision"). The *thalamus* functions as a visual relay center, informing multiple memory areas and other centers of the stimulus. The *superior colliculi* are visual reflex centers, making possible rapid head and body movements in response to a visual threat. Finally, note that the image of the stimulus impinging on the *visual cortex* (K/J) is the reverse of that which was actually seen (J/K). Integration of visual and memory centers at the visual cortex makes possible perception of the image as actually seen.

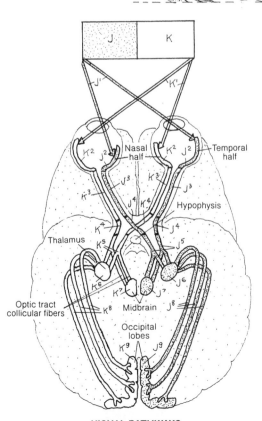

Nasal half

Temporal half

Hypophysis

Thalamus

Optic tract collicular fibers

Midbrain

Occipital lobes

VISUAL PATHWAYS
(Horizontal brain section, schematic)

AUDITORY & VESTIBULAR SYSTEM (1)

CN: Use yellow for Z, and light colors for A, B, G, I, M, N, W, and X. The view of the internal ear is magnified in the upper illustration for coloring purposes. Color your way down the plate, beginning with the diagram at the top.

EXTERNAL EAR-:-
AURICLE A
EXT. AUDITORY MEATUS B
TYMPANIC MEMBRANE C

MIDDLE EAR-:-
MALLEUS (HAMMER) D
INCUS (ANVIL) E
STAPES (STIRRUP) F
AUDITORY TUBE G

INTERNAL EAR-:-
BONY LABYRINTH H
VESTIBULE I
OVAL WINDOW J
SEMICIRCULAR CANAL K
COCHLEA L
SCALA VESTIBULI M
SCALA TYMPANI N
ROUND WINDOW O
MEMBRANOUS LABYRINTH P
SACCULE Q /UTRICLE Q'
ENDOLYMPHATIC DUCT R
SEMICIRCULAR DUCT S
COCHLEAR DUCT T
TECTORIAL MEMBRANE U
ORGAN OF CORTI V
HAIR CELL W
SUPPORTING CELL X
BASILAR MEMBRANE Y
CRANIAL NERVE VIII Z

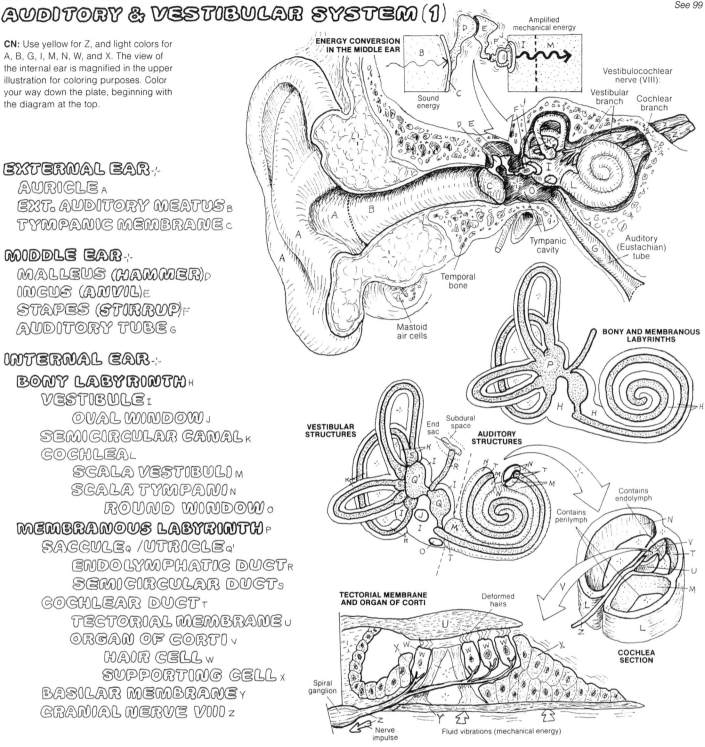

ENERGY CONVERSION IN THE MIDDLE EAR

Amplified mechanical energy

Sound energy

Vestibulocochlear nerve (VIII):
Vestibular branch
Cochlear branch

Tympanic cavity

Temporal bone

Mastoid air cells

Auditory (Eustachian) tube

BONY AND MEMBRANOUS LABYRINTHS

VESTIBULAR STRUCTURES

End sac

Subdural space

AUDITORY STRUCTURES

Contains endolymph

Contains perilymph

COCHLEA SECTION

TECTORIAL MEMBRANE AND ORGAN OF CORTI

Deformed hairs

Spiral ganglion

Nerve impulse

Fluid vibrations (mechanical energy)

The ear is the organ of hearing and equilibrium (auditory and vestibular systems). It is organized into external, middle, and internal parts. The external ear includes the *auricle* (collector of sound energy) and the *external auditory meatus* or canal (a narrow passageway conducting sound energy to the *tympanic membrane*). This membrane, lined externally by skin and internally by respiratory mucosa, converts sound energy into mechanical energy by resonating in response to incoming sound waves.

The middle ear is a small area filled with much structure, including three small bones (*malleus, incus, stapes*) joined together by synovial joints. These ossicles vibrate with movement of the tympanic membrane, and amplify and conduct the mechanical energy imparted to them to the waters of the inner ear at the flexible, water-tight *oval window* (middle ear/inner ear interface). At the anterior-medial aspect of the middle ear cavity, the *auditory tube* runs to the nasopharynx, permitting equilibration of air pressure between nasal cavity (outside) and the middle ear. The internal ear, carved out within the petrous portion of the temporal

bone, consists of a series of interconnecting bony-walled chambers and passageways (*bony labyrinth: vestibule, semicircular canals,* and *cochlea*) filled with perilymph (extracellular-like) fluid. Within the bony labyrinth is a series of interconnecting membranous chambers and passageways (*membranous labyrinth: saccule, utricle, cochlear duct,* and *semicircular ducts*), filled with endolymph (intracellular-like fluid). The *endolymphatic duct*, derived from the *saccule*, ends in a blind sac under the dura mater near the internal auditory meatus (see Plate 25). It drains endolymph and discharges it into veins in the subdural space. Within the coiled, membranous *cochlear duct*, supported by bone and the fibrous *basilar membrane*, a ribbon of specialized receptor (*hair*) *cells* exists integrated with supporting cells, both covered with a flexible, fibrous glycoprotein blanket (*tectorial membrane*). This device (*Organ of Corti*) converts the mechanical energy of the oscillating tectorial membrane scraping against the receptor hair cells into electrical energy. The impulses generated are conducted along bipolar sensory (auditory) neurons of the *VIII cranial nerve*. (Continued on the next plate.)

AUDITORY & VESTIBULAR SYSTEM (2)

CN: Titles with subscripts 1, 2, and 3 require new colors; all other subscripts (A–Z) refer to titles and colors used on the preceding plate, which should be frequently referred to when using those same colors on this plate. (1) Color the numerals with the appropriate color as you follow the sequence of events in the simplified diagram to the right. See the previous plate for the more accurate anatomical structure. (2) Color the parts of the vestibular system concerned with the maintenance of dynamic and static balance.

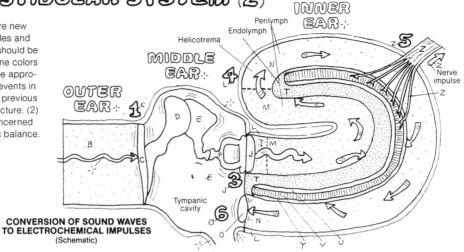

INNER EAR
Perilymph
Endolymph
Helicotrema
Nerve impulse

MIDDLE EAR

OUTER EAR

Tympanic cavity

CONVERSION OF SOUND WAVES TO ELECTROCHEMICAL IMPULSES
(Schematic)

In review: the external ear collects sound waves and rifles them to the *tympanic membrane*, which converts the sound energy into mechanical energy. The linkage of *ossicles* increases the amplitude of the energy and transmits the force to the *oval window* of the bony labyrinth of the inner ear. Vibratory movements of the *stapes* in the window are transmitted to the perilymph of the *vestibule* of the bony labyrinth, creating wave-like motions of the fluid. These waves spread throughout the vestibule, then enter and move through the *scala vestibuli* of the *cochlea* to the helicotrema at the apex of the cochlea (2½ turns) and on around to the *scala tympani*, which

terminates at the *round window*. Here, fluid waves and vibrations are dampened. The fluid motion in the scala vestibuli vibrates the roof of the membranous *cochlear duct*, creating endolymph waves in the cochlear duct. This motion stirs the *tectorial membrane*, which rubs against and bends the hair-like processes of the *receptor (hair) cells*, depolarizing them and inducing electrochemical impulses. These impulses are conducted by the sensory neurons of the cochlear division of the *VIII cranial nerve*. Stimulation of the hair cells from the apex of the cochlea to the base produces a continuum of increasingly high-pitched sound perceptions.

VESTIBULAR SYSTEM / EQUILIBRIUM

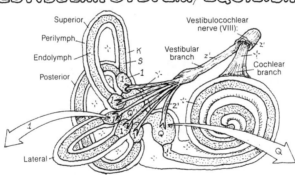

Superior
Perilymph
Endolymph
Posterior
Lateral

Vestibulocochlear nerve (VIII):
Vestibular branch
Cochlear branch

AMPULLA₁
CRISTA
CUPOLA₂
HAIR CELL w'
NERVE FIBER z'
SUPPORTING CELL x'

SACCULE Q **UTRICLE** Q'
MACULA
GELATINOUS LAYER 2
OTOLITH 3
HAIR CELL w'
NERVE FIBER z'
SUPPORTING CELL x'

SEMICIRCULAR CANAL K
SEMICIRCULAR DUCT S

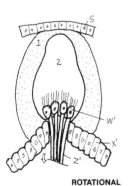

ROTATIONAL MOVEMENT
(Dynamic balance)

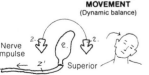

Nerve impulse
Superior

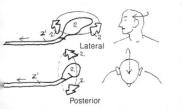

Lateral

Posterior

In review: the vestibular system is located in the inner ear. The bony *semicircular canals* are oriented at 90° to one another. Within these canals are the membranous *semicircular ducts*. Directly communicating with the utricle at one end, each duct terminates at the other end in an *ampulla*. Within the *saccule/utricle* and the ampullae are sensors responsive to fluid (endolymph) movement. Each ampulla has a hillock of cells (*crista* or crest) consisting of receptor (*hair*) and *supporting cells*. The hair-like processes of these receptor cells are embedded in a top-heavy, gelatinous *cupola* (like an inverted cup). Movement of endolymph in response to head turning, and especially rotation, pushes these cupolas, bending the hair cells and causing them to depolarize, generating an electrochemical impulse. The impulses travel out the vestibular part of the VIII nerve to the vestibular nuclei in the lower brain stem. When the body is rotated rapidly, horizontal, oscillatory eye movements occur (nystagmus). These eye movements are mediated by ampullary sensory input to the brain stem. Such movements represent the brain's attempt to maintain spatial orientation (by momentary visual fixation) during head and/or body rotation. Sensations of rotational movement in the absence of body rotation are called vertigo.

Within the utricle/saccule, *hair cells* and their *supporting cells* are covered with a *gelatinous layer* in which are embedded small calcareous bodies (*otoliths*). Movement of the endolymph induces movement of the gelatinous layer against the hair cells, with responses identical to those of the ampullary receptors. Receptor activity in the utricle/saccule is influenced by linear (horizontal and vertical but non-rotational) acceleration of the body. Vestibular receptors have strong neural connections with cranial nerve nuclei concerned with eye movement and with postural motor centers.

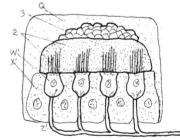

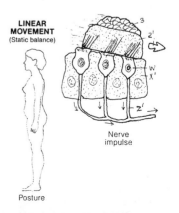

LINEAR MOVEMENT
(Static balance)

Nerve impulse

Posture

TASTE & OLFACTION

CN: Use yellow for H and light colors for A, B, C, G, and I.
(1) Do not color the taste buds in the circumvallate papillae in the modified section at right. (2) In the lowest illustration, color over the neurons within the olfactory bulb.

PAPILLAE

CIRCUMVALLATE A
FUNGIFORM B
FILIFORM C

TASTE BUD D

PORE CANAL E
RECEPTOR CELL F
SUPPORTING CELL G
NERVE FIBER H

Taste receptors (taste buds) are located within the stratified squamous epithelial lining of the sides (moats) of *circumvallate*, *foliate* (not shown), and *fungiform papillae* on the tongue and, to a lesser extent, on the soft palate and lingual side of the epiglottis. They are not seen in the tiny filiform papillae. Each taste bud consists of a number of *receptor cells* and their *supporting cells*. The apex of this oval cell complex faces the moat; here it opens on to the papillary surface via a taste pore or *pore canal*. Dissolved material enters the pore, stimulating the chemoreceptor (gustatory) cells. The impulses generated are conducted along *sensory axons* that reach the brain stem via the VII, IX, and X cranial nerves (recall Plate 83). Taste interpretation occurs at the lower reaches of the sensory cortex (post-central gyrus). Basic tastes (sweet, sour, salt, and bitter) notwithstanding, interpretation of taste, as a practical matter, is a function of smell, food texture, and temperature in association with taste bud sensations.

OLFACTION (SMELL)

OLFACTORY GLAND I
OLFACTORY MUCOSA I'
OLFACTORY NEURON J
OLFACTORY HAIR (CILIA) K
SUPPORTING CELL G'
OLFACTORY BULB H'
OLFACTORY TRACT H²

Olfactory receptors are *olfactory hairs* or cilia (actually modified peripheral processes) of *olfactory bipolar (sensory) neurons* buried in the *olfactory mucosa* at the roof of the nasal cavity. The olfactory mucosa also has tubulo-alveolar *olfactory glands* that function to keep the chemoreceptor endings clean and, along with nasal mucous secretions, dissolve the chemicals that are sensed by these receptors. The olfactory neurons ascend the roof of the nasal cavity, through the cribriform plate of the ethmoid bone, and their central processes synapse with second-order neurons in the *olfactory bulb*. The axons of these neurons form three olfactory bundles (stria) as part of the *olfactory tract*, terminating in the inferior frontal lobe and medial temporal lobe. Here exists the neural basis for olfactory relationships with memory, eating, survival, sex, and other emotional behavior.

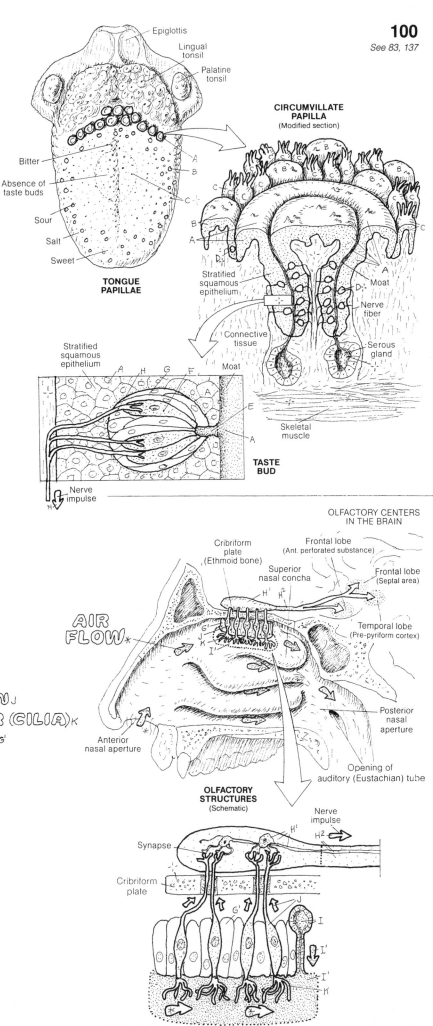

BLOOD & BLOOD ELEMENTS

CN: Color P purple, PP pale purple, O orange, PO pale orange, PB pale blue, R red, and T tan. Except for the latter, these colors match the stains used to observe these cells. First color the cytoplasm of the cell; if you don't have any of the pale colors, leave the cytoplasm background blank. Then stipple the granules with the darker color. The results should create a rough impression of the actual colors.

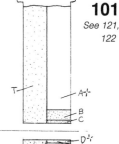

PLASMA (55%)ᴛ
WATER (90%)ₐ₊
PROTEINS (8%)ᵦ
ACIDS, SALTS (2%)ᵪ

RETICULOCYTE

ERYTHROCYTES ʀ
(RED BLOOD CORPUSCLES)

6–8 μm

Erythrocytes (RBCs; approximately 4.5–6.2 million per milliliter of blood in men; 4–5.5 million/ml in women) are formed in the bone marrow. There each cell loses its nucleus and most of its organelles prior to release into the peripheral blood. Recently released immature erythrocytes (reticulocytes) may retain some ribosomal RNA in their cytoplasm; these granules appear dark purple and reticulated when stained. Normally making up about 1% of the RBC population, reticulocytes may increase in number during chronic oxygen lack (e.g., at prolonged high altitude). The circulating RBC (without nucleus or organelles, it is truly a corpuscle and not a cell) is a non-rigid, biconcave-shaped, membrane-lined sac of hemoglobin, a large iron-containing protein. Hemoglobin (12–16 grams/decaliter of blood in women; 14–18 g/dL in men) has a powerful affinity for oxygen and is the principal carrier of oxygen in the body; plasma is the other carrier. Erythrocytes pick up oxygen in the lungs and release it in the capillaries to the tissues/cells. RBCs circulate for about 120 days until defective and are then taken out of the blood and broken down by cells of the spleen.

FORMED ELEMENTS OF THE BLOOD (45%)∗
ERYTHROCYTES (99%)ʀ
THROMBOCYTES (0.6–1.0%)ʟᴘ
LEUKOCYTES (0.2%)ᴅ₊

All constituents of the blood that can be observed as discrete structures with the aid of the light microscope are called *formed elements of the blood*. The rest of the blood is a protein-rich fluid called plasma. When blood is allowed to clot, the cells disintegrate (hemolysis) and a thick yellow fluid called serum emerges. Serum is basically plasma less clotting elements. If whole blood is centrifuged in a test tube, the RBCs will settle to the bottom, the *leukocyte fraction* will form a buffy coat on top of that, and the *plasma*, being the lightest, will take up the upper 55% of the total volume. Packed RBCs in a test tube constitute a *hematocrit* (40–52% of the blood volume in men; 37–47% in women). The difference in blood values between men and women is probably related to iron storage and metabolism differences (men store up to 50% more iron than women). A low hematocrit may be associated with anemia or hemorrhage.

THROMBOCYTES ᴘ
(PLATELETS)

Thrombocytes or platelets (150,000–400,000/ml of blood; 2–5 μm in diameter) are small bits of cytoplasm from giant cells (megakaryocytes) of the bone marrow. Circulating in the blood for a lifetime of 10 days or so, platelets adhere to injured endothelium and play a significant role in limiting hemorrhage (aggregation of platelets, blood coagulation/clotting, and clot removal).

LEUKOCYTES ᴅ₊
(WHITE BLOOD CELLS)

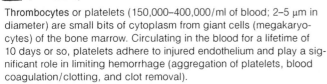

Cytoplasm
PP
Nucleus
Granules

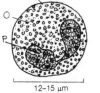
SEGMENTED (MATURE)
12–15 μm

GRANULAR ∗
NEUTROPHIL ₊

Segmented neutrophils (55–75% of the WBC population) arise in the bone marrow and live short lives in the blood and connective tissues (hours–4 days). Immature forms (*band neutrophils*, 1–5%) may be seen in the blood; their numbers often increase in acute infections. Neutrophils rapidly engulf foreign elements/cellular debris; strong enzymes in specific granules and lysosomes destroy them (phagocytosis).

BAND (IMMATURE)

EOSINOPHIL ₊

Eosinophils (1–3% of WBCs) exhibit colorful granules when properly stained. Eosinophils are phagocytic in immune reactions. They are involved in the late-onset phase of asthmatic attacks (subsequent bronchial constriction), possibly enhancing cell injury by increasing cell membrane permeability in the bronchial mucosa to allergic substances. They also appear to limit the expression of mast cell degranulation (histamine release and effects) during allergic reactions.

PO
O
P
12–15 μm

BASOPHIL ₊

Basophils (0–1% of WBCs) contain dark-staining granules. Basophils are known to degranulate in allergic reactions, releasing histamine, serotonin, and heparin. Such degranulation induces contraction of smooth muscle, increases vascular permeability (enhancing the effects of inflammation), and slows down movement of white blood cells in inflammation.

NONGRANULAR ₊
LYMPHOCYTE ₊

PB P
6–18 μm

Lymphocytes (20–45% of WBCs) arise from the bone marrow and reside in the blood as well as the lymphoid tissues (lymph nodes, thymus, spleen, and so on). Lymphocytes generally consist of about 20% B cells (short-lived cells from the bone marrow, concerned with humoral immunity, transformation into plasma cells, and the secretion of antibodies or immunoglobulins) and 70% T cells (long-lived cells from the thymus; may be cytotoxic, helper, or suppressor cells associated with cell-mediated immunity). Lymphocytes with neither B or T surface antigens (less than 5%) are called natural killer cells.

MONOCYTE ₊

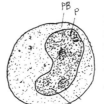
PB P
12–20 μm

Monocytes (2–8% of WBCs) arise in the bone marrow, mature in the blood (about eight hours), then leave the circulation to enter the extracellular spaces as macrophages. They are critical to the functioning of the immune system, as they present antigen to the immune cells, secrete substances in immune reactions, and destroy antigens (see glossary). They phagocytose cellular and related debris in wound healing, bone formation, and multiple other cellular activities where breakdown occurs.

SCHEME OF BLOOD CIRCULATION

CN: Use blue for A, purple for B, red for C, and very light colors for D and E. (1) Color the titles for systemic and pulmonary circulation, the two figures, and the borders bracketing the large illustration. Also color purple (representing the transitional state between oxygenation and deoxygenation) the two capillaries, demonstrating the difference between capillary function in the lungs and that in the body. (2) Begin in the right atrium of the heart and color the flow of oxygen-poor blood (A) into the lungs. After coloring the pulmonary capillary network (B), color the oxygen-rich blood (C) that re-enters the heart and is pumped into and through the systemic circuit.

OXYGEN-POOR BLOOD A
CAPILLARY BLOOD B
OXYGEN-RICH BLOOD C

Circulation of blood begins with the heart, which pumps blood into arteries and receives blood from veins. *Arteries* conduct blood *away* from the heart regardless of the amount of oxygen (oxygenation) in that blood. *Veins* conduct blood toward the heart, regardless of the degree of oxygenation of the blood. *Capillaries* are networks of extremely thin-walled vessels throughout the body tissues that permit the exchange of gases and nutrients between the vessel interior (vascular space) and the area external to the vessel (extracellular space). Capillaries receive blood from small arteries and conduct blood to small veins.

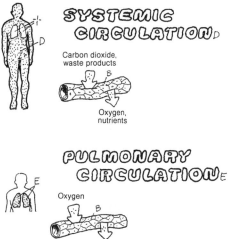

SYSTEMIC CIRCULATION D

Carbon dioxide, waste products

Oxygen, nutrients

PULMONARY CIRCULATION E

Oxygen

Carbon dioxide

There are two circuits of blood flow: (1) the pulmonary circuit, which conveys blood from the right side of the heart to the lungs and fresh blood back to the left side of the heart, and (2) the systemic circuit, which conveys blood from the left heart to the body tissues and returns blood to the right heart. The color red is used universally for *oxygenated blood,* and the color blue is used for *oxygen-poor blood.*

Clearly, not all arterial blood is fully oxygenated (in the pulmonary circulation, arteries conduct poorly oxygenated blood to the lungs), and not all venous blood is oxygen deficient (pulmonary veins conduct oxygenated blood to the heart).

Capillary blood is mixed; it is largely oxygenated on the arterial side of the capillary bed, and it is largely deoxygenated on the venous side, as a consequence of delivering oxygen to and picking up carbon dioxide from the tissues it supplies.

One capillary network generally exists between an artery and a vein. There are exceptions: the portal circulation of the liver involves two sets of capillaries between artery and vein (Plate 119); the hypophyseal portal system involves two capillary networks between artery and vein (Plate 152); and the renal vascular system has a glomerulus and a peritubular capillary plexus between artery and vein (Plate 150).

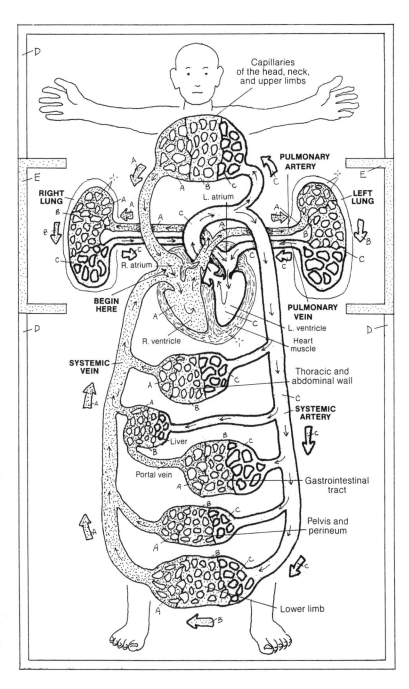

Capillaries of the head, neck, and upper limbs

PULMONARY ARTERY

RIGHT LUNG

LEFT LUNG

L. atrium

R. atrium

BEGIN HERE

PULMONARY VEIN

L. ventricle

Heart muscle

R. ventricle

SYSTEMIC VEIN

Thoracic and abdominal wall

SYSTEMIC ARTERY

Liver

Portal vein

Gastrointestinal tract

Pelvis and perineum

Lower limb

BLOOD VESSELS

CN: Use red for A, purple for B, blue for C, and very light colors for D, F, and H. (1) Complete the upper left diagram, beginning with the large arteries. (2) Color the blood vessels and their titles at the bottom of the plate. Note that the vas and nervus vasorum in the fibrous tissue layer (H) are not colored. (3) In the diagram of venous valve action, the blood in both vein and artery is colored gray.

LARGE ARTERIES A
Elastic conducting vessels

LARGE VEINS c²
Capacitance or reservoir vessels

Heart

MEDIUM A'
Muscular, distributing vessels

MEDIUM c'

SMALL (ARTERIOLE) A²
Resistance vessels

SMALL (VENULE) c

CAPILLARIES B
Larger capillaries are sinusoids

Large arteries (elastic or conducting arteries), such as the aorta or common carotid, contain multiple layers of elastic tissue. They are roughly the size of a finger. *Medium arteries* (muscular, distributing arteries), averaging the size of a pencil, are generally named (e.g., brachial). Diminutive branches of medium arteries are called small arteries (*arterioles*); unnamed, they control the flow of blood into capillary beds (resistance vessels). *Capillaries* are unnamed simple endothelial tubes supported by thin fibrous tissue. Microscopic in dimension, some capillaries are larger (sinusoids) or more specialized than others.

Veins get progressively larger as they get closer to the heart. Veins have tributaries; except in portal circulations, they do not have branches. Venules (small veins) are formed by the merging of capillaries and are basically of the same construction. *Venules* merge to form *medium veins*, and these are the tributaries of *large veins* (capacitance or reservoir vessels). Certain specialized large veins, as in the skull, are called sinuses. The walls of these veins are thinner than those of their arterial counterparts, and their lumens are generally larger. Large veins can stretch significantly, becoming virtual reservoirs of blood.

All vessels demonstrate a simple squamous epithelial (endothelial) lining (tunica interna) supported by a thin layer of fibrous tissue (not shown). Most medium veins of the neck and extremities have a series of small pockets formed from the endothelial layer. These valves are paired and point in the direction of blood flow. Though offering no resistance to blood flow, they will bend into and close off the lumen of the vein when the flow of blood is reversed. Valves resist gravity-induced blood pooling, especially in the lower limb vessels. Venous flow here is enhanced by the contraction of skeletal muscles, whose bulges give an anti-gravity boost to the movement of blood. The *internal elastic lamina*, a discrete layer only in medium-sized arteries, assists in maintaining blood pressure; this tissue is more diffuse in other vessels. The *tunica media* consists of concentrically arranged smooth muscle fibers. It is well developed in medium arteries, least developed in veins. Medium arteries use this layer in distributing blood from one field to another. In arterioles, reduced to only one or two layers, the smooth muscle can literally block blood flow into capillary fields. The *external elastic lamina* exists as a discrete layer only in muscular arteries. The *tunica externa* (adventitia) is fibrous tissue contiguous with the fascial layer in which the vessel is located; within this tunica, much smaller nutrient vessels (vasa vasorum) and motor nerves (nervi vasorum) are found. In very specialized situations, the structure of small vessels may be specially adapted—e.g., the glomerulus, in Plate 149.

VESSEL STRUCTURE *
TUNICA INTERNA
ENDOTHELIUM D
INTERNAL ELASTIC LAMINA E
TUNICA MEDIA
SMOOTH MUSCLE F
EXTERNAL ELASTIC LAMINA G
TUNICA EXTERNA
FIBROUS TISSUE H

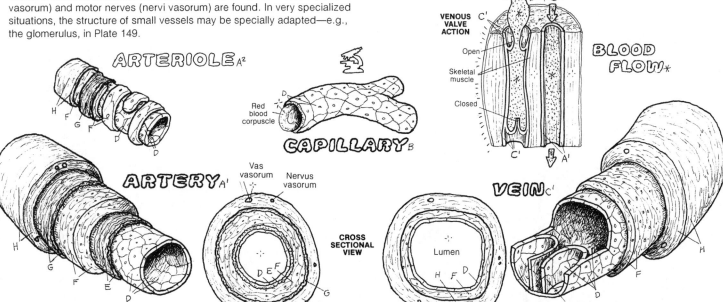

VENOUS VALVE ACTION
Open
Skeletal muscle
Closed

BLOOD FLOW *

ARTERIOLE A²

CAPILLARY B
Red blood corpuscle

ARTERY A'
Vas vasorum
Nervus vasorum
CROSS SECTIONAL VIEW

VEIN c'
Lumen
Valve

MEDIASTINUM, WALLS & COVERINGS OF THE HEART

MEDIASTINUM REGIONS:
SUPERIOR A
INFERIOR:
ANTERIOR B
MIDDLE C
POSTERIOR D

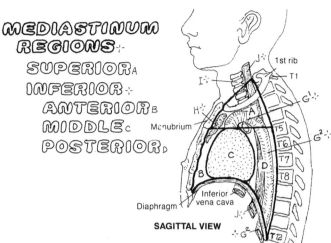

SAGITTAL VIEW

CN: Use blue for F, red for G, and your lightest colors for A–D. (1) Begin with the regions of the mediastinum at upper left and color over all the structures within the dark outline. (2) Color the major structures within the mediastinum in the anterior view. Note that the lungs, not being in the mediastinum, remain uncolored. Note that the thymus, which can be seen in the sagittal view, has been deleted here to show the great vessels covered by it. (3) Finally, color the walls of the heart and layers of pericardium at lower left. The pericardial cavity has been greatly exaggerated for coloring. It is normally only a potential space.

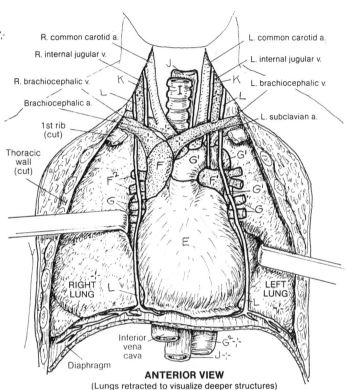

ANTERIOR VIEW
(Lungs retracted to visualize deeper structures)

STRUCTURES:
PERICARDIUM-LINED HEART E
GREAT VESSELS:
SUPERIOR VENA CAVA F
PULMONARY TRUNK F¹
PULMONARY ARTERY F²
PULMONARY VEIN G
AORTIC ARCH G¹
THORACIC AORTA G²:

THYMUS H:
TRACHEA I
ESOPHAGUS J
VAGUS NERVE K
PHRENIC NERVE L

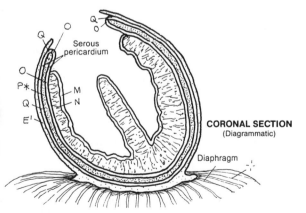

Serous
pericardium

CORONAL SECTION
(Diagrammatic)

Diaphragm

WALLS OF THE HEART / PERICARDIUM:
ENDOCARDIUM M
MYOCARDIUM N
VISCERAL PERICARDIUM O
PERICARDIAL CAVITY P*
PARIETAL PERICARDIUM Q
FIBROUS PERICARDIUM E¹

The **mediastinum** (median septum or partition) is a highly populated region between and excluding the lungs. A variety of passageways, nerves, and vessels enter, pass through, and exit the mediastinum. For descriptive purposes, the mediastinum is divided into the subdivisions (regions) illustrated. The *superior mediastinum* is remarkable for the array of *great vessels of the heart* and the *trachea, esophagus,* and *vagus* and *phrenic nerves.* At the level of the T4–T5 vertebrae (superior/inferior mediastinal border), the trachea bifurcates into the main bronchi (see Plate 133) posterior to the great vessels, and the *aorta* makes its arch. The *posterior mediastinum* includes the inferior continuation of the esophagus embraced by a fine network of vagal nerve fibers, the thoracic duct (see Plate 121), and the descending (*thoracic*) aorta. The floor of the mediastinum is the diaphragm, penetrated by the thoracic aorta, esophagus, and inferior vena cava.

The **heart wall** consists of an inner layer of simple squamous epithelium (*endocardium*) overlying a variably thick *myocardium* (cardiac muscle). External to the myocardium is a three-layered sac (*pericardium*). The innermost layer of this sac is the *visceral pericardium* (epicardium), clothing the heart. At the origin of the aortic arch, this layer turns (reflects) outward to become the *parietal pericardium* (imagine a fist clutching the edges around the opening of a paper bag; now push the fist into the closed bag still clutching the edges; as you do so, note that your fist becomes surrounded by two layers of the paper bag, yet is not inside the bag itself). The relationship of your fist to the two layers of the bag is the relationship of the heart to the visceral and parietal pericardium. The cavity of the bag is empty—the fist is not in the bag (if you did it right!). Similarly, the *pericardial cavity* between the two pericardial layers is empty as well, except for serous fluid that makes for friction-free movement of the heart in its sac.

The **fibrous pericardium** is the outer surface of the parietal pericardium; it is fibrous and fatty and is strongly attached to the sternum, the great vessels, and the diaphragm. It keeps the twisting, contracting, squeezing heart within the middle mediastinum.

CHAMBERS OF THE HEART

CN: Use blue for A–A⁴, red for H–H⁴, and your lightest colors for B, C, I, and J. All dotted arrows (A⁴) receive a blue color; all clear arrows (H⁴) receive a red color. (1) Begin with the arrows A⁴ above the title list and above the superior vena cava (A) in the illustration at upper right and color the structures in the order of the title list (A–H³). (2) Color the circulation chart at lower right, beginning with the arrow A⁴ leading into the right atrium (numeral 1). Color the numerals in order from 1 to 4 and related arrows. Do not color the chambers or the vessels in this drawing at lower right.

A⁴

SUPERIOR VENA CAVA ₐ
INFERIOR VENA CAVA ₐ'
RIGHT ATRIUM ʙ

RIGHT VENTRICLE ᴄ
 A-V TRICUSPID VALVE ᴅ
  CHORDAE TENDINEAE ᴇ
  PAPILLARY MUSCLE ꜰ

PULMONARY TRUNK ₐ²
 PUL. SEMILUNAR VALVE ɢ
 PUL. ARTERY ₐ³

—H⁴
PULMONARY VEIN ʜ
LEFT ATRIUM ɪ

LEFT VENTRICLE ᴊ
 A-V BICUSPID (MITRAL) VALVE ᴅ'
  CHORDAE TENDINEAE ᴇ'
  PAPILLARY MUSCLE ꜰ'

ASCENDING AORTA ʜ'
 AORTIC SEMILUNAR VALVE ɢ'
 AORTIC ARCH ʜ²
 THORACIC AORTA ʜ³

The heart is the muscular pump of the blood vascular system. It contains four cavities (chambers): two on the right side (pulmonary heart), two on the left (systemic heart). The pulmonary "heart" includes the right atrium and right ventricle. The thin-walled *right atrium* receives poorly *oxygenated blood* from the *superior* and the *inferior vena cava* and from the coronary sinus (draining the heart vessels). The thin-walled *left atrium* receives richly *oxygenated blood* from pulmonary veins. Atrial blood is pumped at a pressure of about 5 mm Hg into the *right and left ventricles* simultaneously through the atrioventricular orifices, guarded by the 3-cusp *tricuspid valve* on the right and the 2-cusp *bicuspid valve* on the left. The cusps are like panels of a parachute, secured to the *papillary muscles* in the ventricles by tendinous *chordae tendineae*. These muscles contract with the ventricular muscles, tensing the cords and resisting cusp over-flap as ventricular blood bulges into them during ventricular contraction (systole). The right ventricle pumps oxygen-deficient blood to the lungs via the *pulmonary trunk* at a pressure of about 25 mm Hg (right ventricle), and the left ventricle pumps oxygen-rich blood into the *ascending aorta* at a pressure of about 120 mm Hg simultaneously. This pressure difference is reflected in the thicker walls of the left ventricle compared to the right. The pocket-like *pulmonary and aortic semilunar valves* guard the trunk and aorta, respectively. As blood falls back toward the ventricle from the trunk/aorta during the resting phase (diastole), these pockets fill, closing off their respective orifices and preventing reflux into the ventricles.

**ANTERIOR VIEW
OF HEART CAVITIES
AND GREAT VESSELS**

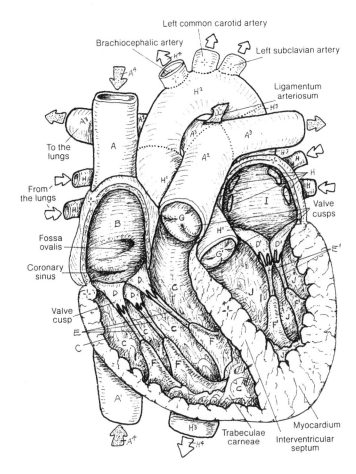

Left common carotid artery
Brachiocephalic artery
Left subclavian artery
Ligamentum arteriosum
To the lungs
From the lungs
Valve cusps
Fossa ovalis
Coronary sinus
Valve cusp
Myocardium
Trabeculae carneae
Interventricular septum

**CIRCULATION
THROUGH
THE HEART**

OXYGEN-RICH BLOOD ʜ⁴
OXYGEN-POOR BLOOD ₐ⁴

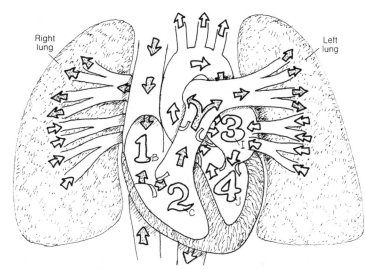

Right lung

Left lung

CARDIAC CONDUCTION SYSTEM & THE ECG

CN: Use blue for D and red for E. Use a very light color for B so that the patterns of dots identifying the segments (B–B³) of the ECG remain visible after coloring. (1) Begin at upper right and color the four large arrows identifying the atria (A²) and ventricles (B³), as well as their titles. The atria and ventricles are not to be colored. (2) In the middle of the page, color the stages of blood flow through the heart, and related letters. These stages relate to voltage changes in the ECG below. (3) Color the ECG and related letters, starting at the left and working to the right. The parts of the ECG are related to the activity of the conduction system or related myocardial activity. (4) Color the horizontal bar below the time line.

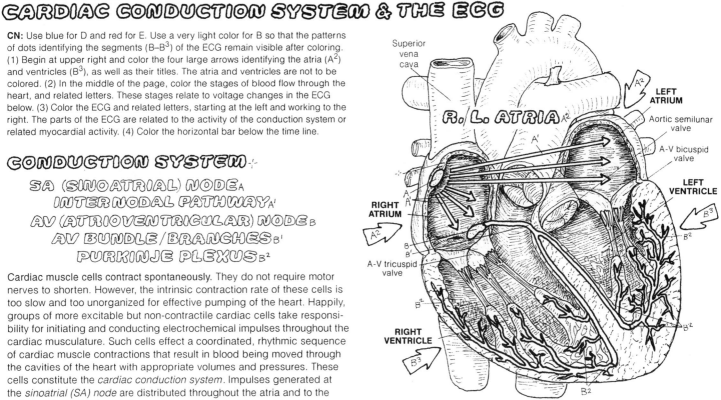

CONDUCTION SYSTEM

SA (SINOATRIAL) NODE A
INTERNODAL PATHWAY A'
AV (ATRIOVENTRICULAR) NODE B
AV BUNDLE/BRANCHES B'
PURKINJE PLEXUS B²

Cardiac muscle cells contract spontaneously. They do not require motor nerves to shorten. However, the intrinsic contraction rate of these cells is too slow and too unorganized for effective pumping of the heart. Happily, groups of more excitable but non-contractile cardiac cells take responsibility for initiating and conducting electrochemical impulses throughout the cardiac musculature. Such cells effect a coordinated, rhythmic sequence of cardiac muscle contractions that result in blood being moved through the cavities of the heart with appropriate volumes and pressures. These cells constitute the *cardiac conduction system*. Impulses generated at the *sinoatrial (SA) node* are distributed throughout the atria and to the *atrioventricular (AV) node* by way of non-discrete *internodal pathways*. Impulses travel from the AV node, down the *AV bundle and its branches*, to the *Purkinje plexus* of cells embedded in the ventricular musculature.

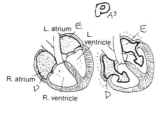

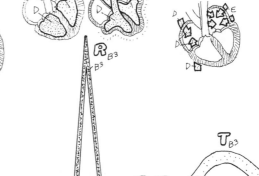

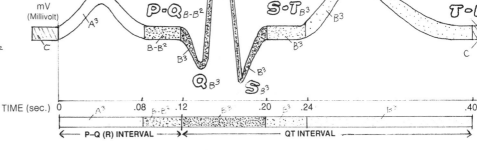

BLOOD FLOW
OXYGEN-POOR D
OXYGEN-RICH E

ELECTRO CARDIOGRAM (ECG)

P WAVE A³
P-Q (P-R) SEGMENT B-B²
QRS COMPLEX B³
S-T SEGMENT B³
T WAVE B³
T-P SEGMENT C

mV (Millivolt)

TIME (sec.) 0   .08   .12   .20   .24   .40

P-Q (R) INTERVAL   QT INTERVAL

The cardiac conduction system generates voltage changes about the heart, some of which can be monitored, assessed, and measured by *electrocardiography (ECG)*. An ECG is essentially a voltmeter reading. It does not measure hemodynamic changes. Electrodes are placed on a number of body points on the skin. Recorded data (various waves of varying voltage over time) are displayed on an oscilloscope or a strip of moving paper. The shape and direction of wave deflections are dependent upon the spatial relationship of the electrodes (leads) on the body surface.

When the SA node fires, excitation/depolarization of the atrial musculature spreads out from the node. This is reflected in the ECG by an upward deflection of the resting (isoelectric) horizontal line (*P wave*). This deflection immediately precedes the contraction of the atrial musculature and filling of the ventricles. The *P-Q interval* (P-R interval in the absence of a Q wave) reflects conduction of excitation from the atria to the Purkinje cell plexus in the ventricular myocardium. Prolongation of this interval beyond .20 seconds may reflect an AV conduction block. The *QRS complex* reflects depolarization of the ventricular myocardium. This complex of deflections immediately precedes ventricular contraction, wherein blood is forced into the pulmonary trunk and ascending aorta. The *S-T segment* reflects a continuing period of ventricular depolarization. Myocardial ischemia may induce a deflection of this normally horizontal segment. The *T wave* is an upward, prolonged deflection and reflects ventricular repolarization (recovery), during which the atria passively fill with blood from the vena cavae and pulmonary veins. The QT interval, corrected for heart rate (QTc), reflects ventricular depolarization and repolarization. Prolongation of this segment may suggest abnormal ventricular rhythms (arrhythmias). In a healthy heart at a low rate of beat, the P-Q, S-T, and T-P segments all are isoelectric (horizontal).

CARDIOVASCULAR SYSTEM **107**

CORONARY ARTERIES & CARDIAC VEINS

CN: Use your brightest colors for A, D, and L. (1) When coloring the arteries, include the broken lines that represent vessels on the posterior surface of the heart. (2) Do the same with the veins. (3) Color the artery in front of the plaque in the circled view; color the vessel after the plaque a lighter shade of the same color or do not color it at all.

CORONARY ARTERIES:

RIGHT CORONARYA
MUSCULAR BRANCHA'
MARGINAL BRANCHB
**POSTERIOR INTERVENTRICULAR
(DESCENDING) BRANCH**C
LEFT CORONARYD
**ANTERIOR INTERVENTRICULAR
(DESCENDING) BRANCH**E
MUSCULAR BRANCHE'
CIRCUMFLEX BRANCHF

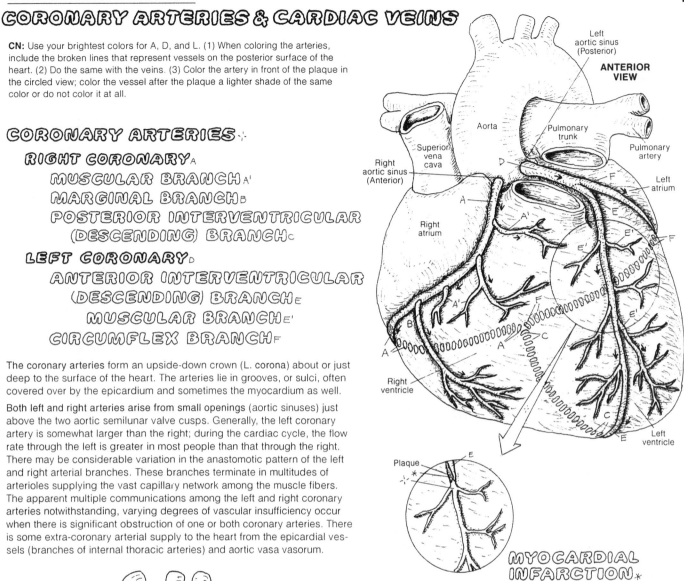

The coronary arteries form an upside-down crown (L. corona) about or just deep to the surface of the heart. The arteries lie in grooves, or sulci, often covered over by the epicardium and sometimes the myocardium as well.

Both left and right arteries arise from small openings (aortic sinuses) just above the two aortic semilunar valve cusps. Generally, the left coronary artery is somewhat larger than the right; during the cardiac cycle, the flow rate through the left is greater in most people than that through the right. There may be considerable variation in the anastomotic pattern of the left and right arterial branches. These branches terminate in multitudes of arterioles supplying the vast capillary network among the muscle fibers. The apparent multiple communications among the left and right coronary arteries notwithstanding, varying degrees of vascular insufficiency occur when there is significant obstruction of one or both coronary arteries. There is some extra-coronary arterial supply to the heart from the epicardial vessels (branches of internal thoracic arteries) and aortic vasa vasorum.

MYOCARDIAL INFARCTION *

Damage to the intimal layer of coronary arteries can occur with lipid deposition or inflammation. Platelet aggregation at these sites contributes to the formation of plaque (cell material, lipid, platelet, fibrin). Plaque builds up within the vessels, forming thrombi that occlude the vessels in progressively greater degrees. Significantly reduced blood flow to the myocardium (ischemia) can cause sharp pain (angina) to the chest, back, shoulder, and arm as well as permanent damage to the myocardium (infarction).

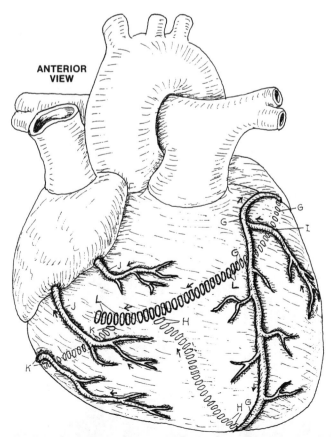

CARDIAC VEINS:
GREAT CARDIAC V.G
MIDDLE CARDIAC V.H
MARGINAL V.I
ANTERIOR CARDIAC V.J
SMALL CARDIAC V.K
CORONARY SINUSL

The cardiac veins travel with the coronary arteries, but incompletely. Vast anastomoses of veins occur throughout the myocardium; most drain into the right atrium by way of the coronary sinus. The anterior cardiac veins conduct blood directly into the right atrium. Other small veins may drain directly into the right atrium as well. Some deep (arteriosinusoidal) veins drain directly into the atria and ventricles. Extracardiac venous drainage can also occur through the vasa vasorum of the vena cavae.

ARTERIES OF THE HEAD & NECK

CN: Use red for A and dark or bright colors for B and L. (1) Begin with the brachiocephalic (A) and the right subclavian (B) and its branches. Color the broken lines that represent deeper vessels. (2) Color the anterior view revealing the absence of the brachio-cephalic artery on the left side. (3) Color the arrows pointing to the four sites where the arterial pulse may be palpated.

BRACHIOCEPHALIC A

RIGHT SUBCLAVIAN B
INTERNAL THORACIC c
VERTEBRAL d
THYROCERVICAL TRUNK e
INFERIOR THYROID f
SUPRASCAPULAR g
TRANSVERSE CERVICAL h
COSTOCERVICAL TRUNK i
DEEP CERVICAL j
HIGHEST INTERCOSTAL k

RIGHT COMMON CAROTID L
INTERNAL CAROTID m
OPHTHALMIC n
EXTERNAL CAROTID o
SUPERIOR THYROID p
LINGUAL q
FACIAL r
OCCIPITAL s
MAXILLARY t
ALVEOLAR BRANCHES: INF. u SUP. u'
MIDDLE MENINGEAL v
POSTERIOR AURICULAR w
SUPERFICIAL TEMPORAL x
TRANSVERSE FACIAL y

PULSE SITES

The subclavian artery is the major source of blood to the upper limb, and it contributes vessels to the lateral and posterior neck and shoulder. On the right, the artery springs from the brachiocephalic; on the left, the artery comes directly off the aortic arch, as does the common carotid (see below). The *vertebral artery* dives deep into the neck to enter the transverse foramen of the 6th cervical vertebra. It supplies vessels to the spinal cord, brain stem, and cerebellum. The *thyrocervical trunk* arises just medial to the anterior scalene muscle (see Plate 48) and immediately gives off its branches, the destinations of which are obvious by name. The subclavian artery ends and the axillary artery begins at the lateral border of the first rib.

The common carotid artery ascends the neck ensheathed with the internal jugular vein and vagus nerve (not shown). Between the hyoid bone and the upper thyroid cartilage, the artery bifurcates into *internal* and *external carotid arteries*. The internal carotid passes into the skull, gives off the *ophthalmic artery* to the orbital region, and joins the circulus arteriosus (Plate 109). The external carotid artery and its branches supply all of the visceral, musculoskeletal, and dental structures of the head and neck less the brain and orbit. The external carotid divides into *maxillary* and *superficial temporal arteries*. The maxillary artery is a major source of blood to the deep skull cavities, the orbit, teeth, the muscles of mastication, and the dura mater (*middle meningeal artery*). The middle meningeal artery on the dura mater immediately deep to the temporal bone is a potential site of rupture with a hard fall on the side of the head (epidural hematoma).

LEFT SUBCLAVIAN ARTERY B'
LEFT COMMON CAROTID A. L'

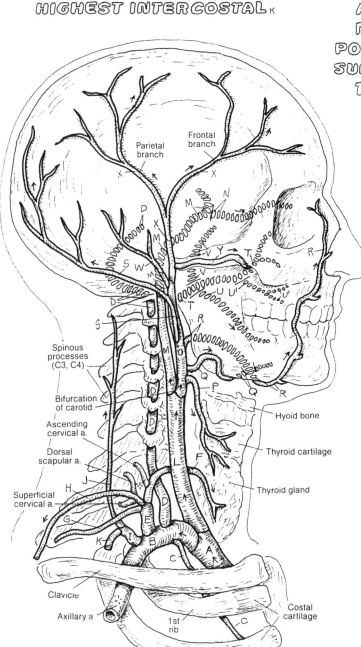

Parietal branch
Frontal branch
Spinous processes (C3, C4)
Bifurcation of carotid
Ascending cervical a.
Dorsal scapular a.
Superficial cervical a.
Hyoid bone
Thyroid cartilage
Thyroid gland
Clavicle
Axillary a
1st rib
Costal cartilage

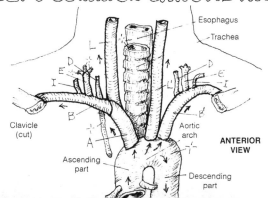

Esophagus
Trachea
Clavicle (cut)
Aortic arch
Ascending part
Descending part

ANTERIOR VIEW

ARTERIES OF THE BRAIN

CN: Use red for A and dark or bright colors for F and G. (1) Begin with the diagram at upper right, coloring both A and F. (2) Then color the diagram of the circulus arteriosus at left, beginning with the internal carotid artery (A). (3) Then color the vessels in the central illustration, relating them to the branches seen on the lateral and medial surfaces of the cerebrum (at the bottom of the plate).

INTERNAL CAROTID A
ANTERIOR CEREBRAL B
ANTERIOR COMMUNICATING C
MIDDLE CEREBRAL D
POSTERIOR COMMUNICATING E

VERTEBRAL F
BASILAR G
CEREBELLAR (3) H
POSTERIOR CEREBRAL I
ANTERIOR SPINAL J

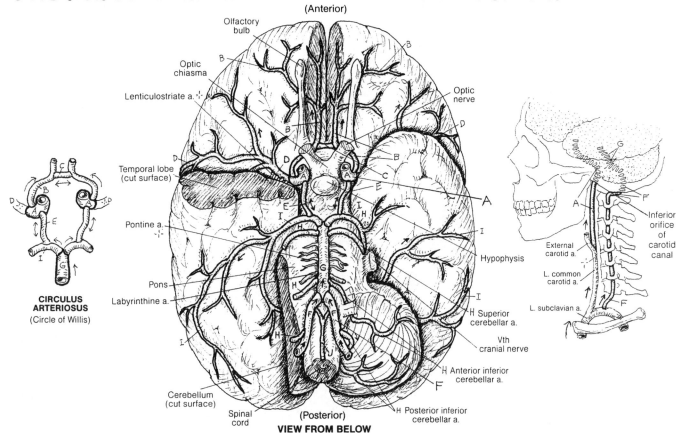

CIRCULUS ARTERIOSUS
(Circle of Willis)

(Anterior)
Olfactory bulb
Optic chiasma
Lenticulostriate a.
Optic nerve
Temporal lobe (cut surface)
Pontine a.
Pons
Labyrinthine a.
Hypophysis
H Superior cerebellar a.
Vth cranial nerve
H Anterior inferior cerebellar a.
Cerebellum (cut surface)
Spinal cord
(Posterior)
VIEW FROM BELOW
H Posterior inferior cerebellar a.

External carotid a.
L. common carotid a.
L. subclavian a.
Inferior orifice of carotid canal

The paired internal carotid arteries ascend the carotid canals (the inferior orifices of which are shown in Plate 25) to arrive in the middle cranial fossa just lateral to the optic chiasma. There each divides into *anterior* and *middle cerebral arteries*. Small lenticulostriate arteries come off the middle cerebral at right angles, supplying the basal ganglia. These "stroke arteries" are commonly the ruptured vessels in intracerebral hemorrhage, often resulting in at least partial paralysis of the limb muscles on the side of the body contralateral to the hemorrhage. Note the distribution of the anterior and middle cerebral arter-

ies on the surface of the cerebrum. Note in the central illustration how the brain stem is supplied by vessels arising indirectly from the vertebral arteries, and how occlusion of one of these vessels, e.g., labyrinthine arteries, might be manifested. Note that the single direct connection of the vertebral system and the carotid system is by way of the posterior communicating artery. Yet there is considerable variation in the components of the arterial circle as seen angiographically, including anomalies and severely narrowed vessels. More emphasis on the branchings of the carotid and vertebral arteries is recommended.

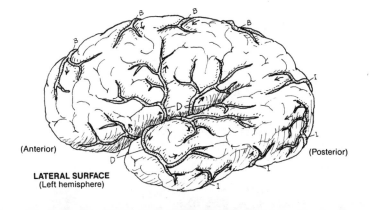

(Anterior)   (Posterior)

LATERAL SURFACE
(Left hemisphere)

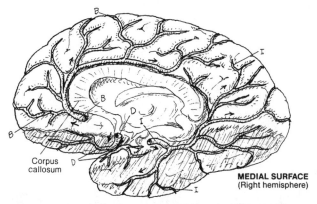

Corpus callosum

MEDIAL SURFACE
(Right hemisphere)

ARTERIES & VEINS OF THE UPPER LIMB

PULSE POINTS *

Common carotid a.

Clavicle

Acromion

Coracoid process

Humerus

Internal jugular v.

External jugular v.

CN: Use red for brachiocephalic artery (A) and blue for brachiocephalic vein (I). You can repeat colors except red and blue. Color arterial pulse point arrows gray. Broken lines represents the posterior surface.

Arteries of the upper limb generally lie deep among or within muscle fascicles. The pulse of a few of the larger ones can be palpated (see large arrows). In the event of hemorrhage, the arterial flow can be checked or slowed distally by the application of compression at these sites. Collateral routes of blood flow exist at many levels, a limb-saving reality in the event of the cessation of flow in a large artery. Note the collateral routes between arm and forearm. The arteries to the hand and wrist have been simplified for coloring.

Middle collateral a.

Recurrent interosseous a.

Radius

Ulna

◀ ARTERIES

BRACHIOCEPHALIC A

SUBCLAVIAN B

AXILLARY C

  SUPERIOR THORACIC D

  THORACO-ACROMIAL & BRANCHES E

  LATERAL THORACIC F

  SUBSCAPULAR G

  ANT./POST. CIRCUMFLEX HUMERAL H

BRACHIAL I

  PROFUNDA BRACHII & BRANCH J

  SUPERIOR ULNAR COLLATERAL K

  INFERIOR ULNAR COLLATERAL L

RADIAL M

  RADIAL RECURRENT N

ULNAR O

  ANT. ULNAR RECURRENT P

  POST. ULNAR RECURRENT Q

  COMMON INTEROSSEOUS R

SUPERFICIAL PALMAR ARCH S

  COMMON PALMAR DIGITAL T

DEEP PALMAR ARCH U

ANTERIOR VIEW
(Right limb)

VEINS ▶

DORSAL DIGITAL A & NETWORK A'

BASILIC B

MEDIAN V. OF FOREARM C

CEPHALIC D

MEDIAN CUBITAL E

BRACHIAL F

AXILLARY G

SUBCLAVIAN H

BRACHIOCEPHALIC I

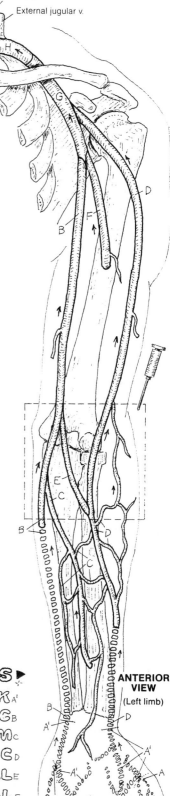

ANTERIOR VIEW
(Left limb)

The veins of the upper limb are variable in number and pattern. There are two sets of veins: deep and superficial. The deep set follows the arteries and is identically named—e.g., radial artery, radial vein. Often traveling in pairs (venae comitantes), *the deep veins of the hand, forearm, and lower arm are not shown on this plate.* The broken lines represent superficial veins on the posterior aspect of the forearm. At the elbow, the veins within the boxed area often are sites for blood sampling and administration of intravenous medication.

ARTERIES OF THE LOWER LIMB

PULSE POINTS ✱

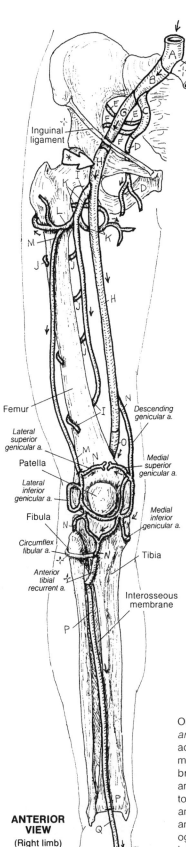

L. common iliac a.

Inguinal ligament

Femur

Lateral superior genicular a.

Patella

Lateral inferior genicular a.

Fibula

Circumflex fibular a.

Anterior tibial recurrent a.

Descending genicular a.

Medial superior genicular a.

Medial inferior genicular a.

Tibia

Interosseous membrane

ANTERIOR VIEW (Right limb)

CN: Use red for A. (1) Work both views of the lower limb simultaneously. Color the parts of the genicular artery (N) as it forms a pattern around the knee joint. Also part of that anastomosis, but not to be colored, are the circumflex fibular and anterior tibial recurrent arteries. (2) The foot in the posterior view is plantar flexed, with the sole showing. (3) Color the arrows at the four pulsation points.

ABDOMINAL AORTA A
RIGHT COMMON ILIAC B
INTERNAL ILIAC C
OBTURATOR D
SUPERIOR GLUTEAL E
INFERIOR GLUTEAL F
EXTERNAL ILIAC G
FEMORAL H
PROFUNDA FEMORIS I
PERFORATING BRANCHES J
MEDIAL CIRCUMFLEX FEMORAL K
LATERAL CIRCUMFLEX FEMORAL L
DESCENDING BRANCH M
GENICULAR N
POPLITEAL O
ANTERIOR TIBIAL P
DORSALIS PEDIS Q
ARCUATE R
DORSAL METATARSAL S
DORSAL DIGITAL T
POSTERIOR TIBIAL U
FIBULAR V
MEDIAL PLANTAR W
LATERAL PLANTAR X
PLANTAR ARCH Y
PLANTAR METATARSAL Z
PLANTAR DIGITAL 1.

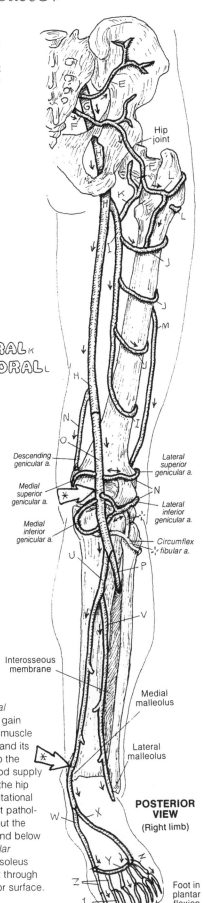

Hip joint

Descending genicular a.

Medial superior genicular a.

Medial inferior genicular a.

Lateral superior genicular a.

Lateral inferior genicular a.

Circumflex fibular a.

Interosseous membrane

Medial malleolus

Lateral malleolus

POSTERIOR VIEW (Right limb)

Foot in plantar flexion

Only the major arteries are shown here. The artery of the thigh (*femoral artery*) pierces the medial muscular compartment (adductor canal) to gain access to the back of the knee and leg. Because of the considerable muscle mass in the posterior thigh, the *profunda femoris artery* is quite large and its branches are extensive. Note how the *circumflex arteries* contribute to the anastomosis about the hip (femoral head and neck, hip joint). The blood supply to the hip joint area can be compromised by congenital anomalies in the hip anastomoses as well as degenerative and traumatic processes. Gravitational and weightbearing factors also can be important mediators in hip joint pathology. The *genicular arteries* form a significant anastomotic pattern about the knee joint and can be an important source of blood flow to the knee and below with an obstruction of the *popliteal artery*. The *posterior tibial* and *fibular arteries* run in a fascial compartment deep to the gastrocnemius and soleus muscles. The *anterior tibial artery* exits the posterior leg compartment through the interosseous membrane and descends on the membrane's anterior surface.

AORTA & BRANCHES

CN: Use red for A and light, bright colors for B, I, J, and L. (1) Color all posterior intercostal arteries (K) under the thoracic aortic section, even though only a few are labeled. (2) In the abdomen, note the location of the inferior phrenic, suprarenal, middle sacral, internal iliac, and external iliac arteries, none of which will be colored. Also note the inferior vena cava (stippled), which echoes the shape of the abdominal aorta.

AORTIC ARCH A
　　CORONARY B
　　BRACHIOCEPHALIC C
　　COMMON CAROTID D
　　SUBCLAVIAN E
　　　　INTERNAL THORACIC F
　　　　MUSCULOPHRENIC F¹
　　　　SUPERIOR EPIGASTRIC F²
　　COSTOCERVICAL TRUNK G
　　　　HIGHEST INTERCOSTAL H

THORACIC AORTA A¹
　　BRONCHIAL I
　　ESOPHAGEAL J
　　POSTERIOR INTERCOSTAL K

ABDOMINAL AORTA A²
　　CELIAC TRUNK L
　　　　LEFT GASTRIC M
　　　　SPLENIC N
　　　　COMMON HEPATIC O
　　SUPERIOR MESENTERIC P
　　RENAL Q
　　TESTICULAR / OVARIAN R
　　LUMBAR S
　　INFERIOR MESENTERIC T
　　COMMON ILIAC U

The branches of the aortic arch are unpaired except for the coronary arteries/sinuses. The branches of the *thoracic aorta* are paired. The *bronchial arteries* supply the lung tissues with oxygenated blood. The visceral branches of the aorta may be paired (e.g., gastric) or not (e.g., sup. mesenteric). Parietal branches of the abdominal aorta (*lumbar arteries*) are segmental and bilateral, in register with the segmental and bilateral posterior intercostals and subcostals of the thoracic aorta.

There are 9 intercostal spaces anteriorly (not shown) and 11 posteriorly (because of the floating ribs 11 and 12, not shown). These spaces are supplied by both *anterior* and *posterior* intercostal arteries. The posterior intercostals for spaces 1 and 2 are branches of the *highest* (supreme, superior) *intercostal artery* from the costocervical trunk. The posterior intercostals for spaces 3–11 come directly off the thoracic aorta. The subcostal arteries (not shown) run inferiorly to the 12th rib, posterior to the diaphragm. The paired first lumbar arteries (not shown) leave the abdominal aorta immediately below the diaphragm. The anterior intercostals arise from the *internal thoracic artery* for the first 6 intercostal spaces; then the latter artery bifurcates into *superior epigastric* and *musculophrenic* branches. Branches of the superior epigastric artery anastamose with branches of the inferior epigastric artery (not shown), a potentially significant collateral pathway. The musculophrenic artery supplies the lower 3 intercostal spaces.

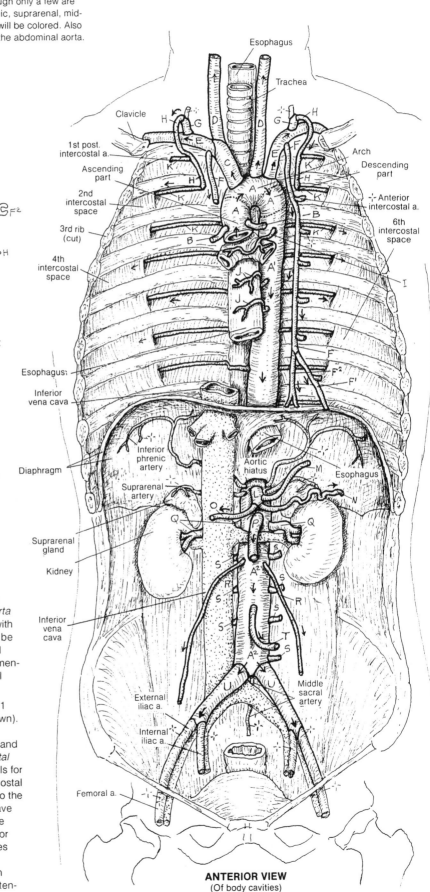

ANTERIOR VIEW
(Of body cavities)

ARTERIES TO GASTROINTESTINAL TRACT & RELATED ORGANS

CN: Use red for A, and use the same colors for the celiac trunk (B), superior mesenteric artery (L), and inferior mesenteric artery (Q) that you used for those structures on Plate 112, where they had different subscripts. (1) Begin with the large illustration. Note that the two pancreaticoduodenal arteries (H, H^1) receive the same color. (2) Color the upper illustration, which summarizes the three sources of blood supply to the digestive system.

AORTA$_A$
CELIAC TRUNK$_B$
HEPATIC: COMMON$_C$/LEFT$_{C'}$/RIGHT$_{C^2}$
RIGHT GASTRIC$_D$
GASTRODUODENAL$_E$
R. GASTROEPIPLOIC$_F$
L. GASTROEPIPLOIC$_G$
PANCREATICODUODENAL (SUP.)$_H$
CYSTIC$_I$
LEFT GASTRIC$_J$
SPLENIC$_K$

The celiac trunk, the first single visceral artery off the abdominal aorta, is a very short vessel that divides immediately into arteries to the liver, spleen, and stomach. Only the major branches of these three arteries are shown here. Note the anastomotic pattern of arteries to the stomach. The blood supply to the pancreas can be better appreciated in Plate 156.

SUPERIOR MESENTERIC$_L$
PANCREATICO-DUODENAL (INF.)$_{H'}$
MIDDLE COLIC$_M$
RIGHT COLIC$_N$
ILEO-COLIC$_O$
BRANCHES TO SMALL INTESTINE$_P$

The superior mesenteric artery supplies most of the small intestine, the head of the pancreas, cecum, and ascending colon, and part of the transverse colon. It travels in the common mesentery. Notice the collateral circulation between the celiac and superior mesenteric arteries in the curve of the duodenum. The superior and inferior mesenteric arteries also interconnect via a marginal artery that runs along the length of the large intestine and is fed by both these arteries. The arteries to the ileum/jejunum (cut short) run in the common mesentery.

INFERIOR MESENTERIC$_Q$
LEFT COLIC$_R$
SIGMOID BRANCHES$_S$
SUPERIOR RECTAL$_T$

The inferior mesenteric artery supplies the transverse colon down to the rectum. Its branches lie, for the most part, behind the peritoneum (retroperitoneal); the vessels to the sigmoid colon run in the sigmoid mesocolon. Note the anastomoses between branches of the *superior rectal artery* and those of the middle and inferior rectal arteries.

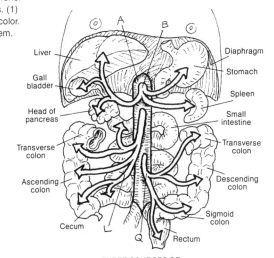

THREE SOURCES OF
DIGESTIVE BLOOD SUPPLY

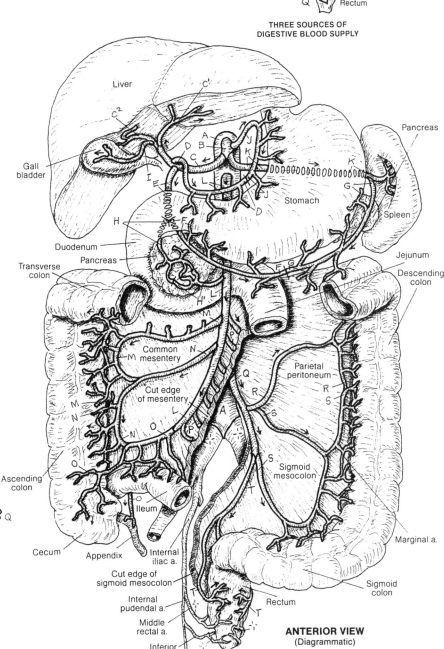

ANTERIOR VIEW
(Diagrammatic)

ARTERIES OF THE PELVIS & PERINEUM

CN: Use a light color for A. (1) Color the medial views of both pelves simultaneously. Only the disposition of reproductive organs and their vessels varies in these two views. (2) Color both halves of the dissected perineum seen from below. The names of the male vessels can be seen in the medial view.

INTERNAL ILIAC ₐ
POSTERIOR TRUNK ₐ₁
ILIOLUMBAR ʙ
SUPERIOR GLUTEAL ᴄ
LATERAL SACRAL ᴅ
ANTERIOR TRUNK ₐ₂
UMBILICAL (FETAL) ᴇ
SUP. VESICAL / A. TO VAS DEF. ꜰ
OBTURATOR ɢ
UTERINE ʜ
VAGINAL ɪ
INFERIOR VESICAL ᴊ
MIDDLE RECTAL ᴋ
INFERIOR GLUTEAL ʟ

The internal iliac artery supplies the pelvis and perineum, with some collaterals from the inferior mesenteric and femoral arteries. Its branches are usually organized into posterior (parietal) and anterior (visceral) divisions. The vascular pattern here is variable; the one shown is characteristic. From the posterior trunk, the *superior gluteal* passes through the greater sciatic foramen above piriformis. The *inferior gluteal* and *pudendal arteries*, from the anterior trunk, depart the pelvis through the lesser sciatic foramen below piriformis. Proximal to the formation of these latter two vessels, the anterior trunk of the internal iliac gives off four branches in both male and female. The first is the *superior vesical* (arising from the proximal part of the fetal umbilical artery; when the umbilical cord is cut, the distal part of the artery atrophies, forming the medial umbilical ligament; the remaining umbilical artery becomes the superior vesical artery, supplying the upper bladder and ductus deferens). The second is the *obturator artery* to the medial thigh region. The third is the *uterine artery*; in the male, it is the *inferior vesical artery*. The *vaginal artery* comes off the uterine artery. The arteries to the prostate and seminal vesicles (not shown) come off the *inferior vesical artery*. The fourth is the *middle rectal artery*.

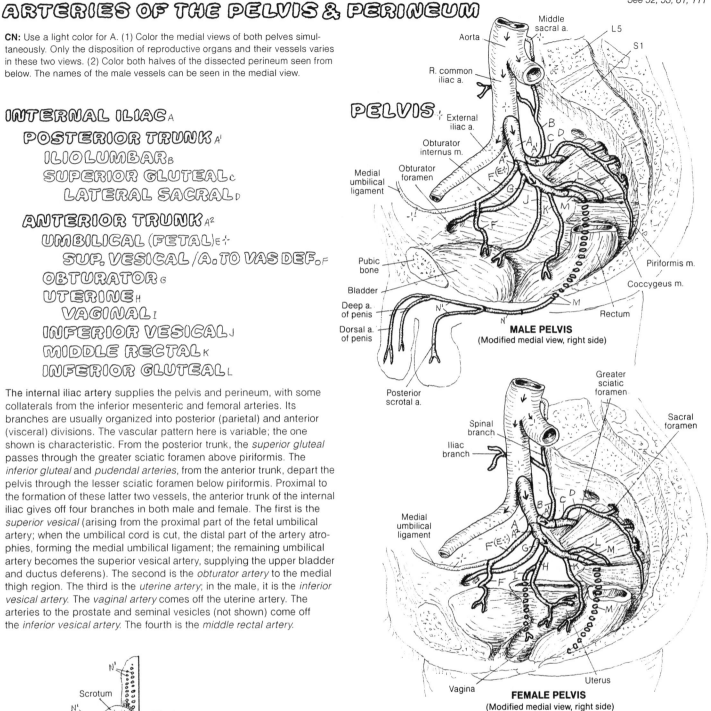

PELVIS

Middle sacral a. — Aorta — L5 — S1 — R. common iliac a. — External iliac a. — Obturator internus m. — Obturator foramen — Medial umbilical ligament — Pubic bone — Bladder — Deep a. of penis — Dorsal a. of penis — Piriformis m. — Coccygeus m. — Rectum

MALE PELVIS
(Modified medial view, right side)

Posterior scrotal a.

Greater sciatic foramen — Sacral foramen — Spinal branch — Iliac branch — Medial umbilical ligament — Vagina — Uterus

FEMALE PELVIS
(Modified medial view, right side)

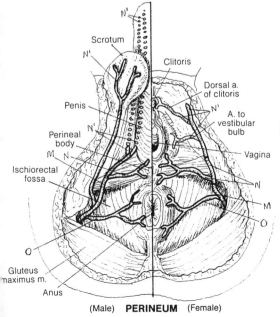

Scrotum — Clitoris — Dorsal a. of clitoris — A. to vestibular bulb — Penis — Perineal body — Vagina — Ischiorectal fossa — Gluteus maximus m. — Anus

(Male) PERINEUM (Female)

PERINEUM

PUDENDAL ᴍ
PERINEAL ɴ / BRANCHES ɴ₁
INFERIOR RECTAL ᴏ

The external genital structures are supplied by the *internal pudendal arteries*, which pass through the pudendal (fascial) canal alongside the inferior pubic ramus. The arteries (and nerves of the same name) enter the perineum just medial to the ischial tuberosities. The *perineal arteries* are significant in that they provide the vascular basis for erection of the penis and clitoris in sexual stimulation. Superficial and deep external pudendal arteries (not shown; branches of the proximal femoral artery) supply the skin of the genital structures.

REVIEW OF PRINCIPAL ARTERIES

CN: Use red for A. The arteries are shown bilaterally in the limbs only. Note that the figure is in the anatomical position, palms facing forward. (1) Using the preceding artery plates for reference where necessary, start with A and color in the order listed. As you color each artery, write its name next to the appropriate letter/number on each answer line. Use a pencil (you may change your mind).

(See appendix A in the back of the book for answers)

A _____

ARTERIES OF THE UPPER LIMB
B _____
C _____
D _____
E _____
F _____
G _____
H _____
I _____
J _____

ARTERIES OF THE HEAD AND NECK
K _____
L _____
M _____

ARTERIES OF THE CHEST
A _____
A^1 _____
N _____
O _____
P _____
Q _____
R _____
S _____

ARTERIES OF THE ABDOMEN AND PELVIS
A^2 _____
T _____
U _____
V _____
W _____
X _____
Y _____
Z _____
1 _____
2 _____

ARTERIES OF THE LOWER LIMB
3 _____
4 _____
5 _____
6 _____
7 _____
8 _____
9 _____
10 _____
11 _____
12 _____
13 _____
14 _____

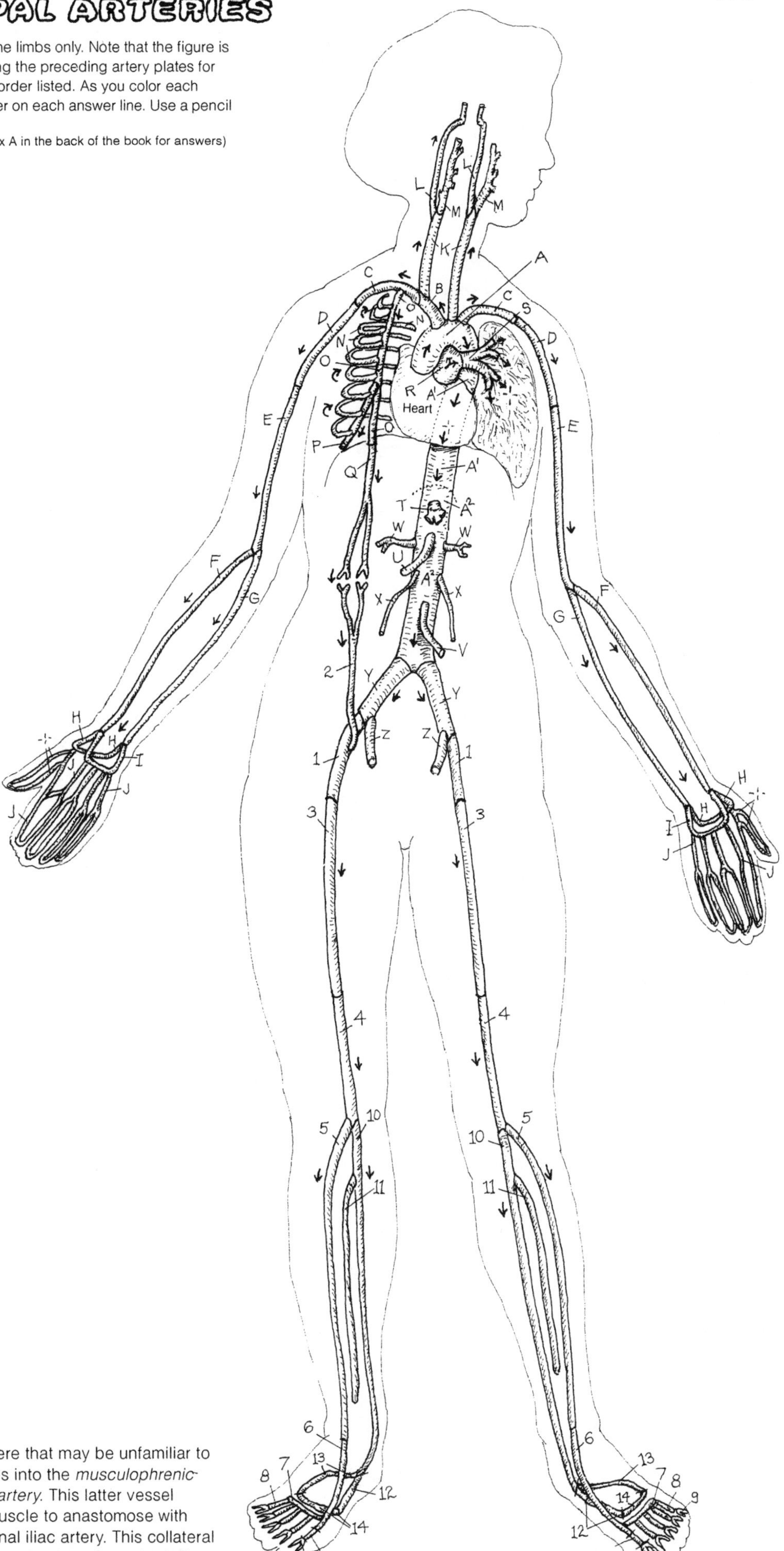

There is one pair of anastomotic vessels shown here that may be unfamiliar to you: note that the *internal thoracic artery* bifurcates into the *musculophrenic-artery* and the more medial *superficial epigastric artery*. This latter vessel descends inthe sheath of the rectus abdominis muscle to anastomose with the inferior epigastric artery, a branch of the external iliac artery. This collateral path from the heart (indirectly) to the lower limb represents an important alternate route for blood in the event of a seriously obstructing aortic aneurysm.

VEINS OF THE HEAD & NECK

CN: Note the order of titles and their indentations. We begin with titles of tributaries, indenting them above the vein with which they merge or join. This order is in the direction of flow. It will hold for all plates on the veins. Use lighter colors for the sinuses (A–K), represented in the lateral view by broken lines. (1) Begin with the venous sinuses. When coloring the falx and tentorium gray, color lightly over the vessels contained within (A, B, D, and E). Do not color the superior cerebral veins that join the superior sagittal sinus (A). The occipital sinus (K) is shown only in the lateral view.

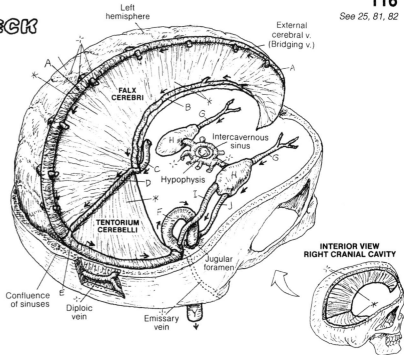

INTERIOR VIEW
RIGHT CRANIAL CAVITY

SINUSES OF DURA MATER :-
SUPERIOR SAGITTAL SINUS A
INFERIOR SAGITTAL SINUS B
GREAT CEREBRAL V. C
STRAIGHT SINUS D
TRANSVERSE SINUS E
SIGMOID SINUS F
SUPERIOR OPHTHALMIC V. G
CAVERNOUS SINUS H
SUPERIOR PETROSAL SINUS I
INFERIOR PETROSAL SINUS J
OCCIPITAL SINUS K

The veins draining the brain are tributaries of large venous channels (cranial dural venous sinuses) in the dura mater. The external cerebral ("bridging") veins drain the cerebral surface and merge with the *superior sagittal sinus*. They are known to rupture when excessive inertial loads are imposed on the brain (subdural hematoma). Two internal cerebral veins form the *great cerebral vein*; these drain the deeper hemispheres (subcortical areas). The *confluence of merging sinuses* (*occipital, straight*, and *superior sagittal*) is variable. The *cavernous sinus* offers significant collateral drainage of blood from the brain. Other collateral veins include the diploic and emissary veins.

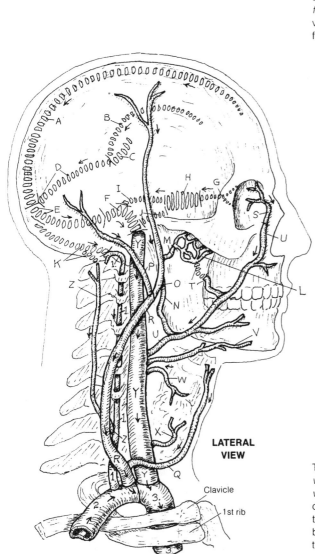

LATERAL VIEW

Clavicle

1st rib

VEINS OF HEAD & NECK :-
PTERYGOID PLEXUS L
MAXILLARY M
RETROMANDIBULAR N
SUPERFICIAL TEMPORAL O
POSTERIOR AURICULAR P
ANTERIOR JUGULAR Q
EXTERNAL JUGULAR R
ANGULAR S
DEEP FACIAL T
FACIAL U
LINGUAL V
SUPERIOR THYROID W
MIDDLE THYROID X
INTERNAL JUGULAR Y
DEEP CERVICAL Z
VERTEBRAL 1.
RIGHT SUBCLAVIAN 2.
RIGHT BRACHIOCEPHALIC 3.

The internal jugular veins drain the venous sinuses, assisted by the *angular veins* and the *pterygoid plexus*. The tributaries of the internal/*external jugular veins* are variable. The internal jugular travels in the same sheath with the common/internal carotid arteries and the vagus nerve. The arrangement of tributaries of the brachiocephalic vein is similar bilaterally, except the left brachiocephalic vein is longer than the right (see next plate). Recall that there is only one brachiocephalic artery (Plate 112).

See 119, 147

CAVAL & AZYGOS SYSTEMS

CN: Use blue for the superior and inferior venae cavae (H, H^1). Note that a large segment of the latter has been deleted to reveal the azygos vein (N). Use bright colors for the first posterior intercostal (D) and internal thoracic (F) veins, both of which drain into the brachiocephalic.

SUPERIOR VENA CAVAL SYSTEM ✣

SUPERIOR THYROID A
MIDDLE THYROID B
INTERNAL JUGULAR C
1ST POSTERIOR INTERCOSTAL D
INFERIOR THYROID E
INTERNAL THORACIC F
R. & L. BRACHIOCEPHALIC G
SUPERIOR VENA CAVA H

AZYGOS SYSTEM ✣

POSTERIOR INTERCOSTAL D'
SUPERIOR INTERCOSTAL I
LUMBAR J
ASCENDING LUMBAR K
HEMIAZYGOS (ACCESSORY) L
HEMIAZYGOS M
AZYGOS N

INFERIOR VENA CAVAL SYSTEM ✣

COMMON ILIAC O
TESTICULAR / OVARIAN P
RENAL Q
HEPATIC R
INFERIOR VENA CAVA H'

The superior vena cava drains the head, neck, and upper limbs. In addition, it drains the intercostal and lumbar regions by way of a remarkable but irregular and variable collection of veins called the azygos system. In conjunction with the veins of the vertebral canal (vertebral venous plexus, Pl. 77), the azygos system (*azygos, accessory hemiazygos*, and *hemiazygos veins*) provides a means of returning blood to the heart from the lower limb and trunk in the event of obstruction of the inferior vena cava.

The azygos and hemiazygos veins are derived from the merging of the *ascending lumbar* and subcostal veins. The anterior intercostal veins (not shown) drain into the *internal thoracic veins*. The inferior vena caval (and azygos) systems have no major tributaries draining the gastrointestinal tract, gall bladder, and pancreas. These viscera have their own venous (portal) system (see Plate 119). However, note that the liver is drained by *hepatic veins* into the inferior vena cava (IVC) just below the diaphragm. Note that the left *testicular/ovarian veins* merge with the inferior vena cava on the right but with the renal vein on the left. Testicular varicoceles are more common on the left, owing in part to the venous back pressure created at this 90° angle intersection between the renal and testicular veins.

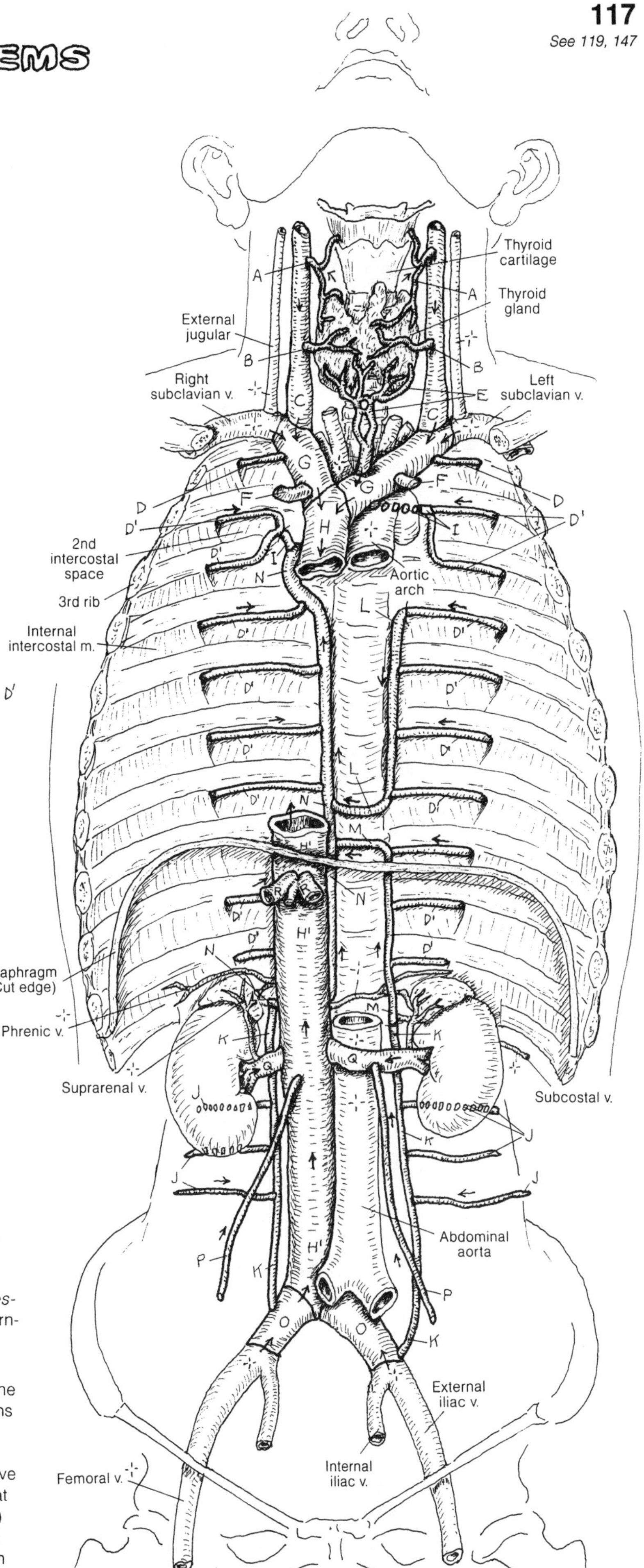
Thyroid cartilage
Thyroid gland
External jugular
Right subclavian v.
Left subclavian v.
2nd intercostal space
3rd rib
Internal intercostal m.
Aortic arch
Diaphragm (Cut edge)
Phrenic v.
Suprarenal v.
Subcostal v.
Abdominal aorta
External iliac v.
Internal iliac v.
Femoral v.

ANTERIOR VIEW
(of body cavities)

VEINS OF THE LOWER LIMB

CN: Use blue for P and light colors for A–C, the deep veins. Use dark colors for the superficial veins (Q–V); they have been drawn with darker outlines. (1) Begin with the deep veins and work both views simultaneously. (2) After completing the superficial veins, color the main ones in each of the small illustrations, but not the fine lines representing their tributaries.

DEEP VEINS:-

PLANTAR DIGITAL A / METATARSAL A'
DEEP PLANTAR VENOUS ARCH B
MED. C / LAT. PLANTAR C'
POSTERIOR TIBIAL D
DORSAL E
ANTERIOR TIBIAL F
POPLITEAL G
LAT. H / MED. CIRCUMFLEX FEMORAL H'
PROFUNDA FEMORIS I
FEMORAL J
EXTERNAL ILIAC K
SUPERIOR L / INFERIOR GLUTEAL L'
OBTURATOR M
INTERNAL ILIAC N
RIGHT COMMON ILIAC O
INFERIOR VENA CAVA P

SUPERFICIAL VEINS:-

DIGITAL Q / METATARSAL Q'
DORSAL VENOUS ARCH R
LATERAL MARGINAL S
MEDIAL MARGINAL T
GREAT SAPHENOUS U
SMALL SAPHENOUS V

The flow of blood in the deep veins of the lower limb is generally an uphill course. In concert with gravity, prolonged horizontal positioning of the legs (and other conditions) can result in slowed flow (stasis) in the deep veins, producing venous distention and inflammation (phlebitis). Formation of clots may follow (deep vein thrombosis), resulting in thrombophlebitis. In these conditions, fragments of clots may become detached and released into the venous circulation (embolism). The emboli continue up the venous pathway of ever-increasing size, easily pass into the right heart, and are pumped into the progressively smaller vessels of the lung, where they become stuck (pulmonary embolism).

While deep veins generally follow the arteries (venae comitantes), superficial veins do not. They travel with cutaneous nerves in the superficial fascia; many are easily visualized in the limbs. The blood in these long veins has to overcome gravity for a considerable distance, and their valves (Pl. 103) often come under weightbearing stress. Happily, there exist a number of communicating vessels (perforating veins, or perforators, not shown) between superficial and deep veins, permitting runoff into the deep veins, significantly offsetting the effect of incompetent valves. Incompetent valves lead to pooling of blood and swelling in the lower superficial veins, with potential inflammation. In the chronic condition, the *saphenous veins* and their tributaries can become permanently deformed and dysfunctional (varicosities).

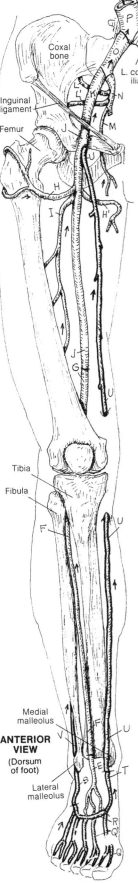

Coxal bone
L. common iliac v.
Inguinal ligament
Femur
Tibia
Fibula
Medial malleolus

ANTERIOR VIEW
(Dorsum of foot)

Lateral malleolus

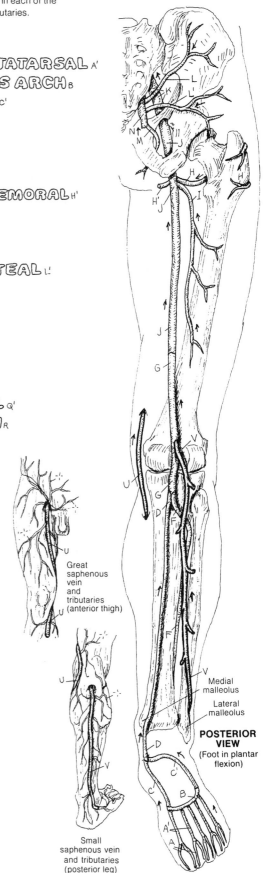

Great saphenous vein and tributaries (anterior thigh)

Small saphenous vein and tributaries (posterior leg)

Medial malleolus
Lateral malleolus

POSTERIOR VIEW
(Foot in plantar flexion)

HEPATIC PORTAL SYSTEM

CN: Use blue for I and a dark color for J. (1) Color the veins draining the intestines, pancreas, gall bladder, and spleen. There are both left and right gastro-epiploic (D, D¹) and gastric (G, G¹) veins. For the darkly outlined directional arrows adjacent to blood vessels, use the color of the blood vessel. (2) After coloring gray the inferior vena cava (✳) and its tributaries (✳¹), the tributaries of the superior vena cava (✳²), and their directional arrows, color the three large arrows (✳³) identifying anastomotic sites (include the esophageal veins passing posterior to the heart).

HEPATIC PORTAL SYSTEM ₋ₗ
SUPERIOR RECTAL ₐ
INFERIOR MESENTERIC ʙ
PANCREATIC c
L. GASTRO-EPIPLOIC ᴅ
SPLENIC ᴇ
R. GASTRO-EPIPLOIC ᴅ'
SUPERIOR MESENTERIC ꜰ
R. GASTRIC ɢ
L. GASTRIC ɢ'
CYSTIC ʜ
PORTAL ᴛ
HEPATIC V. ⱼ & TRIBUTARIES ⱼ'
INF. VENA CAVA ✳ / TRIBUTARY ✳¹
TRIB. OF SUP. VENA CAVA ✳²

ANASTOMOSES SITE ✳³ ⇨

Capillaries of the gastrointestinal tract, gall bladder, pancreas, and spleen are drained by tributaries of the *hepatic portal vein*. Within the liver, branches of this vein (like those of an artery) discharge blood into capillaries (sinusoids) surrounded by liver cells. These cells remove digested (molecular) lipids, carbohydrates, amino acids, vitamins, and iron from the sinusoids and store them, alter their structure, and/or distribute them to the body tissues (and, in the case of unnecessary molecules or degraded remains of toxic substances, the kidneys). The distribution process begins with the selective release of molecular substances from the liver cells into the small *tributaries* of the three *hepatic veins*. The hepatic veins join the inferior vena cava (IVC) immediately below the diaphragm.

The portal system of veins is so called because it *transports* nutrients and other molecules from the first capillary network in the intestines directly to the second capillary network (sinusoids) of the liver without going through the heart first.

In certain diseases of the liver, the formation of scar tissue blocks flow through the sinusoids. As a result, blood begins to back up in the portal system. In the chronic condition, the portal vein and its tributaries enlarge significantly. The blood, seeking the path of least resistance, finds collateral routes to the heart, enlarging them (varices). These routes (noted by arrows) occur by way of anastomoses between veins of the portal system and veins of the inferior caval, superior caval, azygos, and vertebral venous systems. Other anastomoses exist via a number of retroperitoneal veins and the umbilical vein (not shown). In the latter vessel, irregular varicose veins can appear on the surface of the abdominal wall (caput Medusae; see glossary).

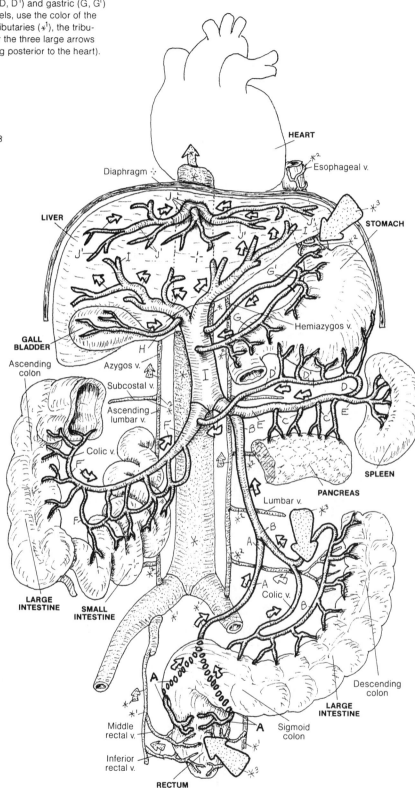

**PORTAL VEIN
AND ITS TRIBUTARIES**
(Anterior view, diagrammatic)

REVIEW OF PRINCIPAL VEINS

CN: Use blue for K and K[1]. Superficial veins of the limbs are on the left and deep veins are shown on the right. Only a few are shown bilaterally. Palms are facing forward. (1) Using the preceding vein plates for reference where necessary, start with A (right hand) and color in the order listed. As you color each vein write the name of the vein in pencil. After completing the superficial veins in the limbs, color through the deep veins, starting at the hand/foot. (2) Recall that the deep veins travel with the arteries of the same name.

(See appendix A in the back of the book for answers)

VEINS OF THE UPPER LIMB
A _____   H _____
B _____   I _____
C _____   J _____
D _____   K _____
E _____   L _____
F _____   M _____
G _____   N _____

VEINS OF THE HEAD AND NECK
O _____
P _____

VEINS OF THE CHEST
Q _____
R _____
S _____
T _____

VEINS OF THE LOWER LIMB
U _____
V _____
W _____
X _____
Y _____
Z _____
1 _____
2 _____
3 _____
4 _____
5 _____
6 _____
7 _____
8 _____
9 _____

VEINS OF THE PELVIS AND ABDOMEN
10 _____   16 _____
11 _____   17 _____
12 _____   18 _____
13 _____   19 _____
14 _____   20 _____
15 _____   21 _____

SUPERFICIAL VEINS

DEEP VEINS

Heart

Lung

Liver

Thoraco-epigastric vein

Superficial veins on the posterior aspect

DORSAL VEINS

PLANTAR VEINS

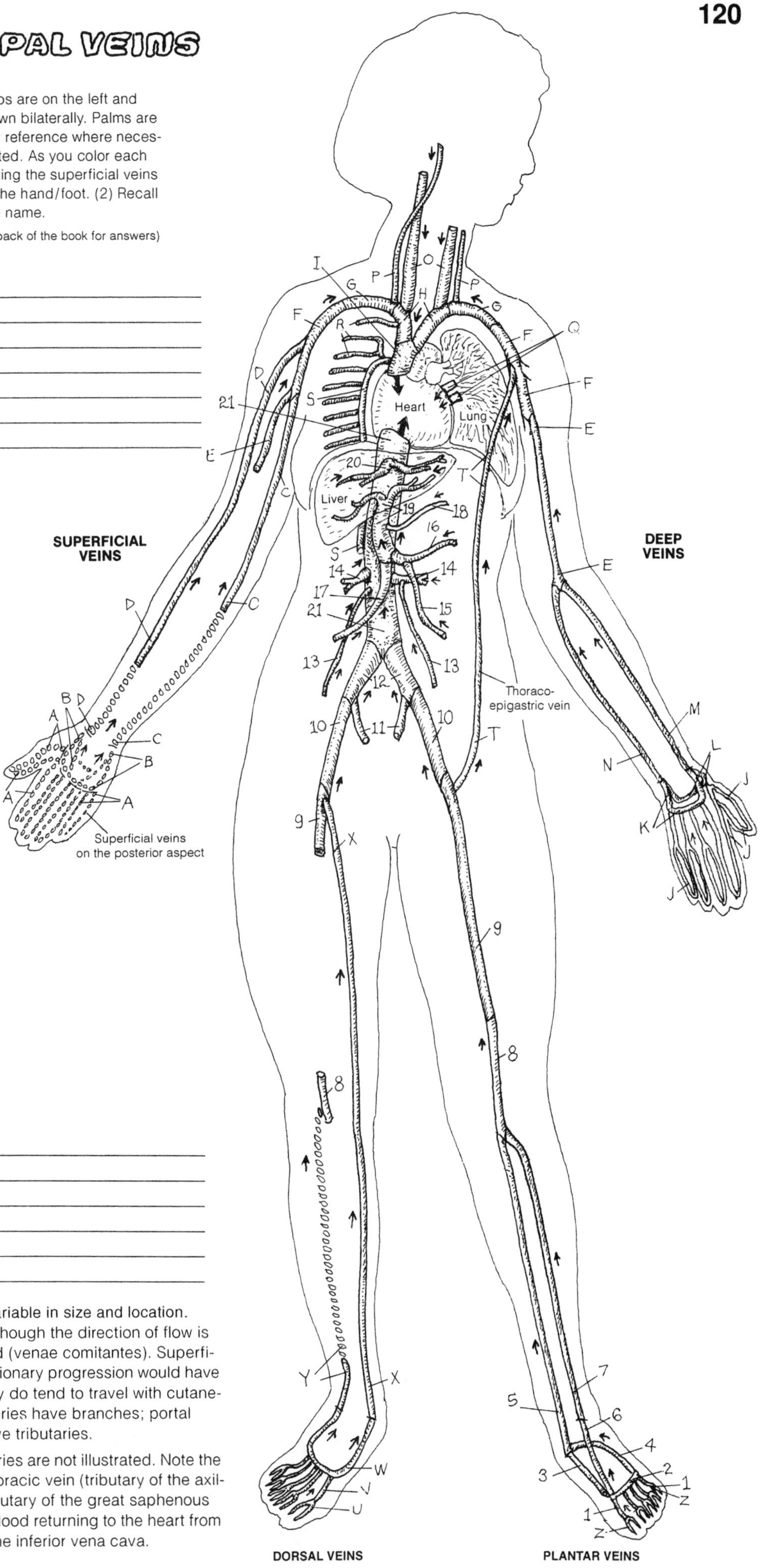

As all phlebotomists know, veins are extremely variable in size and location. Deep veins run with arteries of the same name (though the direction of flow is opposite); in the limbs, the veins are often paired (venae comitantes). Superficial veins have no companion arteries (our evolutionary progression would have been severely shortened if this were not so); they do tend to travel with cutaneous nerves. In review, it should be clear that arteries have branches; portal veins also have branches, but all other veins have tributaries.

Note that the internal thoracic vein and its tributaries are not illustrated. Note the *thoracoepigastric vein* (P) between the lateral thoracic vein (tributary of the axillary vein) and the superficial epigastric vein (tributary of the great saphenous vein). It is a potential route of collateral flow for blood returning to the heart from the lower limb in the event of an obstruction of the inferior vena cava.

LYMPHOCYTE CIRCULATION

CN: Use blue for H, red for I, purple for L, and green for M. (1) Color over the light lines representing peripheral (superficial) lymph vessels (A). (2) Color each large step numeral in the diagram below with the related titles. In the bottom diagram, do not color the lymphocytes circulating in and between the blood and lymph capillaries.

SUPERFICIAL DRAINAGE:
PERIPHERAL LYMPH VESSELS A
CERVICAL NODE B
AXILLARY NODE B¹
INGUINAL NODE B²

DEEP DRAINAGE:
LYMPHATIC TRUNK C
CYSTERNA CHYLI D
THORACIC DUCT E
RIGHT LYMPH DUCT F

The body is about 60% fluid (by volume), which fills cells, vessels, and spaces. Fluid requires circulation. Some of the fluid of the blood and some lymphocytes leave the circulatory system and enter the tissue spaces. Some of this fluid (lipids) and lymphocytes (lymph) are recovered by thin-walled vessels (*lymphatic capillaries*) that form in the loose connective tissue spaces. Unlike the closed-loop blood capillary networks, these tiny vessels are closed at one end. They merge to form progressively larger lymphatic vessels that drain into large veins in the neck. These vessels constitute the lymphatic system. Certain lymphatic vessels enter and leave lymph-filtering stations called *lymph nodes*.

Region-draining *lymph trunks* converge into a dilated lymph sac (*cysterna chyli*) lying deep to the abdominal aorta on the first lumbar vertebra. The *thoracic duct* begins at the upper end of the sac, ascends the anterior surface of the vertebral column, and drains into the left subclavian vein at its junction with the internal jugular vein. The *right lymph duct* terminates similarly on the opposite side. It drains the dotted area.

LYMPHOCYTE CIRCULATION:
GENERATIVE ORGAN G
VENOUS BLOOD H
ARTERIAL BLOOD I
LYMPHOID TISSUE J
PERIPHERAL TISSUE K
CAPILLARY NETWORK L
LYMPH VESSEL M
LYMPH NODE M¹

Lymphocytes are among the principal cells of the immune system. The circulation scheme reveals the primary pathway for the dissemination of lymphocytes from their *generative organs* (*bone marrow, thymus*) into the *lymphoid tissues* and organs as well as organs and tissues in general (*peripheral tissues*). Such a circulation pattern provides for maximum exposure of lymphocytes to microorganisms and subsequent body defense operations (immune responses).

Formed and developed in the bone marrow and thymus (1), lymphocytes leave with the *venous blood* to enter the circulation. By way of *arterial blood* (2), lymphocytes enter the *capillary networks* of the lymphoid tissues (3) and other peripheral tissues (4). The lymphocytes may remain in or migrate from the lymphoid organs/peripheral tissues, entering blood capillaries or lymph vessels. From lymph capillaries, the lymphocytes flow with the lymph fluid into regional lymph nodes (5). Here they may become resident or they may depart the node and merge with other lymph vessels to join the lymph ducts (6) that connect with the blood circulatory system.

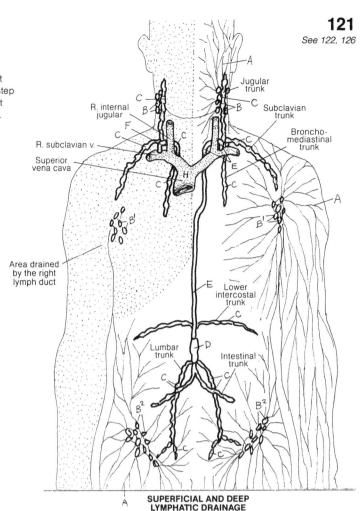

SUPERFICIAL AND DEEP LYMPHATIC DRAINAGE

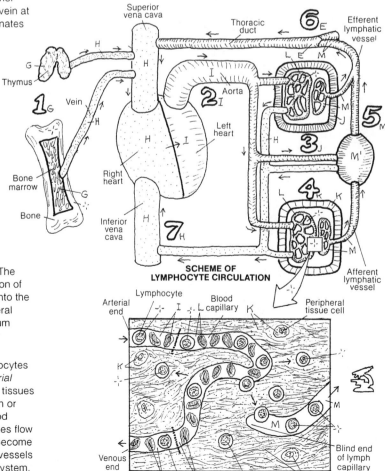

SCHEME OF LYMPHOCYTE CIRCULATION

INTRODUCTION

CN: Use green for D, the same colors for bone marrow (A) and thymus (B) used on Plate 121. (1) Structures depicting mucosal-associated lymphoid tissue (E) are generalizations; more accurate representations can be seen on Plate 127. (2) The three lymphocyte types have identifying letters drawn into their nuclei. Color over the entire cell in all cases. (3) The various types of cells appearing in this section will generally be identified by more descriptive letters/labels (e.g., PC = plasma cell). Try to use the same light color for each type wherever it appears on plates 122–128.

PRIMARY ORGANS :
BONE MARROW A
THYMUS B

The lymphoid system, the anatomical component of the immune system, functions in defense against microorganisms entering the body as well as the destruction of cells or cell parts no longer recognizable as "self." Lymphoid tissues and organs are predominantly collections of lymphocytes and related cells (see below), often supported by a meshwork of reticular fibers and cells.

The red bone marrow and thymus are primary lymphoid organs. The *bone marrow* contains the precursors of all lymphocytes and disburses lymphocytes into the circulation. It consists largely of great varieties of blood cells in various stages of maturation, phagocytes, reticular cells and fibers, and fat cells. Some of the lymphocytes mature and undergo structural and biochemical revision (differentiation) in the bone marrow to become B lymphocytes. Large lymphocytes enter the circulation from the bone marrow and function as natural killer cells. Some partly differentiated lymphocytes migrate via the blood to the thymus. There they become T cells and differentiate further. Those cells then re-enter the circulation and migrate to secondary lymphoid organs.

The *thymus* is located in the superior and anterior (inferior) mediastinum. It receives uncommitted lymphocytes from the bone marrow The thymus is actively engaged in T lymphocyte proliferation and differentiation during embryonic and fetal life as well as the first decade of extrauterine life. The thymus begins to undergo degeneration (involution) after puberty.

SECONDARY ORGANS :
SPLEEN C
LYMPH NODE D
MUCOSAL ASSOCIATED
LYMPHOID TISSUE (M.A.L.T.) E
TONSILS / ADENOIDS F
APPENDIX G

Secondary lymphoid organs are structures predominantly populated by lymphocytes that migrated from the primary lymphoid organs. The structural arrangement of these organs ranges from encapsulated, complex structures, like the *spleen* and *lymph nodes*, to a diffuse disposition of lymphocytes throughout the loose connective and epithelial tissues of the digestive system, if not all open cavities. These secondary organs represent satellite sites for lymphocytic activation when challenged by antigens. The *spleen* processes incoming blood. Its lymphocytes and phagocytes react rapidly to the presence of microorganisms and aged red blood corpuscles. *Lymph nodes* screen lymph from incoming (afferent) lymphatic vessels, much in the same manner as the spleen processes blood. Partly encapsulated, nodular masses of lymphoid tissue (tonsils and adenoids) guard the pharynx, marking incoming microorganisms for destruction. Unencapsulated, variably sized, nodular masses (follicles) of lymphocytes occur throughout the mucosal layers of open cavities (primarily the digestive tract), as do more diffuse distributions of lymphocytes. These unencapsulated follicles and lymphocyte collections constitute mucosal-associated lymphoid tissues (M.A.L.T.); in the intestines, they may be called "gut-associated lymphoid tissues" (G.A.L.T.). The vermiform appendix harbors multiple lymphoid follicles in its mucosa. The density of lymphocytes and follicles of lymphocytes in all these groups varies with the degree of immune responsivity required.

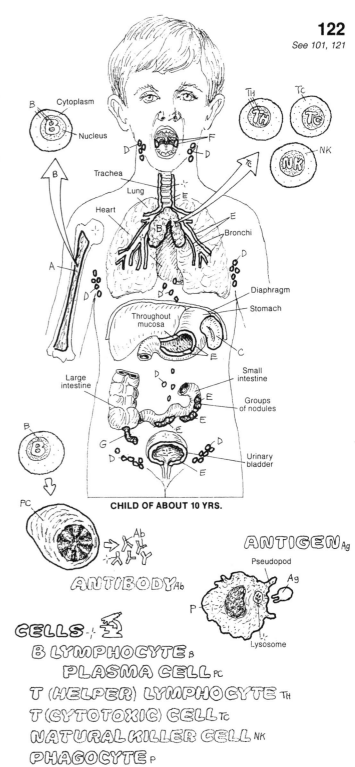

CHILD OF ABOUT 10 YRS.

ANTIBODY Ab

ANTIGEN Ag

CELLS :
B LYMPHOCYTE B
PLASMA CELL PC
T (HELPER) LYMPHOCYTE TH
T (CYTOTOXIC) CELL Tc
NATURAL KILLER CELL NK
PHAGOCYTE P

Activated B lymphocytes (B = bone marrow–derived) differentiate along specific lines, one of which becomes plasma cells. Plasma cells secrete protein molecules called antibodies into tissue fluids. Antibodies interact with and destroy antigens, a term restricted to those molecules (free or attached to/are part of cells and microorganisms) that elicit activation of the B cells.

Early T lymphocytes (T = thymus-derived) differentiate into one of a number of cells, including helper (TH), cytotoxic (Tc), and memory cells (not shown). Activated by antigen stimulation, TH cells stimulate and regulate specific and nonspecific immune operations against cells, without necessarily being assisted by B cells. Thus, they are concerned with cell-mediated immunity. Tc cells kill cells targeted by other T cells or lymphokines. Natural killer (NK) cells are neither B nor T cells. They are not activated by other cells or lymphokines (they kill naturally). In association with Tc cells, they destroy tumor cells and virus-infected cells primarily. Phagocytes are cells that destroy antigen by phagocytosis. They function as antigen-presenting cells (APC) for T cells; T cells, in turn, activate phagocytes.

NATURAL & ACQUIRED IMMUNITY

CN: Use pink for IR and the same colors used on Plate 122 where possible. Radial lines surrounding a cell indicate activation. All elements shown have been magnified and schematized for coloring.

MICROORGANISM A
NATURAL IMMUNITY :-
ANATOMIC BARRIER ABa
COMPLEMENT C
PHAGOCYTE P
INFLAMMATORY RESPONSE IR

Immunity is an anatomic and physiologic state of security against disease. *Natural* immunity exists independent of any specific microorganismal interaction with a lymphocyte. Shortly before birth and following, one progressively *acquires* a specific immunity following each lymphocyte's encounter with antigen and resulting activation. Phagocytes participate in both natural and acquired immunity; lymphocytes participate in acquired immunity and enhance natural immunity.

Natural immunity operates indiscriminately against *microorganisms* and degenerated cells/cell parts. *Anatomic barriers* (1), such as skin or mucous membranes, physically resist microorganismal invasion. *Phagocytes* approach their prey from the blood (2) or connective tissues (3), engulf them (4, phagocytosis), and destroy them with lysosomal enzymes (5). *Complement* (certain soluble proteins in body fluids) bind to microorganisms, enhancing their phagocytosis. Tissue irritation, e.g., disruption by a splinter, induces an inflammatory response that involves both natural and acquired immunity.

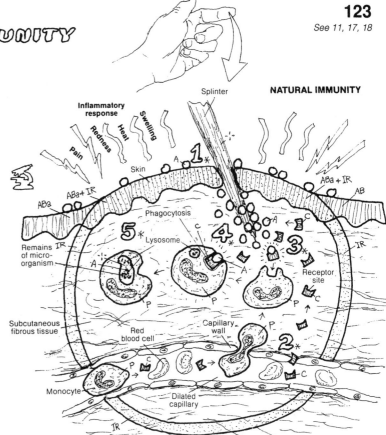

ACQUIRED IMMUNITY :-

Acquired immunity involves diverse but specific lymphocyte responses to the presence of *antigen*. A specific lymphocytic reaction to antigens (immune response) is characterized by the activation and proliferation of lymphocytes followed by the destruction of antigens. Two kinds of acquired immunity are possible, based on lymphocyte types: *humoral immunity* and *cellular immunity*. Inherent in both kinds of immunity are specificity and diversity of response, retention of cellular memory of antigen, and the ability to recognize self from non-self among the body's proteins.

ANTIGEN Ag
HUMORAL IMMUNITY B-
B LYMPHOCYTE B
MEMORY CELL BM
PLASMA CELL PC
ANTIBODY Ab

INFECTED CELL IC
CELLULAR IMMUNITY T-
T LYMPHOCYTE T
MEMORY CELL TM
HELPER CELL (TH) TH
CYTOTOXIC CELL (Tc) TC

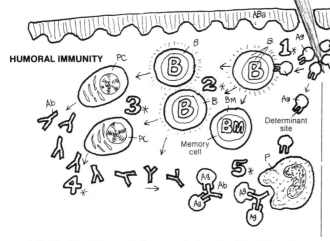

Humoral (fluid-related) immunity is characterized by *B lymphocytes* being activated by antigen (Ag) (1), proliferating, forming *memory cells (BM)*, secreting *antibody* (Ab) (2), and forming *plasma cells (PC)* (3), which secrete *antibody* (4). Antibodies are complex proteins formed in response to a specific antigen and attached to it at the antigenic determinant site (5), facilitating its phagocytosis.

Cellular immunity is characterized by *T lymphocytes (T)* being activated by antigens attached to antigen-presenting cells: *phagocytes (P)* (1).

Most T cells differentiate into *helper T lymphocytes (TH)* and *cytotoxic T lymphocytes (Tc)*. Helper T lymphocytes (2) enhance humoral immunity by activating B cells, augmenting the inflammatory response, activating phagocytes with stimulating factors (lymphokines), and forming *memory cells (TM)*. Cytotoxic T lymphocytes (3) bind to and destroy *infected cells* and form memory cells. Memory cells recognize specific structural characteristics of the antigens encountered ("memory") and facilitate rapid immune responses on subsequent exposure to those antigens.

THYMUS & RED MARROW

CN: Use red for G, blue for H, green for J. (1) Begin with red marrow (A) at the bottom of the plate. Then color the red marrow (K) portion of the newborn's arm bone. Color the thymus section and the diagrammatic description of lymphocyte maturation in the thymus. Note that the borders of the diagram represent the cortex and medulla of the thymus. Then color the schematic overview of thymic function.

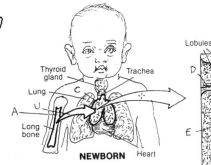

NEWBORN

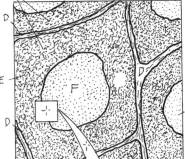

THYMUS SECTION

THYMUS c

FIBROUS SEPTA d
CORTEX e
　UNDIFFERENTIATED LYMPHOCYTE u
　IMMATURE T LYMPHOCYTE i
MEDULLA f
　MATURE T LYMPHOCYTE t
ARTERIAL VESSEL g
VENOUS VESSEL h
SINUSOID h'
LYMPH VESSEL j

The thymus seeds the entire body with T lymphocytes, the protagonists of cellular immunity. It consists of two lobes of glandular tissue in the anterior and superior mediastinum. The thymus is functional and relatively large in the late fetus/newborn (15 gms), continues to grow and function during the pre-teen years, and declines in size and activity in the following years.

The functional thymus consists of microscopic lobules partitioned by *fibrous septa* containing blood vessels. Each lobule has an outer *cortex* dense with lymphocytes and a much less dense central *medulla*. The *epithelial cells* of the lobule form a structurally supporting "reticular" network. Distinctive concentric rings of keratinized epithelial cells (Hassal's corpuscles) are seen in the medulla; although associated with degenerative signs in aging, they may support lymphocyte differentiation. *Arterial vessels* bring *undifferentiated lymphocytes* into the cortex. The cells migrate into the medulla, showing signs of differentiating into T cells. In the inner cortex, the cells are largely *immature* (but committed) *T cells*. The medulla contains largely *mature T cells*. These cells leave the thymus by venules (*venous vessels*) to enter the systemic circulation. Some T lymphocytes enter *lymph vessels* destined for mediastinal lymph nodes and beyond. The thymus also produces a number of factors (hormones) stimulating lymphocyte differentiation.

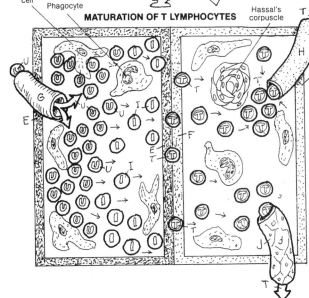

MATURATION OF T LYMPHOCYTES

SCHEMATIC OVERVIEW

RED MARROW a

LYMPHOCYTE STEM CELL l
GROWTH FACTOR gf
B LYMPHOCYTE b
PRE-T CELL tp
NATURAL KILLER CELL nk

Red marrow (recall Plate 20) is densely populated with a great variety of blood cells in various stages of development. The supporting framework of marrow is reticular fibers and cells. Fed by arterioles from the nutrient artery of the bone, the capillaries within the marrow are enlarged to the extent of being small sinuses (*sinusoids*). They reveal transient cytoplasmic "pores" for the immediate passage of cells into the circulation. Among the developing blood cells are *lymphocyte precursors*. These are stimulated to divide by certain *growth factors*. The progeny of these cells are mostly small and some large *lymphocytes*. B lymphocyte (B cell) maturation, natural killer cell (large lymphocyte) development, and pre–T cell development occur in the bone marrow. These lymphocytes enter the sinusoids and the venous outflow to be distributed bodywide.

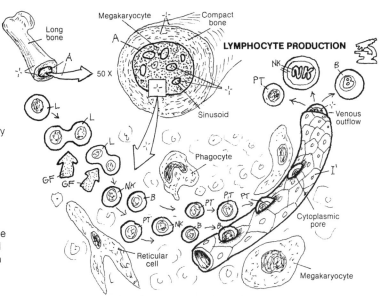

LYMPHOCYTE PRODUCTION

SPLEEN

CN: Use red for F and blue for H. Continue with the same colors as were used on preceding plates for the cells. (1) When coloring the schematic representation of spleen structures, note the underlying brackets that designate the structures fitting within the white pulp and red pulp regions. (2) Do not color the venous sinuses (G), in order to keep visible the gaps and highly branched reticular cells (lower illustration).

SPLEEN A
- CAPSULE A'
  - TRABECULA C
- WHITE PULP D
  - LYMPHOID FOLLICLE D'
- RED PULP E

BLOOD VESSELS -:-
- ARTERY F
  - ARTERIOLE F'
- VENOUS SINUSOID G-:-
  - VENULE H
- VEIN H'

CELLS -:-
- T LYMPHOCYTE T
- B LYMPHOCYTE B
- MITOTIC LYMPHOCYTE ML
- PHAGOCYTE P
- PLASMA CELL PC

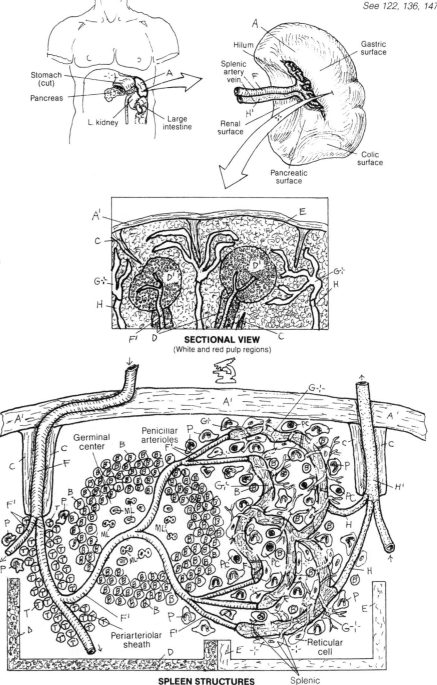

SECTIONAL VIEW
(White and red pulp regions)

SPLEEN STRUCTURES
(Schematic)

The soft, blood-filled, dark purple spleen lies posteriorly in the upper left abdominal quadrant, just above the left kidney, at about the level of the 11th and 12th ribs. It is generally about the size of your closed fist. The *capsule* of the spleen projects inward extensions (*trabeculae*) that support the organ and incoming/outgoing vessels. The microscopic view of the spleen is complicated by the endless sea of lymphocytes and phagocytes and, in this organ, red blood cells.

Small, downstream branches of the splenic artery travel in the fibrous trabeculae; branching *arterioles* become enveloped in lymphocytes (periarteriolar sheath) and branch among *lymphoid follicles*. These follicles, the arterioles, and their cellular sheaths constitute the *white pulp*. The follicles enlarge with antigenic stimulation; large *mitotic lymphocytes* (in various stages of cell division) begin to appear in the central part of each follicle (germinal center) following stimulation, creating a zone less dense than the surrounding, cell-packed area. As the straight (penicillar) arterioles leave the white pulp, they lose their muscular tunics to open into venous sinuses surrounded by phagocytes. These sinuses appear to have gaps amidst the irregular strands of splenic cords (highly branched reticular cells, B cells, and plasma cells). Phagocytes hanging out among the branches engage aged red blood corpuscles that slip out of the sinuses. The sinuses and splenic cords constitute the red pulp. The sinusoids drain into venules, the tributaries of trabecular veins. These form the tributaries of the splenic veins.

Antibody production and phagocytosis are major activities of the spleen. As blood flows into the arterioles of the white pulp, antigen is greeted by phagocytes and myriad T cells, setting off cellular immune responses in the periarteriolar sheaths. Snaking into the follicles, the vessels are surrounded by B cells. Once activated by the presence of antigen, B cells transform into plasma cells, and antibody is produced (humoral immunity). Systemic infection markedly increases the output of lymphocytes, causing palpable splenic enlargement (splenomegaly). Removal of the spleen (splenectomy) is not a benign event. Absent splenic tissue, the body may have reduced immune capabilities.

LYMPH NODE

CN: Use red for M, blue for N, and green for O (if you have additional greens, use them on J–L); continue using the same colors for the various cells. (1) Color the circular insets identifying the dominant cell in the regions of the lymph node. In the paracortex, note the small circles representing venules (N¹).

LYMPH NODE A
CAPSULE A¹
TRABECULA C
RETICULAR NETWORK D
CORTEX E
LYMPH FOLLICLE F
GERMINAL CENTER G
MEDULLA H
PARACORTEX I

LYMPH VESSELS ·⁄·
AFFERENT LYMPH VESSEL J
LYMPH SINUS K
EFFERENT LYMPH VESSEL L

BLOOD VESSELS ·⁄·
ARTERY M
VEIN N / VENULE N¹

LYMPHOID CELLS ·⁄·
PHAGOCYTE P
T LYMPHOCYTE T
B LYMPHOCYTE B
MITOTIC LYMPHOCYTE ML
PLASMA CELL PC
LYMPH O
ANTIGEN Ag

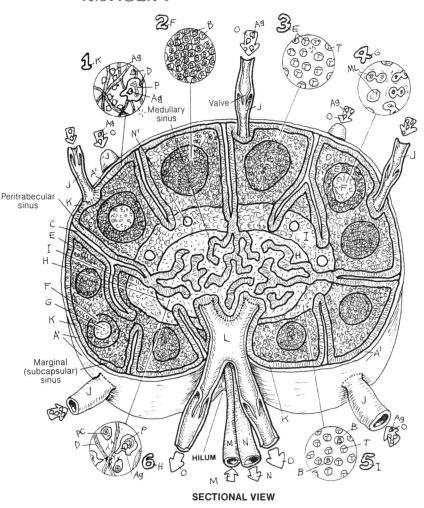

Peritrabecular sinus

Valve
Medullary sinus

Marginal (subcapsular) sinus

HILUM

SECTIONAL VIEW

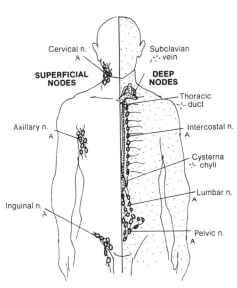

Cervical n.
Subclavian vein
SUPERFICIAL NODES
DEEP NODES
Thoracic duct
Axillary n.
Intercostal n.
Cysterna chyli
Lumbar n.
Inguinal n.
Pelvic n.

The lymph node has a fibrous *capsule* from which *trabeculae* invade the organ, dividing it incompletely into compartments. Fine reticular fibers and cells spread out from the trabeculae to form a thicket of interwoven branches throughout the node (*reticular network*). This intricate weave of fibers supports the dense populations of lymphocytes throughout the node. Lymph percolates through parts of the reticular network called *lymph sinuses* (only the marginal sinuses appear to be endothelial lined). The reticular fibers in these sinuses (1) form a spatial framework from which *phagocytes* can readily engage *antigens* in the lymph flow.

The node interior is characterized by an outer *cortex* and an inner *medulla*. The cortex reveals a group of particularly dense masses of *B lymphocytes* (2, *lymphoid follicles*) existing among a more sparse array of largely *T lymphocytes* in the interfollicular areas (3). In the presence of significant amounts of antigen, the follicles develop *germinal centers*; here are seen *mitotic lymphocytes* in varying degrees of mitosis (4). The outer part of the medulla (*paracortex*) has more diffuse arrangements of phagocytes, T cells, and some B cells (5). The endothelial cells of the venules in the paracortex are specialized and provide lymphocyte homing receptors that influence the localization of T and B cells within the node. The medulla contains a

concentrated array of interconnecting sinuses, with phagocytes and *plasma cells* in significant numbers (6).

Lymph enters the nodes by afferent vessels with valves controlling unidirectional traffic. As the lymph meanders through the throngs of reticular fibers in the sinuses, phagocytes pick off the antigens and present them to the T cells in the interfollicular areas. Activated B cells in the follicles, facilitated by helper (T$_H$) cells, transform into plasma cells and memory cells. The plasma cells and B cells secrete antibody with receptors that bind a portion of the antigen (antigenic determinant). Binding of antibody to the antigen facilitates destruction of the antigen. Major stimulus promotes the formation of germinal centers. Further immune activity occurs in the paracortical and inner medullary areas. Lymph leaves the medullary sinuses and the node by way of the *efferent vessels*. Lymphocytes also enter the node by small *arteries*; these cells can migrate into the sinuses from the venules while the remaining blood leaves the node by *veins*.

In summary, the lymph node is the site of both humoral-mediated (B cell) and cell-mediated (T cell) immune responses to antigens in the lymph. Palpable enlargement of cervical lymph nodes during an upper respiratory infection, for example, gives testimony to the existence of such mechanisms operative in the face of microorganismal invasion.

MUCOSAL ASSOCIATED LYMPHOID TISSUE (MALT)

CN: Use green for C and the same cell colors as you used previously. (1) Begin with the representations of normal and inflamed tonsils, using pink and red colors, respectively. Include the circular insets identifying the dominant cells within the follicle and germinal center.

PRIMARY FOLLICLE A
GERMINAL CENTER B
EFFERENT LYMPH VESSEL C

LYMPHOID CELLS -/-
MITOTIC LYMPHOCYTE ML
PHAGOCYTE P
B LYMPHOCYTE B
T LYMPHOCYTE T
PLASMA CELL PC
ANTIBODY Ab
ANTIGEN Ag

Unencapsulated lymphoid tissue abounds throughout the epithelial and connective tissues of the body. This tissue may be represented by mobile loose or dense collections of lymphocytes and/or phagocytic cells. Single or multiple follicles (nodules) of organized lymphoid tissue, as seen in lymph nodes and the spleen, also fit this category. These nodules may disappear and then form anew in response to antigenic challenge. These lymphoid cells/nodules in the mucosal (and sub-mucosal) layers of the viscera, spanning several systems, constitute "mucosal-associated lymphoid tissue" (M.A.L.T.). In the "gut," such tissue is called "gut associated lymphoid tissue" (G.A.L.T.). Skin-associated lymphoid tissue (S.A.L.T.) has also been described (see Langerhan's cell in Plate 18). Here some examples of M.A.L.T. are presented.

TONSIL D / INFLAMED TONSIL E

Masses of primary lymphoid follicles surrounding infoldings of the pharyngeal mucosa are called *tonsils*. Tonsils have no definitive lymph sinuses; however, lymph capillaries can be seen draining into *efferent lymphatic vessels*. With antigenic stimulation, inflammation of the tonsil commonly occurs (tonsillitis). An inflamed tonsil is swollen, red (often with streaks of vessels ranging across the mucosal surface), and painful. As the follicles are activated, germinal centers are formed, numbers of *B* and *T lymphocytes* increase, *phagocytes* and *plasma cells* appear, and *antibody* is produced. Considered a culturally accepted "rite of passage" in the past, tonsillectomies are accomplished now only for good cause (obstructed airways, chronic infections). Tonsils/adenoids (M.A.L.T.) respond quickly to the presence of microorganisms, either marking them for presentation to T cells or undertaking their destruction directly by means of antibodies.

PEYER'S PATCHES -/-

Aggregates of lymphoid follicles in the submucosa of the distal ileum are called Peyer's patches. Seen sporadically throughout the intestine, lymphoid follicles are more concentrated here. With antigenic stimulation, these follicles increase in much the same manner as tonsils.

VERMIFORM APPENDIX F

The vermiform appendix is a thin, tubular extension of the cecum (large intestine). It contains a number of lymphoid follicles that extend from the submucosa up to the epithelial lining of the mucosa. The mucosa of the appendix experiences fairly frequent insults (tomato and chile seeds, popcorn kernels, and ingested foreign matter), and inflammatory events are fairly common (appendicitis). The structure swells, reddens, and is often quite painful. Classical immune responses occur (formation of germinal centers, and so on). Because of the thin walls of the appendix, inflammations induced by acute infections can rupture through to the peritoneum (peritonitis). Surgical removal of the appendix (appendectomy) is common. There is no evidence that depressed immune activity occurs following appendectomy.

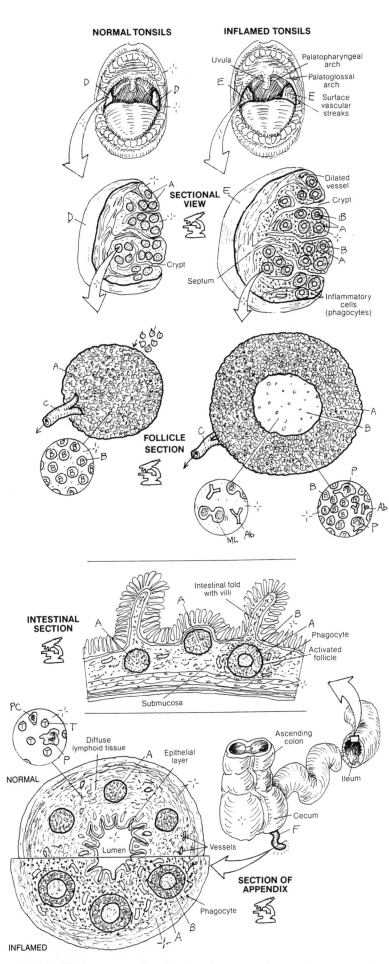

HIV-INDUCED IMMUNOSUPPRESSION

CN: Do not color this plate until you have colored the previous plates on the immune system (use the earlier colors for the cells on this plate). See the glossary for explanation of new terms. (1) Begin with HIV infection and follow the numbered text as you work down the page. The three cells at the top represent different stages of a single cell. (2) Dotted outlines represent destroyed structures. (3) Downward pointing, outlined arrows indicate a reduction in number or negative reaction; upward arrows equal increase in number or positive effect.

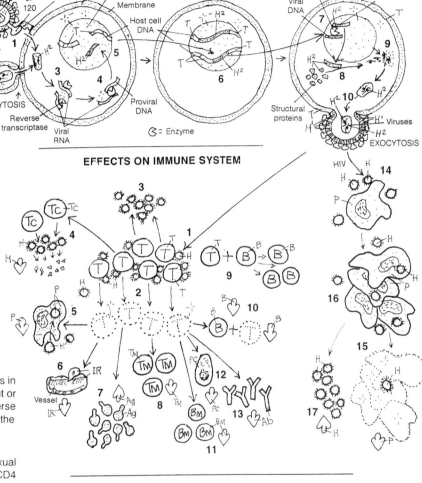

HIV INFECTION:-

T LYMPHOCYTE/PARTS T

HIV H

ENVELOPE H¹

CORE/PARTS H²

EFFECTS ON IMMUNE SYSTEM:-

T LYMPHOCYTE T

CYTOTOXIC T CELL Tc

PHAGOCYTE P

INFLAMMATORY RESPONSE IR

ANTIGEN Ag

T MEMORY CELL Tm

B CELL B

B MEMORY CELL Bm

PLASMA CELL PC

ANTIBODY Ab

Human immunodeficiency virus (HIV) is a retrovirus that causes acquired immunodeficiency syndrome (AIDS), a disease process in which most parts of the immune system are affected at one point or another. A retrovirus is an RNA virus containing the enzyme reverse transcriptase, which converts RNA to DNA. This plate illustrates the immunosuppressive effects of the disease process.

Free HIV or HIV-infected cells are passed from one person to another by body fluids. Such transfer occurs predominant by sexual contact or by injection. Once in the body, HIV infects cells with CD4 receptors (certain phagocytes and T cells) at a rapid rate, generally inducing fatigue, fever, sore throat, and palpably swollen lymph nodes. Following a decline of symptoms, a period of latency (usually 2+ years) follows. HIV replication appears relatively subdued during this time, and there may be no symptoms. But replication of HIV continues in the lymphoid tissues and wherever mobile, infected phagocytes take them. At some point, when the viral load is highly elevated and the competence of the immune system is sufficiently compromised, a variety of non-HIV/non-HIV-infected microorganisms proliferate in the body without adequate immune challenge. These "opportunistic infections" may be viral, bacterial, fungal, or protozoal in origin and can occur simultaneously. These infections ultimately kill the host. Current regimens of treatment can reduce the viral load and improve the quality of life, but do not eliminate the HIV infection.

In our example here, HIV is free and blood-borne. Note that (1) the CD4 receptor of the *lymphocyte* is attached to the outer envelope glycoprotein (GP) 120 of an HIV. Fusion occurs, mediated by the co-receptor chemokine. The envelope is discarded as the *viral core* enters the cell (2). The nucleocapsid is ruptured (3), freeing the viral RNA and the enzyme reverse transcriptase into the cell cytoplasm. The viral RNA is transcribed by the reverse transcriptase into proviral DNA (4). The proviral DNA flows into the nucleus of the cell and integrates with the host cell DNA (5). Once integrated (6), RNA polymerase transcribes the viral DNA segments into viral mRNA

OPPORTUNISTIC INFECTIONS

(7). The viral mRNA translates large, virus-specific proteins that are broken down by protease enzymes into structural proteins (8). Viruses are formed (9) from these proteins and then discharged from the cell by exocytosis (10). The viral envelope is derived from the cell membrane. The discharged HIV is infectious.

Freed into the circulation, HIV attaches to CD4+ T cells (1). Some of the T cells are killed by the infection (2) and some survive, pumping out a continuum of HIV (3). Some infected *T cells* induce *cytotoxic T cells* to destroy the virus (4). T cell destruction results in reduced T cell–stimulated *phagocytosis* (5), reduced inflammatory responses (6), increased *antigen* (7), and decreased T memory cells (Tm) (8). *T and B cells* (9) working together enhance B cell activity. With reduction of T cells, enhanced B cell activity (10) decreases, as do *B memory cells* (Bm) (11), *plasma cells* (12), and *antibodies* (13). HIV also infects *phagocytes* (14), killing them individually or in fused cell masses (syncytia) (15). Surviving, infected phagocytes (16) produce a continuum of viruses (17). The global effect of all these events, and more, is progressive immunosuppression and onset of opportunistic infections.

OVERVIEW OF THE SYSTEM

CN: Use red for L and light colors throughout. (1) Begin with the structures of the respiratory system. (2) Color the cross section of the trachea (D), including the respiratory mucosa (I). (3) Color the magnified section of the mucosa in the lowest view.

NASAL CAVITY_A
PHARYNX_B
LARYNX_C
TRACHEA_D
PRIMARY BRONCHI_E
BRONCHIAL TREE_F
R. LUNG,_G L. LUNG_{G'}
DIAPHRAGM_H

The respiratory tract conducts air to the respiratory units of the lungs where it can readily be absorbed by the blood, and it removes carbon dioxide–laden air from the air cells and exhausts it to the external atmosphere. It develops and refines sounds into potentially intelligible vocalization, and it helps maintain the acid-base balance of the blood by blowing off excess acid in the form of carbon dioxide. Nowhere in the body does the outside world, with all its creatures of microscopic dimension, have such easy access to the protected interior cavities of the body as it does at the air/blood interfaces of the lung. The respiratory tract has both air-conducting and respiratory (gas exchange) parts.

The air conduction tract includes an upper (*nasal cavity, pharynx, larynx*) and a lower tract (*trachea, primary bronchi and bronchial tree*). The upper tract is lined with respiratory mucosa, except in the lower pharynx where it has a stratified squamous epithelial surface. Except for the nose and pharynx, the skeleton of the respiratory tract is cartilaginous down to the smallest airways (bronchioles), where the cartilage is replaced by smooth muscle. The parts associated with gaseous exchange are the smallest bronchioles and alveoli (respiratory units), which take up much of the *lung*'s volume.

The muscular diaphragm provides much of the force necessary for inspiration and expiration of air. One-quarter of that force is generated by the intercostal muscles moving the ribs.

RESPIRATORY MUCOSA_I
PSEUDOSTRATIFIED
COLUMNAR EPITHELIUM_J
LAMINA PROPRIA_K
BLOOD VESSEL_L
GLAND_M

The mucosa of the respiratory tract is largely *pseudostratified columnar* and (in the bronchioles) cuboidal *epithelia* with mucus-secreting *goblet* (unicellular gland) *cells* and *cilia*. Here excreted mucus traps foreign particulate matter, inhaled air is hydrated (mixed with water), putting oxygen in solution, and the air is heated from underlying vessels. The epithelial cells are supported by a loose fibrous, *glandular, vascular lamina propria*, replete with fibroblasts and cells of the lymphoid system. Deep to this connective tissue layer is the supporting tissue (bone in the nasal cavity, muscle in the pharynx, hyaline cartilage in the trachea, larynx, and bronchi, smooth muscle in the bronchioles, and thin fibers supporting the air cells).

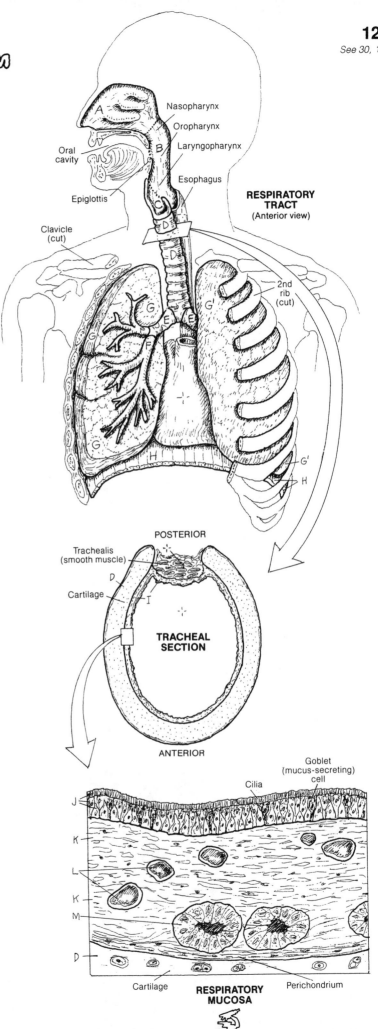

RESPIRATORY TRACT (Anterior view)

Nasopharynx
Oropharynx
Laryngopharynx
Esophagus
Oral cavity
Epiglottis
Clavicle (cut)
2nd rib (cut)

POSTERIOR
Trachealis (smooth muscle)
Cartilage
TRACHEAL SECTION
ANTERIOR

Goblet (mucus-secreting) cell
Cilia
Cartilage
Perichondrium
RESPIRATORY MUCOSA

EXTERNAL NOSE, NASAL SEPTUM & NASAL CAVITY

CN: Use very light colors for H and I. (1) Begin with the upper illustration. (2) Color the nasal septum and its structure in the nasal cavities diagram. (3) Color the elements of the lateral wall of the nasal cavity and relations in the lowest illustration.

EXTERNAL NOSE
NASAL BONE A
CARTILAGE OF NASAL SEPTUM B
LATERAL NASAL CARTILAGE C
ALAR CARTILAGE D
FIBRO-FATTY TISSUE E

NASAL SEPTUM
CARTILAGE OF NASAL SEPTUM B
ALAR CARTILAGE D
PERPENDICULAR PLATE
    OF ETHMOID BONE F
VOMER BONE G

NASAL CAVITY & RELATIONS
NASAL BONE A
FRONTAL BONE H
SPHENOID BONE I
CRIBRIFORM PLATE OF ETHMOID F'
VESTIBULE OF NOSE D'
SUPERIOR CONCHA J
MIDDLE CONCHA K
INFERIOR CONCHA L
HARD PALATE M
SOFT PALATE N
LATERAL WALL O*

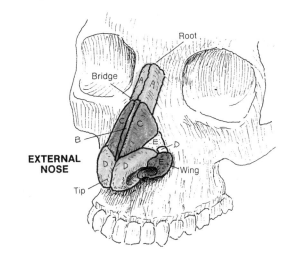

EXTERNAL NOSE

(labels: Root, Bridge, Wing, Tip)

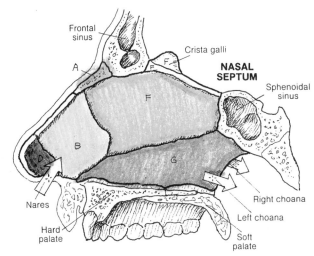

NASAL SEPTUM

(labels: Frontal sinus, Crista galli, Sphenoidal sinus, Right choana, Left choana, Soft palate, Hard palate, Nares)

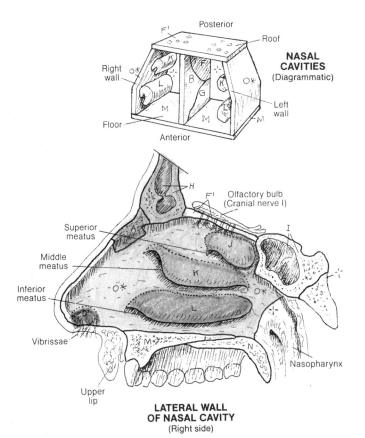

NASAL CAVITIES
(Diagrammatic)

(labels: Posterior, Roof, Right wall, Left wall, Floor, Anterior)

LATERAL WALL OF NASAL CAVITY
(Right side)

(labels: Olfactory bulb (Cranial nerve I), Superior meatus, Middle meatus, Inferior meatus, Vibrissae, Upper lip, Nasopharynx)

The nose is a largely cartilaginous affair external to the skull proper. Its orifices (nares, or nostrils) open into the nasal cavity of the skull, which is a bony tunnel divided by a partly cartilaginous *nasal septum*. The nasal cavity opens into the muscular pharynx through two bony-walled posterior apertures called choanae. The nose, situated as it is in front of the face, often receives the brunt of a facial impact. In such an event, it is not unusual for the *cartilage of the nasal septum* (septal cartilage) to break off from the *perpendicular plate of the ethmoid*. This "deviated septum" may obstruct air flow through the narrowed half of the cavity. The skin-lined *vestibule* of the nose has long hairs (vibrissae) that serve to discourage entrance of foreign bodies. The nasal cavity is carpeted with a mucosal lining characterized by ciliated epithelial cells that secrete mucus and whose cilia sweep small particulate matter down into the nasopharynx. The bony *conchae* (so called because of their resemblance, in frontal section, to the conch shell) increase the surface area of the nasal cavity, significantly boosting the local temperature and moisture content. The *inferior concha* on each side is attached to the ethmoid bone by an immovable joint (suture); the *superior and middle conchae* are part of the ethmoid bone. The spaces under the conchae (meatuses) are open to paranasal sinuses (air-filled cavities), the subject of the next plate. Note that the roof of the nasal cavity (*cribriform plate*) transmits the olfactory nerve fibers; resting on or near this plate are the frontal lobes of the brain. Note that the floor of the nasal cavity is the *palate*, which is also the roof of the oral cavity. The *soft palate* is a muscular extension of the bony palate and plays a role in swallowing.

PARANASAL AIR SINUSES

CN: Use the same colors for the bones A and B, and conchae F, G, and H, that were used for those structures on Plate 130. (1) Color the sinus drainage sites in the lateral wall of the nasal cavity. Include the edges of the conchae which have been cut away to reveal the meatuses and related drainage sites. (2) Color the coronal section. Note that it is a composite view, showing openings into the nasal cavity that do not appear in any one single coronal plane. Even so, this view cannot show the relations of the sphenoid sinus and opening, nor the mastoid air cells and the auditory tube. (3) Color the lower drawings. Note that nasolacrimal duct and the duct of the frontal sinus are shown on one side only.

AIR SINUSES ⫶
FRONTAL A
SPHENOID B
ETHMOID C
MAXILLARY D
MASTOID E

NASAL CONCHAE ⫶
SUPERIOR F
MIDDLE G
INFERIOR H

OPENING OF AUDITORY TUBE I
NASOLACRIMAL DUCT J
NASAL SEPTUM K
NASAL CAVITY L*

The skull has a number of cavities in it. You are familiar with some of them (mouth, nose, external ear, orbits), but perhaps not so familiar with others. The frontal, sphenoid, maxillary, ethmoid, and temporal bones have variably sized cavities, all of which directly or indirectly communicate with the nasal cavity. These are the *paranasal air sinuses*, to be distinguished from the venous sinuses of the dura mater. These air sinuses serve to lighten the skull and they add timbre to the voice. They are lined with respiratory-type epithelium, which is continuous with the epithelium of the nasal cavity. The mucus secretions from these epithelial linings pass down canals and enter the nasal cavity just under the conchae (meatuses). Their specific drainage sites are indicated by the arrows. Should these passageways become blocked by inflammation and swelling, pressure builds within the sinuses to a point where considerable pain can be experienced (sinusitis, sinus headache). Agents that constrict the blood vessels (decongestants) help to reduce the swelling and reestablish proper drainage. The *mastoid air cells*, in the mastoid process of the temporal bone, drain into the middle ear (tympanic) cavity, communicating by way of the auditory (pharyngotympanic) tube with the nasopharynx just posterior to the nasal cavity. The *nasolacrimal duct* receives secretions from the lacrimal gland, which functions to keep the covering (conjunctiva) of the eye globe moist. Tears drain into slits at the medial aspect of the eyelids, which open into sacs that narrow into the nasolacrimal ducts. These ducts pass downward along the lateral walls of the nasal cavity and open into the meatus of the inferior concha on each side—and that is why one blows one's nose after crying.

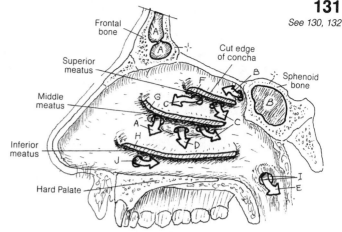

SINUS DRAINAGE SITES
(Right lateral wall of nasal cavity, nasal conchae removed)

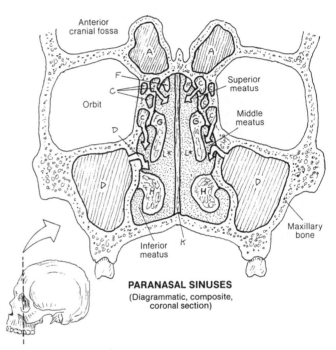

PARANASAL SINUSES
(Diagrammatic, composite, coronal section)

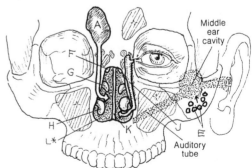

PARANASAL SINUS AND DUCTS

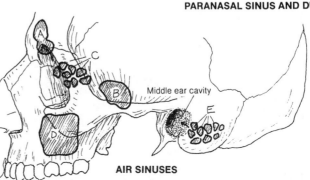

AIR SINUSES

PHARYNX & LARYNX

CN: Use dark or bright colors for N, O, and Q. (1) Begin with the over-view diagram in the upper right corner. (2) Complete the large composite sagittal section (do not color the arrows representing air flow). Take note of the surrounding structure as a frame of reference (not to be colored). (3) Color all six laryngeal views simultaneously.

PHARYNX A
NASOPHARYNX B
PHARYNGEAL TONSIL C
OROPHARYNX D
PALATINE TONSIL E
LARYNGOPHARYNX F

The pharynx is an incomplete tube of mostly skeletal (constrictor) muscle and fibrous tissue, appearing to hang from the edges of the choanae (posterior nasal apertures) at the base of the skull. Posteriorly, it is supported by fascia in front of the sphenoid bone and the upper six cervical vertebrae. It is the posterior and inferior continuation of the *nasal cavity*; it is open to the *oral cavity* anteriorly. Inferiorly, it continues as the *esophagus* behind and the *larynx* in front. Most of pharynx is lined with stratified squamous epithelium, except the nasopharynx (respiratory lining). Coordinated muscular activity in the pharynx underlies the mechanism of swallowing (deglutition).

Masses of partially encapsulated lymphoid tissue incompletely encircle the nasal and oral openings into the pharynx (Waldeyer's ring)—i.e., at the opening of the auditory tube (tubal tonsils), at the roof of the nasopharynx (adenoids), between the palatoglossal and palatopharyngeal pillars (palatine tonsils; see Plate 137), and at the posterior tongue (lingual tonsils). See tonsil function in Plate 127.

HYOID BONE G
LARYNX H
LARYNGEAL CAVITY H'
EPIGLOTTIS I
THYROID CARTILAGE J
THYROHYOID MEMBRANE K
CRICOID CARTILAGE L
CRICOTHYROID LIGAMENT M
ARYTENOID CARTILAGE N
CORNICULATE CARTILAGE O
VESTIBULAR FOLD P
VOCAL FOLD Q
RIMA GLOTTIS R *

The larynx provides a mechanism for sound production, manipulation of sound waves, and protection from inadvertent aspiration (inhaling) of solid matter. The larynx is supported by a framework of hyaline cartilage connected by ligaments. Although associated with the larynx, the hyoid bone is not a laryngeal structure.

The thyroid cartilage is composed of two laminae that together are V-shaped when seen from above. The *arytenoid cartilages* articulate with the top of the cricoid, pivoting on it. The *vocal folds* are mucosa-lined ligaments stretching between thyroid and arytenoid cartilages. They are abducted/adducted by the movement of the arytenoid cartilages. In breathing they are abducted; in coughing, they are momentarily fully adducted (closing the *rima*), permitting intrathoracic pressure to build; opened rapidly by abduction of the folds, the rima experiences hurricane-force winds from the depths of the respiratory airway (explosive cough). During phonation, the vocal folds are generally adducted, varying somewhat with pitch and volume. The *vestibular folds* (false vocal folds) are fibrous and move only passively.

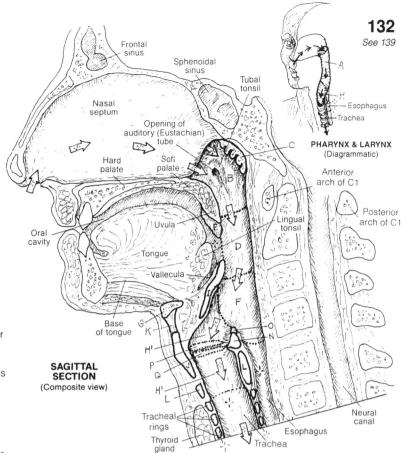

SAGITTAL SECTION
(Composite view)

PHARYNX & LARYNX
(Diagrammatic)

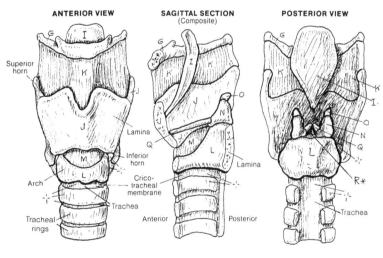

ANTERIOR VIEW   **SAGITTAL SECTION** (Composite)   **POSTERIOR VIEW**

VIEWS OF THE LARYNX

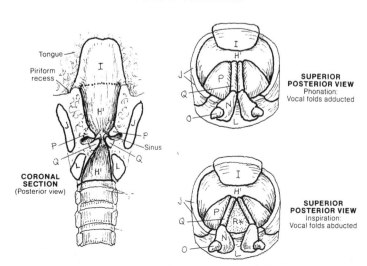

CORONAL SECTION
(Posterior view)

SUPERIOR POSTERIOR VIEW
Phonation:
Vocal folds adducted

SUPERIOR POSTERIOR VIEW
Inspiration:
Vocal folds abducted

LOBES & PLEURAE OF THE LUNGS

CN: Use bright colors for A–E, very light colors for F and G, and a reddish-brown color for H. In all of the illustrations the thickness of the pleurae (F and G) has been enlarged for coloring purposes. (1) Begin with the anterior view. Note that the ribs and intercostal muscles have been removed (see Plate 50). Sections of the pleurae have been stripped away and separated. The potential pleural space is between these layers; in the coronal and cross sections, this space is drawn as a dark line and not as a structure to be colored. Similarly, the title is left uncolored. A small section of the parietal pleura has been cut and pulled away to reveal the underlying visceral pleura and a portion of the costodiaphragmatic recess below the lung superficial to the diaphragm. (2) Color the coronal view, noting the left crus of the diaphragm and the cardiac notch of the left lung. (3) Color the cross section of the lung lobes and pleurae (as seen from above), noting the vertebral level and the roots of the lungs.

LOBES

R. UPPER A   L. UPPER D
R. MIDDLE B   L. LOWER E
R. LOWER C

PLEURAE

VISCERAL PLEURA F
PLEURAL SPACE
PARIETAL PLEURA G

DIAPHRAGM H

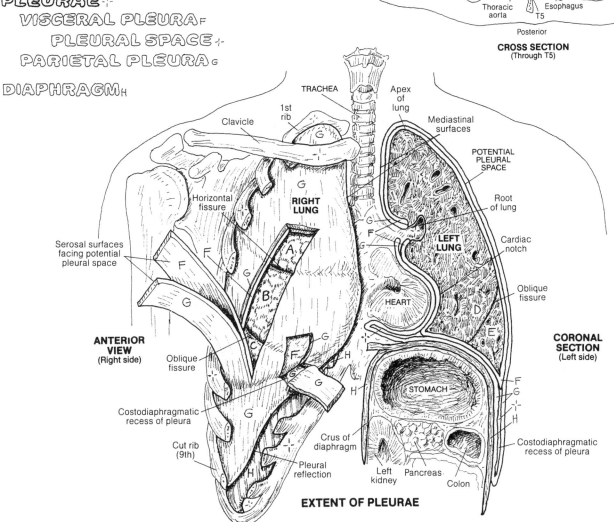

Heart and great vessels — Sternum — 2nd rib
Root of lung — LEFT LUNG — RIGHT LUNG — A — G
G — F
G — F
E — Oblique fissure
Oblique fissure
Thoracic aorta — T5 — Esophagus
Posterior

CROSS SECTION
(Through T5)

TRACHEA — Apex of lung
1st rib — Mediastinal surfaces
Clavicle — POTENTIAL PLEURAL SPACE
G — Root of lung
Horizontal fissure — RIGHT LUNG
G — Cardiac notch
Serosal surfaces facing potential pleural space — F — F — LEFT LUNG
F — A — G
B — HEART — Oblique fissure
ANTERIOR VIEW (Right side) — C — D — **CORONAL SECTION** (Left side)
Oblique fissure — F — E
Costodiaphragmatic recess of pleura — G — F — G — H — STOMACH — F — G — H
Cut rib (9th) — Crus of diaphragm — Costodiaphragmatic recess of pleura
Pleural reflection — Left kidney — Pancreas — Colon
EXTENT OF PLEURAE

The lobes of the lungs are largely enveloped in *visceral pleura*, a thin serosal membrane that turns (reflects) off the lungs at their roots to become the *parietal pleura*, which lines the inner surface of the chest wall, the lateral mediastinum, and much of the diaphragm. These serous membranes are in contact with each other, separated by a thin layer of serous (watery, glycoprotein) fluid. The interface of these membranes is potentially a cavity or space (*pleural space/cavity*). With certain diseases, the space is capable of expanding to accommodate increasing amounts of fluid (pleural effusion) at the expense of the lung, resulting in a reduction of total lung capacity. The serous fluid maintains surface tension between the pleural surfaces (resisting separation of visceral and parietal layers in contact with one another) and prevents frictional irritation between moving pleural membranes. During quiet inhalation, the inferior and anterior margins of the visceral pleura–lined lungs do not quite reach the parietal pleura, leaving a narrow space or recess—i.e., the costomediastinal recess between the rib cage and the mediastinum (not shown), and the costodiaphragmatic recess between rib cage and diaphragm (see coronal section at lower right).

LOWER RESPIRATORY TRACT

TRACHEA   MAIN (PRIMARY) BRONCHUS   LOBAR (SEC.) BRONCHUS
SEGMENTAL (TERTIARY) BRONCHI / BRONCHOPULMONARY SEGMENTS:
1 APICAL   2 POST.   3 ANT.   4 LAT.(R.L.)   4 SUP.(L.L.)
5 MED.(R.L.)   5 INF.(L.L.)   6 SUP.   7 MED. BASAL
8 ANT. BASAL   9 LAT. BASAL   10 POST. BASAL

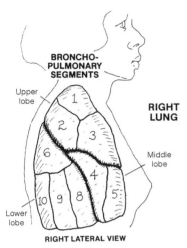

BRONCHO-PULMONARY SEGMENTS
Upper lobe
Middle lobe
Lower lobe
RIGHT LUNG
RIGHT LATERAL VIEW

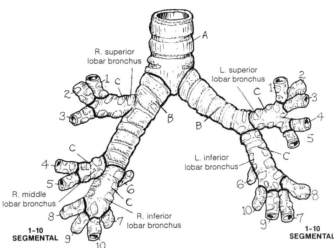

R. superior lobar bronchus
L. superior lobar bronchus
R. middle lobar bronchus
L. inferior lobar bronchus
R. inferior lobar bronchus
1-10 SEGMENTAL
1-10 SEGMENTAL

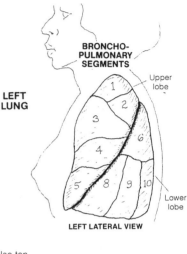

BRONCHO-PULMONARY SEGMENTS
Upper lobe
Lower lobe
LEFT LUNG
LEFT LATERAL VIEW

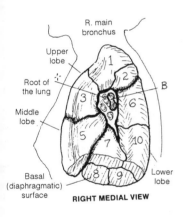

R. main bronchus
Upper lobe
Root of the lung
Middle lobe
Basal (diaphragmatic) surface
Lower lobe
RIGHT MEDIAL VIEW

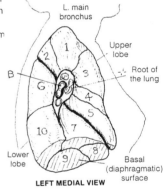

L. main bronchus
Upper lobe
Root of the lung
Lower lobe
Basal (diaphragmatic) surface
LEFT MEDIAL VIEW

CN: Save blue for H, purple for I, and red for J (in the respiratory unit below). (1) Use ten different colors for both lungs, and key those colors to the ten segmental bronchi of each lung. (2) Below, use the same color as above for the 7th segmental bronchus. Use one light color for the alveoli (G¹) and the alveolar sacs (G). Note in the gas exchange diagram that red blood cells in the purple capillary (F) receive three different colors according to their stage of oxygenation.

The lower respiratory tract consists of the *trachea* and the *bronchial tree*, including the respiratory units that are engaged in gaseous exchange. The lungs are divided by connective tissue septa into triangular-shaped, surgically resectable anatomical and functional units called *bronchopulmonary segments*, each served by a segmental bronchus, supplied by a segmental artery, and drained by segmental veins and lymphatics. Segments are of special significance to those interpreting lung sounds by stethoscope (auscultation) or listening to the sounds coming from the lungs when the chest wall is tapped (percussion). By such methods, sites of alveolar dysfunction/disease and levels of abnormal accumulations often can be determined.

BRONCHIOLE D
RESPIRATORY BRONCHIOLE E
ALVEOLAR DUCT F
ALVEOLAR SAC G & ALVEOLUS G'
PULMONARY ARTERIOLE H
CAPILLARY NETWORK I
PULMONARY VENULE J

Within each bronchopulmonary segment, a segmental bronchus branches into several *bronchioles* (less than 1 mm in diameter, supported by smooth muscle instead of cartilage). These bronchioles give off smaller terminal bronchioles, characterized by ciliated cuboidal cells without glands. The terminal bronchioles represent the end of the air-conducting pathway. Each terminal bronchiole divides into two or more *respiratory bronchioles*, characterized by occasional alveolar sacs on their walls. Each respiratory bronchiole supplies a respiratory unit, which is a discrete group of air cells (*alveoli*), arranged in *alveolar sacs*, fed by *alveolar ducts*. Extending from its source bronchiole, each respiratory bronchiole has more and more alveolar sacs, terminating as an alveolar duct opening into alveolar sacs. The walls of the air cells, composed of simple squamous epithelia supported by thin interwoven layers of elastic and reticular fibers, are surrounded by capillaries that arise from pulmonary arterioles and become the tributaries of pulmonary venules. The walls of these capillaries are fused to and structurally similar to those of the alveoli. Oxygen and carbon dioxide rapidly diffuse, on the basis of pressure gradients, through these walls.

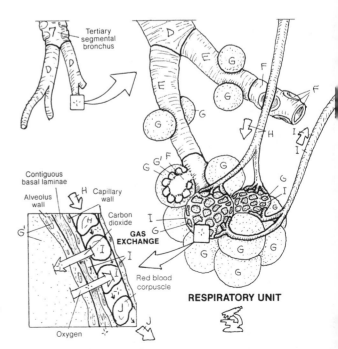

Tertiary segmental bronchus
Contiguous basal laminae
Alveolus wall
Capillary wall
Carbon dioxide
GAS EXCHANGE
Red blood corpuscle
Oxygen
RESPIRATORY UNIT

MECHANISM OF RESPIRATION

CN: Use light colors throughout, except for a bright or dark color for E. (1) Begin with the illustration at far left (inspiration); note that the thoracic wall (A) is shown only in the far right diagram. Color the diaphragm, its location represented by broken lines. (2) Color the expiration illustration and the bucket handle analogy. (3) Finish with the illustration at far right.

THORACIC WALL A
RIB & COSTAL CARTILAGE B
STERNUM C
THORACIC VERTEBRAE D

MUSCLES OF INSPIRATION -/-
DIAPHRAGM E
EXTERNAL INTERCOSTAL F

MUSCLE OF EXPIRATION -/-
INTERNAL INTERCOSTAL G

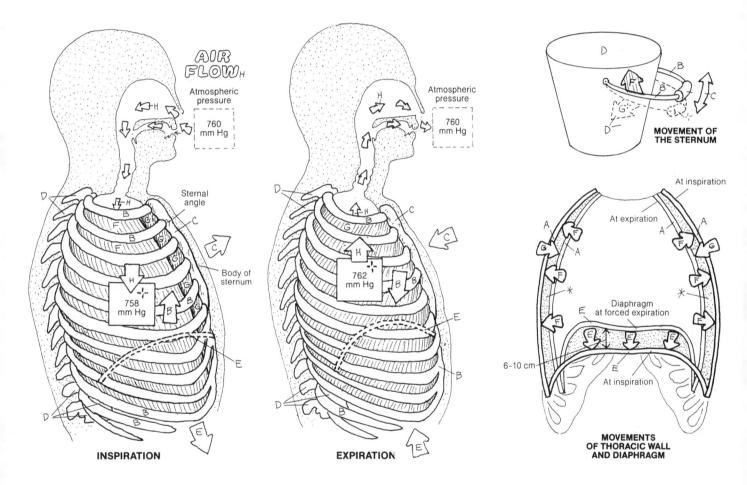

INSPIRATION

EXPIRATION

MOVEMENT OF THE STERNUM

MOVEMENTS OF THORACIC WALL AND DIAPHRAGM

The mechanism of respiration makes possible breathing, which consists of inhalation (inspiration) and exhalation (expiration) phases. The physical principle underlying air movement in/out of the thorax is the inverse relationship of pressure and volume (as one goes up, the other goes down). Volume changes within the *thorax* alter the intrathoracic pressure 1–2 mm Hg above/below atmospheric pressure (outside the body) in quiet breathing—enough of a change to move about 500 ml of air with each breath. The thoracic *diaphragm* accomplishes about 75% of the inspiratory effort, the *external intercostals* 25%. Expiration is largely diaphragm and external intercostal relaxation/stretch, and lung elasticity, with some help from the *internal intercostals*. In inspiration, contraction of the diaphragm flattens the muscle and lowers the floor of the thorax, increasing the vertical dimension of the thoracic cavity. Contraction of the external intercostals

elevates the ribs, swinging the sternal body slightly outward at the sternal angle. This increases the transverse and anteroposterior dimensions of the thoracic cavity. These actions collectively increase the intrathoracic volume, momentarily lowering the pressure within. Given the relatively higher atmospheric pressure outside the head, *air* is induced to enter the respiratory tract to find lower pressure. The action of the bucket handle demonstrates the hinge action at the sternal angle and related rib elevation. In expiration, the relaxed diaphragm forms "domes" over the underlying liver and stomach, decreasing the vertical dimension of the thorax. Recoil/descent of the ribs decreases the transverse and anteroposterior dimensions. The thoracic volume is thus decreased, momentarily increasing the intrathoracic pressure above atmospheric. *Air* escapes to the outside, aided by the natural elastic recoil of the lungs.

CN: When coloring the organs that overlap each other, use your lightest colors for D, E, T, V, and W. Each overlapping portion receives the color of both structures. (1) After coloring the alimentary canal, review the structures before completing the accessory organs. The central section of the transverse colon (J) has been removed to show deeper structures.

ALIMENTARY CANAL⁻ᵢ⁻
ORAL CAVITYₐ
PHARYNX ʙ
ESOPHAGUS c
STOMACH ᴅ
SMALL INTESTINE⁻ᵢ⁻
DUODENUM ᴇ
JEJUNUM ꜰ
ILEUM ɢ
LARGE INTESTINE⁻ᵢ⁻
CECUM ʜ
VERMIFORM APPENDIX ʜᵢ
COLON⁻ᵢ⁻
ASCENDING COLON ɪ
TRANSVERSE COLON ᴊ
DESCENDING COLON ᴋ
SIGMOID COLON ʟ
RECTUM ᴍ
ANAL CANAL ɴ

ACCESSORY ORGANS⁻ᵢ⁻
TEETH o
TONGUE ᴘ
SALIVARY GLANDS⁻ᵢ⁻
SUBLINGUAL ǫ
SUBMANDIBULAR ʀ
PAROTID s
LIVER ᴛ
GALL BLADDER ᴜ
BILE DUCT ᴠ
PANCREAS ᴡ

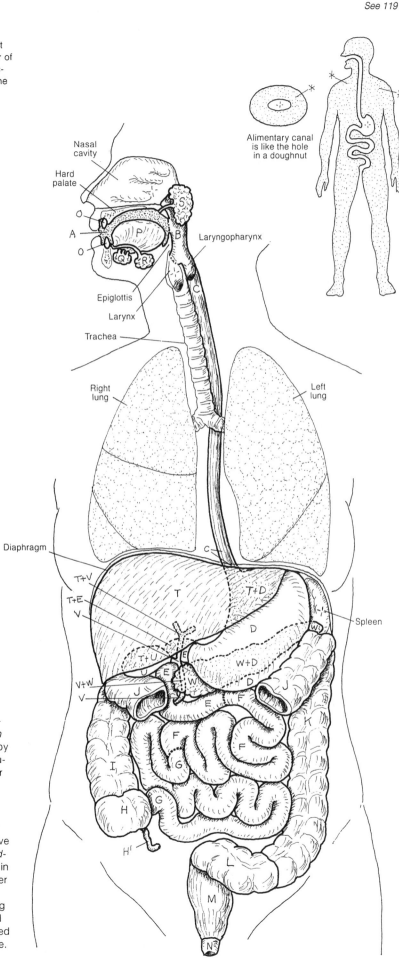

Nasal cavity

Hard palate

Alimentary canal is like the hole in a doughnut

Laryngopharynx

Epiglottis

Larynx

Trachea

Right lung

Left lung

Diaphragm

Spleen

The digestive system consists of an alimentary canal with accessory organs. The canal begins with the *oral cavity*. Here the *teeth* pulverize ingested food while it is softened and partly digested by *salivary gland* secretions. The *tongue* aids in mechanical manipulation of the food and literally flips the food into the fibromuscular *pharynx* during swallowing.

The esophagus moves the bolus along to the *stomach* by peristaltic muscular contractions. Here the bolus is treated to mechanical and chemical digestion, then passed into the highly coiled *small intestine* for more enzymatic and mechanical digestive processes. Bile, produced by the *liver* and stored in the *gall bladder*, is discharged into the *duodenum* by a *bile duct*. Bile assists in the breakdown of fats. Digestive enzymes from the *pancreas* enter the duodenum as well. Nutrients of molecular size are extracted primarily from the lumen of the small intestine, absorbed by lining cells, and transferred to blood and lymph capillaries for eventual delivery to the liver for processing. The large intestine is concerned with absorption of minerals and water (proximal half) and storage. Undigested, unabsorbed material continues to the rectum for discharge through the anal canal and anus.

ORAL CAVITY & RELATIONS

CN: Use pink or red for I and very light colors for N, O, and P. The asterisks preceding titles F, G, and H refer to the footnote under the list of titles, and not the color gray. (1) Color the two upper views of the oral cavity simultaneously. (2) Color the papillae of the tongue with the color of the tongue (I) but not the tongue itself. (3) Color the three salivary glands and the cellular diagram to their right. Note that the lumen, which receives glandular secretions, is not colored as it passes through the various colored structures.

ORAL CAVITY-¡-
TEETH A-¡-
GINGIVA (GUM) B
HARD PALATE C
SOFT PALATE D
UVULA E
* PALATOGLOSSAL ARCH F
* PALATINE TONSIL G
* PALATOPHARYNGEAL ARCH H

TONGUE I
LINGUAL TONSIL J
VALLATE PAPILLAE I¹
FOLIATE P₀ I²
FUNGIFORM P₀ I³
FILIFORM P₀ I⁴

These structures are considered as part of the oropharynx in Plates 132 and 139; as a practical matter, they are part of the oral cavity bordering the oropharynx.

The oral cavity (mouth) is essentially concerned with preparation of food for swallowing. Food is pulverized with *teeth* (presented in next plate), which act on food through chewing (mechanical digestion), made possible by the muscles of mastication and the temporomandibular joint, which permits mouth opening to an interincisor distance of 35–50 mm. Wetting the food is a function of the thousands of mucous and serous glands in the tongue and the mucosa lining the oral cavity. Wetting and enzymatic action also are functions of the salivary glands (discussed below). Mechanical digestion is enhanced by the *papillae* on the surface (dorsum) of the tongue. These provide a site for taste receptors (except filiform papillae) and an abrasive surface, for breaking down food.

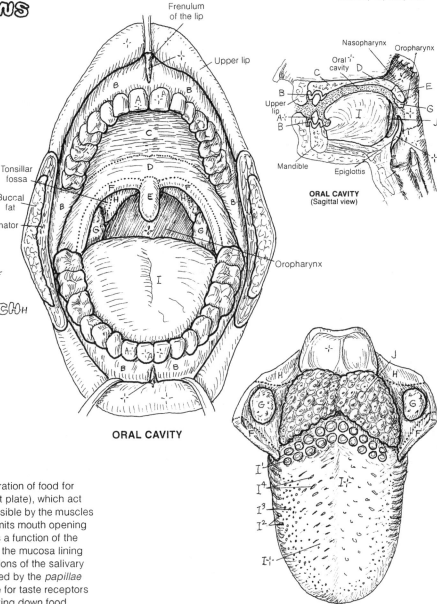

ORAL CAVITY

ORAL CAVITY (Sagittal view)

TONGUE

SALIVARY GLANDS-¡-
SUBLINGUAL K
SUBMANDIBULAR L
PAROTID M

CELLS OF THE PAROTID GLAND

DUCT N
MUCOUS TUBULE O
SEROUS ACINUS P
MYOEPITHELIAL CELL Q

SALIVARY GLANDS

Salivary glands secrete an enzyme-rich fluid into the mouth during periods of eating or anticipated eating. The largest is the *parotid gland*, situated bilaterally in front of and below each external auditory canal, partly overlying the masseter muscle. Its duct arches over the masseter, penetrating the cheek mucosa to enter the oral cavity opposite the upper 2nd molar. Its glandular cells are serous. The smallest of the salivary glands, the mucus-type *sublingual glands*, lie under the tongue below the oral mucosa. The *submandibular glands* are U-shaped and wrap around the mylohyoid muscle (Plate 48). They consist of ducts and mixed glands, primarily mucous.

An example of a mixed (muco-serous) gland is shown here. The serous glands consist of cells that are pyramid-shaped. The cells form rounded, grape-shaped alveoli or acini, whose center forms the *duct*. The more *tubular glands* are mucous-secreting; they are cylinder-shaped, with a central duct. Collections of *serous cells* capping a mucous gland are called serous demilunes (half-moon shaped). Contractile *myoepithelial cells* within the basal laminae of both duct and gland cells are responsible for forcing the secretions into the ducts and out of the glands.

ANATOMY OF A TOOTH

CN: Use yellow for F, red for G, blue for H, and light colors for A, B, and L.
(1) Begin with the anatomy of a tooth. Color gray the titles and arrows/bands arranged vertically. (2) Use only light colors on the teeth below. Note that the identifying letter and number labels are those used by the dental profession.

TOOTH:
ENAMEL_A DENTIN_B
PULP CAVITY_C PULP_E
ROOT CANAL_D
NERVE_F ARTERY_G VEIN_H
CEMENTUM_I
PERIODONTAL LIGAMENT_J
GINGIVA_K ALVEOLAR BONE_L

Crown — A
Neck
Root — B

Note the longitudinal section of a molar; two roots are shown. The core substance of the tooth is *dentin*. Composed of packed microscopic tubules, pain-sensitive, avascular dentin is composed like bone, though more mineralized (70% by weight). Dentin is capped by a 1.5 mm layer of insensitive *enamel*, 95% mineral by weight, less than 1% organic. Enamel consists of microscopic circular rods filled with hydroxyapatite (bone) crystals, and it is the hardest material in the body. The dentin of each tooth has a hollow *pulp cavity*; it extends into each root of the tooth as the root canal. At the apex of each root, an opening (apical or root foramen) permits the passage of *blood vessels and nerves* into/from the alveolar bone. *Pulp* is a well-innervated and vascular loose connective tissue, continuous with the periodontal ligament through the root foramen.

Each tooth has a crown, extending above the *gingiva* (gum line), a *neck* (at the level of the gum; here the enamel ends and abuts the cementum), and one or more *roots* buried in *alveolar bone* of the maxilla (upper teeth) or mandible (lower teeth). Incisors and canine teeth each have a single root canal; premolars and molars may have 1–3 roots. The surface of the crown is characterized by tubercle-like cusps separated by fissures, except for incisiors, which have only a cutting edge. Canines have one cusp (cuspid); premolars have two (bicuspids), and molars have 4 or 5 cusps. Multiple cusps enhance the grinding and abrasive functions of the teeth.

The fibrous *periodontal ligament*, about 0.2 mm thick, interfaces the *cementum* (lining the root of the tooth) and the alveolar bone. The cementum is a highly mineralized substance. Collagen fibers stuck in the cementum penetrate the ligament to insert into the alveolar bone. The gingiva (gum) is a mucous membrane with stratified squamous epithelia that attaches to the enamel by a thickened basal lamina; the lamina propria of the membrane is strongly anchored to the underlying alveolar bone.

ADULT/CHILD DENTITION:
CENTRAL INCISOR 8, 9, 24, 25, E, F, O, P
LATERAL INCISOR 7, 10, 23, 26, D, G, N, Q
CANINE 6, 11, 22, 27, C, H, M, R
1ST PREMOLAR 5, 12, 21, 28
2ND PREMOLAR 4, 13, 20, 29
1ST MOLAR 3, 14, 19, 30, B, I, L, S
2ND MOLAR 2, 15, 18, 31, A, J, K, T
3RD MOLAR (WISDOM) 1, 16, 17, 32

There are naturally 32 teeth in an adult—8 in each of four quadrants (right and left, in both upper and lower dental arches). Two sets of teeth (dentition) develop in a lifetime: deciduous and permanent. The deciduous set (20) are absorbed/shed in early life; the permanent set (32) is not naturally shed. Babies are born with the deciduous dentition submerged under the gingiva, for which breast-feeding mothers are grateful. In general, the deciduous incisors are the first to erupt, at 6 months. The entire deciduous dentition (see boxed area at right) has erupted by 18 months; it is gone by 12 years. The first permanent tooth (1st molar) appears at about 6 years; the last (3rd molar) appears at about 18 years ("wisdom tooth").

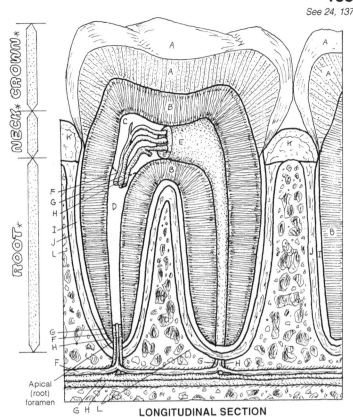

ROOT* NECK* CROWN*

Apical (root) foramen

G H L

LONGITUDINAL SECTION
(1st molar in bone)

Mesial surface
Distal surface
Maxilla
Alveolar bone
Mandible

PERMANENT TEETH IN OCCLUSION
(21 years of age)

Buccal or labial aspect
Cusp
Maxilla
Palatal or lingual aspect
Palatine bone
Fissure

UPPER ARCH
(Right)   (Left)

DECIDUOUS "MILK" TEETH
(3 years of age)

DECIDUOUS AND PERMANENT TEETH
(5 years of age/alveolar walls removed)

LOWER ARCH

Mandible
Mesial surface
Distal surface

PHARYNX & SWALLOWING

CN: Use pink for K. (1) Color the three lower illustrations simultaneously. In the posterior view of the interior of the pharynx, the posterior pharyngeal wall is divided and retracted so you can note the relationship of internal pharyngeal structure to the constrictor muscles (A, B, C) and the subdivisions of the pharynx (D, G, I). Color gray the boluses of food in all views. In the two outer views (below), add the color of the overlying structure to the representation of bolus movement. (2) Follow the text when coloring the deglutition diagrams.

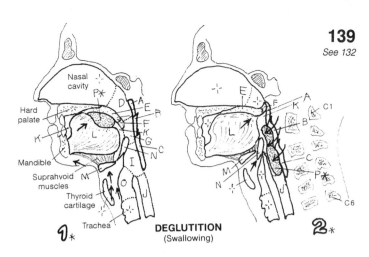

DEGLUTITION
(Swallowing)

MUSCULAR WALL OF PHARYNX⊹
SUPERIOR CONSTRICTOR A
MIDDLE CONSTRICTOR B
INFERIOR CONSTRICTOR C

INTERIOR OF PHARYNX⊹
NASOPHARYNX D
SOFT PALATE E
UVULA F
OROPHARYNX G
PALATOPHARYNGEAL M. H
LARYNGOPHARYNX I

ESOPHAGUS J

RELATED STRUCTURES⊹
ORAL CAVITY K
TONGUE L
HYOID BONE M
EPIGLOTTIS N
LARYNX O

BOLUS OF FOOD P*

Swallowing (deglutition) begins with food in the oral cavity. (1) The *food bolus* is voluntarily pushed upward and backward into the *oropharynx* by the *tongue*. The suprahyoid muscles lift and shorten or lengthen the floor of the mouth by pulling the *hyoid bone* up and forward or backward to one degree or another, depending on the material being moved into the pharynx. You can feel the ascent and descent of the hyoid bone. Place your thumb and index finger around the front of your mid-neck at the level of the palpable hyoid bone; swallow and feel the bone move upward and downward. The bilateral *palatopharyngeal muscles*/folds partially close off the oral cavity from the pharynx, selectively permitting appropriately sized boluses to enter the pharynx and resisting regurgitation into the mouth. Once the bolus is in the oropharynx, the following mechanism is involuntary.

The soft palate is elevated (2), blocking the *nasopharynx* and preventing regurgitation into the nasal cavity. Indeed, the entire pharynx is elevated by the pharyngeal elevator muscles. Simultaneously, the larynx is elevated by infrahyoid and other muscles, and the laryngeal opening is closed by intrinsic laryngeal muscles. During the lifting of the larynx, the epiglottis moves over the laryngeal opening. The *superior and middle constrictor muscles* of the pharynx, assisted by gravity, sequentially contract from above downward, driving the bolus into the laryngopharynx. The contractions of the palatopharyngeal muscles orient the descent of the bolus downward and slightly backward. Contractions of the *inferior constrictor muscle* force the bolus into the esophagus.

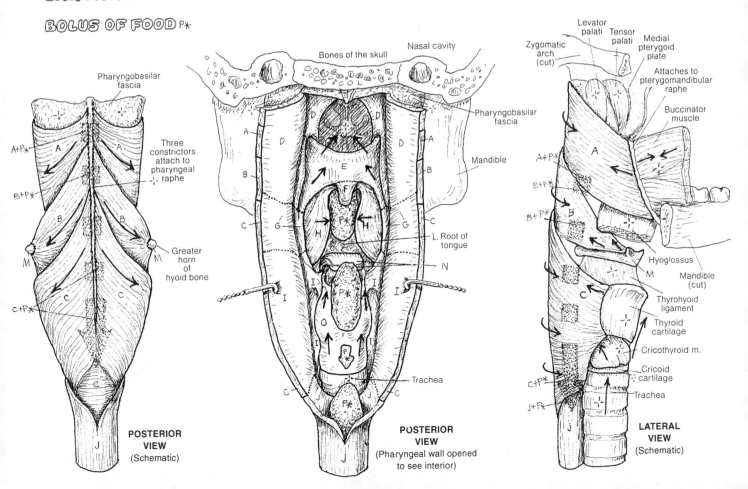

POSTERIOR VIEW
(Schematic)

POSTERIOR VIEW
(Pharyngeal wall opened to see interior)

LATERAL VIEW
(Schematic)

PERITONEUM

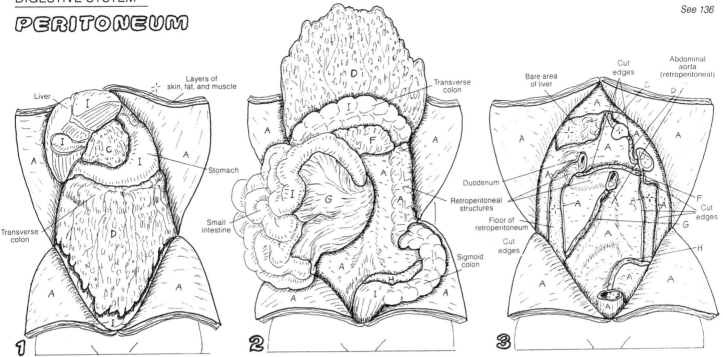

1 With the anterior abdominal wall opened through its deepest (*parietal peritoneal*) layer, the liver, the stomach, and the fatty *greater omentum* are generally all that can be seen with the contents undisturbed. Lifting the liver exposes the *lesser omentum*, a double peritoneal layer between the stomach and liver. It is the anterior wall of the *omental bursa* (E). The greater omentum connects the transverse colon to the stomach.

2 With the greater omentum lifted, the double-layered, *transverse mesocolon* between the transverse colon and parietal peritoneum can be seen. Retracting the intestines to one side reveals the *common mesentery* between most of the small intestine and the parietal peritoneum on the posterior body wall. The sigmoid colon has a mesentery (*sigmoid mesocolon*) as well. Abdominal structures posterior to these mesenteries/omenta are retroperitoneal.

3 The parietal peritoneum of the posterior body wall is seen when all structures except retroperitoneal ones (aorta, inferior vena cava, kidneys, ureters, pancreas, duodenum, ascending/descending colon) are removed. Many nerves and vessels travel in this retroperitoneal space. As organs emerge from the peritoneum, they develop a mesentery to suspend them. The cut layers of several of them can be seen (C, D, F, G, and H).

CN: Use a very light color for A and I. (1) Color the upper three diagrams in numerical order. Note that the digestive organs are covered with visceral peritoneum (I). (2) Color the sagittal view. Use a darker gray or black for the omental bursa (E). The space of the peritoneal cavity (B) has been greatly exaggerated for clarity of peritoneal membranes.

PERITONEAL STRUCTURES
PARIETAL PERITONEUM A
PERITONEAL CAVITY B*
LESSER OMENTUM C
OMENTAL BURSA E•
GREATER OMENTUM D
TRANSVERSE MESOCOLON F
COMMON MESENTERY G
SIGMOID MESOCOLON H
VISCERAL PERITONEUM I

Peritoneum is a serosal membrane of the abdominal cavity. The disposition of the peritoneum is similar to that of the serosal layers around heart (pericardium) and lungs (pleura): peritoneum attached to the body wall is *parietal*; peritoneum attached to the outer visceral wall is *visceral*. Structures deep to the posterior parietal peritoneum are *retroperitoneal*. Peritoneal layers suspending organs are called *mesenteries*; those suspending an organ from another organ are called *omenta* or *ligaments*. When coloring the sagittal view, the continuity of these peritoneal membranes can be appreciated. The cavity of the peritoneum is empty; it can fill with fluid in disease and trauma. The view at right shows intestines separated from one another; in life, they are as close together as strands of coiled wet rope. Vessels/nerves to the intestines and stomach travel in the mesenteries/omenta; they do not penetrate peritoneal layers. The source vessels are retroperitoneal. The *omental bursa* is a peritoneal-lined sac created by rotation of the stomach during fetal life. It is open on the right at the epiploic foramen between the lesser omentum and the parietal peritoneum. Here the omental bursa (lesser sac) communicates with the collapsed, empty peritoneal cavity (greater sac).

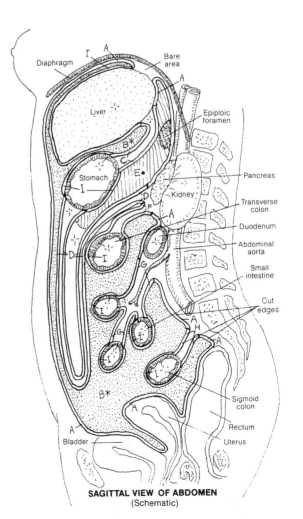

SAGITTAL VIEW OF ABDOMEN
(Schematic)

ESOPHAGUS & STOMACH

CN: Use only light colors. (1) Color the esophagus (A) as it descends through the chest. Color over the dotted parts of the trachea and heart that lie anterior to the esophagus. Color the stomach and the dotted part of the liver anterior to it. (2) Color the names of the digestive products secreted by tile cells shown below.

ESOPHAGUS A

STOMACH REGIONS -:-
CARDIA B
FUNDUS C
BODY D
PYLORUS E

STOMACH WALL K
MUCOSAL SURFACE (RUGAE) F
SUBMUCOSA G
MUSCULARIS EXTERNA -:-
OBLIQUE M. H
CIRCULAR M. I
LONGITUDINAL M. J
SEROSA K'

The esophagus connects the laryngo pharynx and the *cardia* of the stomach. It is a muscular tube whose mucosa is lined with non-keratinized stratified squamous epithelium. At the gastroesophageal junction, the epithelia transition abruptly to simple columnar. The muscle layer is arranged in both longitudinal and circular orientations. In the proximal part of the esophagus, the muscle is striated; in the midesophagus, it is both striated and smooth; in the lower third, the muscle is entirely smooth. The glands of the submucosa secrete mucus, enhancing flow. The esophagus conducts its contents by sequential, rhythmic muscular contractions.

The gastroesophageal junction has an area of specialized circular muscle (lower esophageal sphincter) that permits passage of a food bolus by muscle relaxation during swallowing. The right crus of the diaphragm also contributes fibers to the esophagus (external sphincter) and functions to resist gastroesophageal reflux (reverse flow) during inspiration.

The stomach is the first part of the gastrointestinal tract. The stomach is generally located in the upper left quadrant of the abdomen, although a full stomach can droop into the pelvis. Suspended by peritoneal layers, it receives the esophagus just below the diaphragm. At the duodenal end, the stomach narrows to a muscular pyloric sphincter. The stomach classically has four regions whose shape varies with the quantity of contents. The stomach mechanically manipulates ingested material, acidifies it to enhance protein digestion, secretes proteolytic enzymes (pepsin), and induces secretion of bile from the gall bladder and enzymes from the pancreas, both of which enter the duodenum. Microorganisms do not generally survive these activities.

Note the arrangement of the stomach wall and the various cells that make up the epithelial layer of the mucosa. The epithelial cells are the working cells, providing a cocktail of digestive products, whose main target is protein. The *lamina propria* provides vascular and mechanical support for the *gastric pits*. The muscle layer of the *mucosa* and the external muscle layers produce peristaltic contractions to assist in mechanical digestion and moving the residue of digestion along the tract. The fibrous submucosa supports lymphoid follicles, vessels, and nerves.

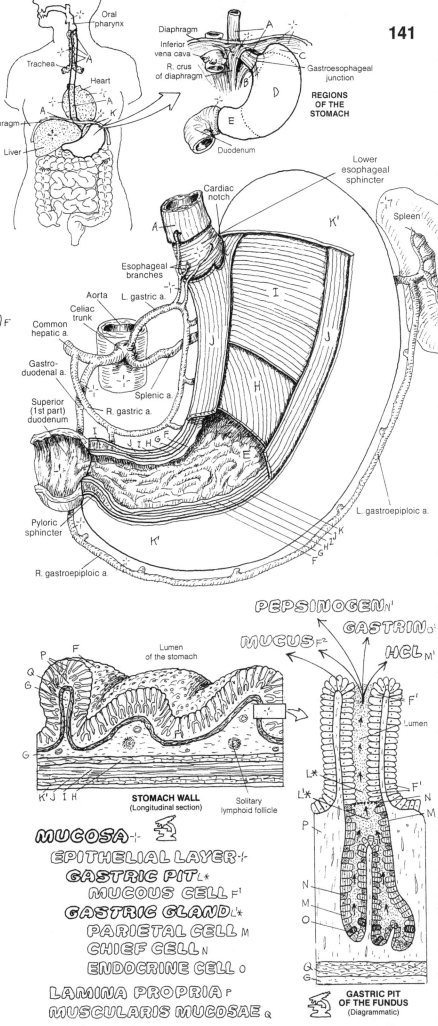

REGIONS OF THE STOMACH

STOMACH WALL
(Longitudinal section)

MUCOSA -:-
EPITHELIAL LAYER -:-
GASTRIC PIT L*
MUCOUS CELL F¹
GASTRIC GLAND L*
PARIETAL CELL M
CHIEF CELL N
ENDOCRINE CELL O
LAMINA PROPRIA P
MUSCULARIS MUCOSAE Q

PEPSINOGEN N'
GASTRIN O'
MUCUS F²
HCL M'

GASTRIC PIT OF THE FUNDUS
(Diagrammatic)

SMALL INTESTINE

CN: Use green for N, red for Q, purple for R, blue for S, yellow for T, and a very light color for H. (1) Begin with the three divisions of the small intestine. (2) Color the parts of the duodenum and the section of duodenal wall. The lamina propria (L) is identified and colored only in the enlarged view of the villi below.

DUODENUM_A
SUPERIOR (1ST) PART_B
DESCENDING (2ND) PART_C
HORIZONTAL (3RD) PART_D
ASCENDING (4TH) PART_E

JEJUNUM_F

ILEUM_G

INTESTINAL WALL_:
PLICA CIRCULARE_H
MUCOSA_:
VILLUS_H' / CRYPT_H²
EPITHELIUM_:
ABSORPTIVE CELL_H³
MUCOUS (GOBLET) CELL_I
ENTEROENDOCRINE CELL_J
PANETH CELL_K
LAMINA PROPRIA_L
MUSCULARIS MUCOSAE_M
LYMPHOID NODULE_N
SUBMUCOSA_O
DUODENAL GLAND_P
ARTERY_Q CAPILLARY_R VEIN_S LACTEAL_N'
PARASYMP. POSTGANG. NEURON_T
MUSCULARIS EXTERNA: CIRC_U / LONG_U'
SEROSA_D'

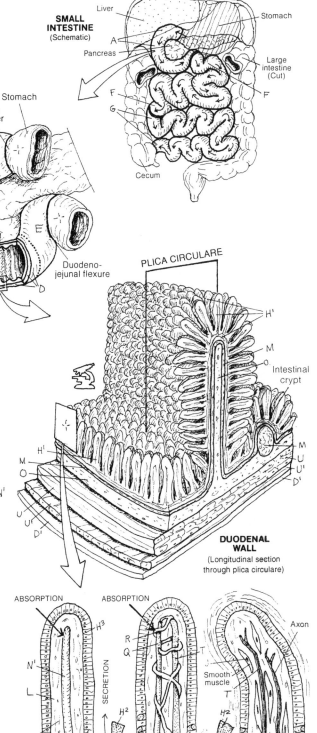

SMALL INTESTINE (Schematic)

Liver
Stomach
Pancreas
Large intestine (Cut)
Cecum

PARTS OF DUODENUM
Stomach
Pyloric sphincter
Pancreas
Plica circulare
Duodeno-jejunal flexure

PLICA CIRCULARE
Intestinal crypt

DUODENAL WALL
(Longitudinal section through plica circulare)

The small intestine is a highly convoluted, thin-walled tube that undertakes much of the chemical and mechanical digestive process and almost the whole of the absorptive process of the entire gastrointestinal tract. The first part of the *duodenum* is suspended by the lesser omentum. The second and third parts are retroperitoneal. The fourth part emerges anteriorly to become embraced by the common mesentery, pulled upward/suspended by a band of smooth muscle at the duodenojejunal junction. The *jejunum* is highly coiled, suspended by the common mesentery between the peritoneal layers through which travel its blood and nerve supply and draining veins. The thinner but longer *ileum* also is suspended by the common mesentery. It opens into the cecum of the large intestine.

The internal or luminal surface of the small intestine, especially the jejunum, consists of a continuous series of circumferential (ring-like) folds (*plicae circulares*) composed of *mucosal* and *submucosal* tissue. Myriad conical, finger-like projections (*villi*) and deep tubular glands (*intestinal crypts*) characterize the mucosal surface. Simple columnar epithelia, mostly goblet-shaped mucous cells and absorptive cells, line the villi and crypts. In the crypts, the cells are secretory and produce a watery medium, enhancing uptake of minerals and nutrients. *Enteroendocrine* cells secrete a number of hormones that encourage glandular secretion (e.g., cholecystokinin and secretin). Potentially phagocytic *paneth cells* secrete lysozymes into the broth in the deep crypts. This digestive enzyme destroys bacterial cell walls. The loose fibrous, vascular *lamina propria* supports the lacteal-, blood vessel-, and axon-containing villi and the glands of the crypts. The *submucosa* supports large blood/lymph vessels and the cell bodies/axons of *parasympathetic neurons*. Both submucosa and lamina propria contain masses of lyphoid nodules (Peyer's patches; recall Plate 127). Specialized epithelial M or membranous cells at the epithelium-lymphoid nodule interface play a role in taking antigen to immune-reactive lymphocytes. In the duodenum, the glands of Brünner of the submucosa secrete a bicarbonate-containing mucus that neutralizes the hydrochloric acid entering from the stomach.

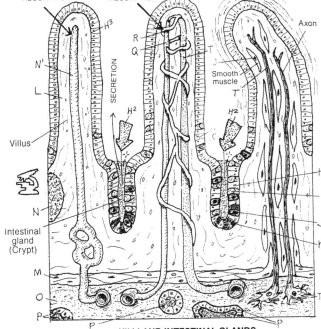

ABSORPTION
ABSORPTION
Axon
SECRETION
Smooth muscle
Villus
Intestinal gland (Crypt)

VILLI AND INTESTINAL GLANDS
(Structures common to all villi are shown separately)

LARGE INTESTINE

CN: By using the same colors for the parts of the intestinal wall you used on the preceding plate, you can demonstrate the similarity between the structures of the two intestines. The epithelium/mucous glands (N) should receive the same color as the villi (H[1]) of Plate 142, and the serosa in both plates should receive the same color. Use a very light color for B. (1) Begin with the section above.

LARGE INTESTINE
CECUM A
ILEOCECAL VALVE B
VERMIFORM APPENDIX C
ASCENDING COLON D
TRANSVERSE COLON E
DESCENDING COLON F
SIGMOID (PELVIC) COLON G
RECTUM H
ANAL CANAL I
INTERNAL ANAL SPHINCTER J
EXTERNAL ANAL SPHINCTER K

TAENIA COLI L
OMENTAL APPENDICES M

INTESTINAL WALL
MUCOSA
EPITHELIUM/MUCUS GLANDS N
LAMINA PROPRIA O
MUSCULARIS MUCOSAE P
SUBMUCOSA Q
MUSCULARIS EXTERNA
CIRCULAR M. R
LONGITUDINAL M. L'
SEROSA D'

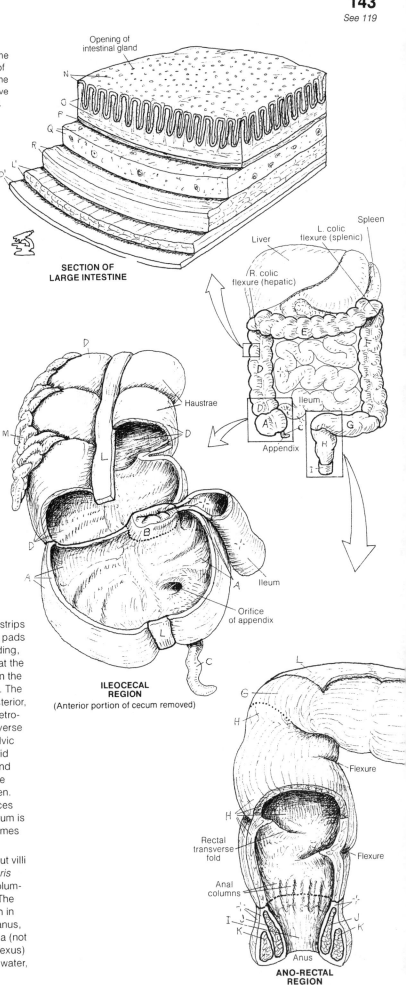

Opening of intestinal gland

SECTION OF LARGE INTESTINE

Spleen
L. colic flexure (splenic)
Liver
R. colic flexure (hepatic)
Ileum
Appendix

Haustrae
Ileum
Orifice of appendix

ILEOCECAL REGION
(Anterior portion of cecum removed)

Flexure
Rectal transverse fold
Flexure
Anal columns
Anus

ANO-RECTAL REGION

The large intestine is characterized by large sacculations (haustrae), strips of *longitudinal muscle* in the muscularis externa (*taenia coli*), and fat pads (*appendices epiploica*) attached to the serosal surface of the ascending, transverse, and descending colon (only). The large intestine begins at the *ileocecal valve* with the *cecum*, usually suspended by a mesentery, in the right lower abdominal quadrant. The function of the valve is not clear. The *vermiform appendix* varies in length (2–20 cm); it may lie anterior, posterior, or inferior to the cecum. The *ascending* and *descending colons* are retroperitoneal; the *transverse colon* is suspended by a mesentery (transverse mesocolon). Note the colic flexures and their relationships. At the pelvic inlet (not shown), the colon turns medially, gains a mesentery (sigmoid mesocolon), and is named the *sigmoid colon*. Variable in its extent and shape, it becomes the *rectum* at the level of the S3 vertebra. Here the haustrae, the appendices epiploica, and the taenia are no longer seen. About 12 cm long, the rectum has a dilated lower part (ampulla). Feces entering the rectum stimulate the desire for defecation; thus, the rectum is not a long-term storage site. As the rectum narrows inferiorly, it becomes the *anal canal* surrounded by *sphincter muscles*.

The wall of the large intestine is characteristic: *mucosal* surface without villi or plicae, underlying vascular *submucosa*, and two-layered *muscularis externa* lined with peritoneal *serosa*. The epithelial lining is simple columnar except in the anal canal where it becomes stratified squamous. The glands are tubular and mucus-secreting. Lymphoid nodules are seen in the lamina propria. At the anorectal junction, about 2 cm above the anus, a remarkably large number of veins can be seen in the lamina propria (not shown). Varicose dilatations of these veins (rectal or hemorrhoidal plexus) are called hemorrhoids. The large intestine functions in absorption of water, vitamins, and minerals as well as the secretion of mucus.

LIVER

CN: Use blue for I, red for J, and yellow for K. Use very light colors for A, B, and L. (1) Color the two upper views simultaneously. (2) Color the group of lobules, and then the enlargement. Begin with the branches of the portal vein (I^1) at the bottom of the section. (3) Begin the overview of blood and bile with the arterial flow.

LOBES
- RIGHT LOBE A
- LEFT LOBE B
- QUADRATE LOBE C
- CAUDATE LOBE D

LIGAMENTS
- CORONARY L. E
- TRIANGULAR L. F
- LESSER OMENTUM G
- FALCIFORM L. H

PORTA
- PORTAL VEIN I
- HEPATIC ARTERY J
- BILE DUCT K

LIVER LOBULE L
- TRIAD *'
    - BRANCH OF PORTAL V. I'
    - BRANCH OF HEPATIC A. J'
    - BILE DUCT K'
- SINUSOID I^2
- HEPATIC CELL L'
- CENTRAL VEIN I^3
- TRIBUTARY OF HEPATIC V. I^4

The liver is the largest gland in the body. Wedge-shaped (rounded upper border, thin, sharp inferior border) when seen from the side, the liver occupies the whole of the upper right quadrant of the abdominal cavity. Weighing about 1.5 kg in health, it can weigh over 10 kg when diseased (chronic cirrhosis). It is relatively large in young children, causing protuberance of the upper abdomen. The liver is enveloped in visceral peritoneum, except for a part of the right posterior surface that is flush against the fascia covering the diaphragm (bare area). The visceral peritoneum around the bare area turns or reflects upward (*coronary ligaments*) on to the diaphragm to become parietal peritoneum. The edges of the coronary ligaments are called the *triangular ligaments*. The two anterior leaves of the coronary ligaments join to become the *falciform ligament*; the two posterior leaves become the *lesser omentum*, which encircles the *porta*, of the liver. Here the hepatic *portal vein* and the *hepatic artery* approach the visceral surface of the liver and branch, and the common bile duct receives the common hepatic duct and cystic duct. The two-layered lesser omentum descends to support the pyloric end of the stomach and the first part of the duodenum. The falciform ligament is continuous with the parietal peritoneum of the anterior abdominal wall.

Connective tissue septa divide the liver cells/tissue (parenchyma) into irregular polyhedral *lobules*. Within each lobule, the cells form radially arranged cords; on two surfaces of these cords are sinusoids that converge onto a more or less *central vein*. At the corners of the lobules are the hepatic artery and portal vein branches, and bile ducts (called a *triad*). The portal vein branches feed into the sinusoids; the hepatic artery branches supply the cells; the *bile ducts* drain the bile ductules formed from tiny canaliculi (not shown) surrounding the cells. The liver cells discharge their products into the sinusoids (except bile) and absorb from the same sinusoids various nutrients and non-nutrients as well. Liver cells store and release proteins, carbohydrates, lipids, iron, and certain vitamins (A, D, E, K); they manufacture urea from amino acids, and bile from pigments and salts; they detoxify many harmful ingested substances. Bile is released from the cells into tributaries of bile ducts. The central veins are tributaries of larger veins that merge to form three *hepatic veins* at the posterior, superior aspect of the liver. These veins join the inferior vena cava just inferior to the diaphragm.

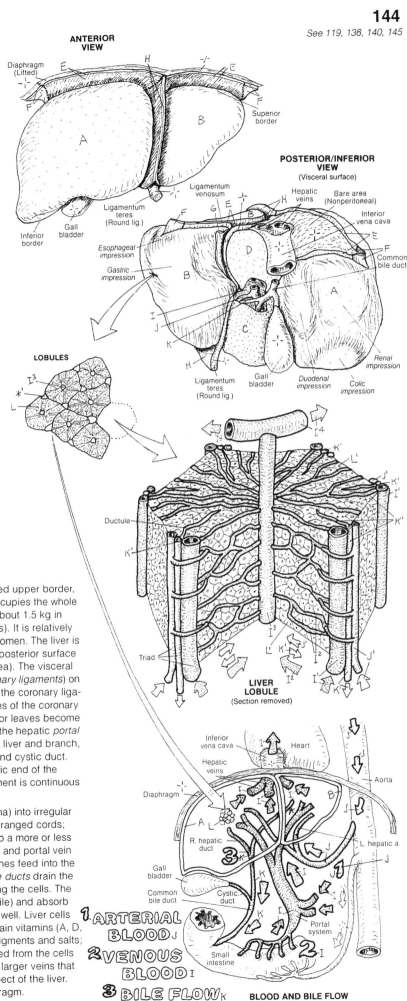

ANTERIOR VIEW

POSTERIOR/INFERIOR VIEW
(Visceral surface)

LOBULES

LIVER LOBULE
(Section removed)

BLOOD AND BILE FLOW
(Schematic)

1 ARTERIAL BLOOD J
2 VENOUS BLOOD I
3 BILE FLOW K

BILIARY SYSTEM & PANCREAS

LIVER (HEPATIC) CELL A
    BILE B
      R. c & L. HEPATIC DUCT c'
        COMMON HEPATIC DUCT D

GALL BLADDER E
      CYSTIC DUCT F

COMMON BILE DUCT G

PANCREAS H
      PANCREATIC DUCT I

HEPATO-DUODENAL AMPULLA J

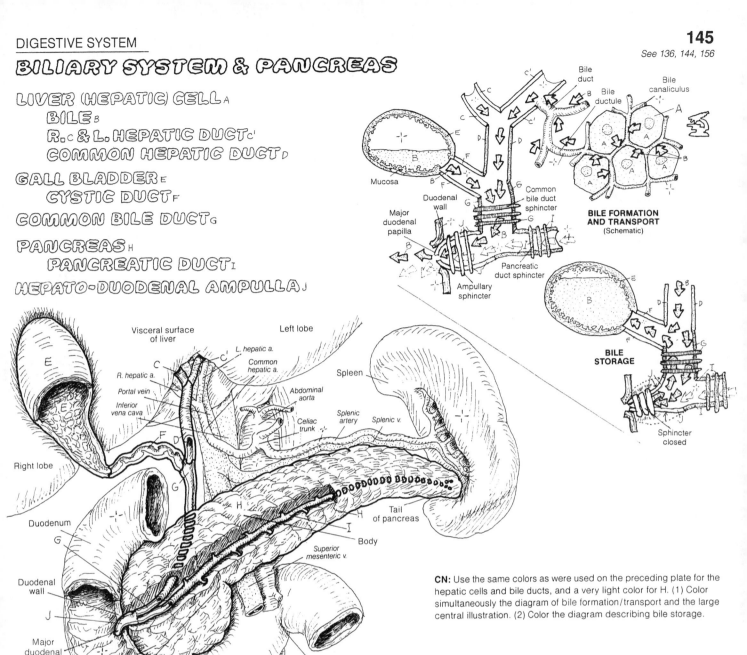

BILE FORMATION AND TRANSPORT
(Schematic)

BILE STORAGE

Sphincter closed

ANTERIOR VIEW
(Stomach removed)

CN: Use the same colors as were used on the preceding plate for the hepatic cells and bile ducts, and a very light color for H. (1) Color simultaneously the diagram of bile formation/transport and the large central illustration. (2) Color the diagram describing bile storage.

The biliary system consists of an arrangement of *ducts* transporting *bile* from the *liver cells* that manufacture it to the *gall bladder* and to the second part of the duodenum.

It is worth repeating: bile is formed in the liver (not the gall bladder). It is a fluid consisting largely of water (97%), with bile salts and pigments (from the breakdown products of hemoglobin in the spleen). Once formed, bile is discharged from liver cells into surrounding *bile canaliculi*. These small canals merge to form bile ductules that join the bile ducts, which travel in company with the branches of the portal vein and hepatic artery. The bile is brought out of the liver by the *right and left hepatic ducts,* which merge at the porta to form the *common hepatic duct.* That duct descends between the layers of the lesser omentum and receives the 4-cm-long *cystic duct* from the *gall bladder.* The gall bladder is pressed against the visceral surface of the right lobe of the liver, covered with visceral peritoneum. The *common bile duct* (or just bile duct) is formed by the cystic and common hepatic duct. About 8 cm long, it descends behind the first part of the duodenum, deep to or through the head of the pancreas. It usually joins with the main *pancreatic duct,* forming an *ampulla* in the wall of the second part of the duodenum. Here the duct opens into the lumen of the duodenum. There can be variations in the union of these two ducts.

The gall bladder serves as a storage chamber for bile discharged from the liver. Bile is concentrated here several times, a fact reflected in the multiple microvilli on the luminal surfaces of the simple columnar epithelial cells that absorb water from the dilute bile. In response to the gastric or duodenal presence of fat, secretion of cholecystokinin is induced, which stimulates the gall bladder to discharge its contents into the cystic duct. Peristaltic contractions of the duct musculature squirt bile into the duodenal lumen through the ampullary sphincter. Bile saponifies and emulsifies fats, making them water soluble and amenable to digestion by enzymes (lipases).

The pancreas is a gland in the retroperitoneum, consisting of a head, neck, body, and tail. Most of the pancreas consists of exocrine glands that secrete enzymes into the pancreatic duct tributaries and on into the duodenum at a rate of about 2000 ml per day. These enzymes are responsible for a major part of chemical digestion in the small intestine (lipases for fat, peptidases for protein, amylases for carbohydrates, and others). Pancreatic secretion is regulated by hormones (primarily cholecystokinin and secretin) from entero-endocrine cells and by the vagus nerves (acetylcholine). The endocrine portion of the pancreas is covered in Plate 156.

URINARY TRACT

URINARY TRACT -:-
- **KIDNEY** A
- **URETER** B
- **URINARY BLADDER** C
- **URETHRA** D
  - **PROSTATIC U. (MALE)** D¹
  - **MEMBRANOUS U. (MALE)** D²
  - **SPONGY U. (MALE)** D³

KIDNEY RELATIONS -:-
- **SUPRARENAL (ADRENAL) GLAND** E
- **LIVER** F
- **DUODENUM** G
- **TRANSVERSE COLON** H
- **SPLEEN** I
- **STOMACH** J
- **PANCREAS** K
- **JEJUNUM** L

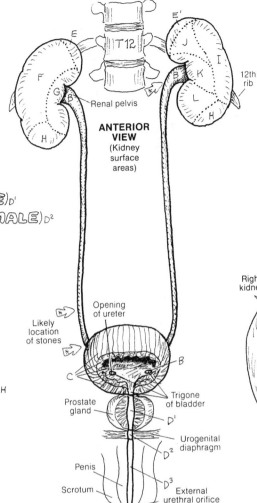

ANTERIOR VIEW
(Kidney surface areas)

T 12

Renal pelvis

12th rib

Likely location of stones

Opening of ureter

Trigone of bladder

Prostate gland

Urogenital diaphragm

Penis

Scrotum

External urethral orifice

CN: Use very light colors for C and E–L.
(1) Color the three views of the urinary tract simultaneously. Note that the kidneys at the top of the plate are to be colored according to areas that are in contact with other organs. Also note that the ureters penetrate the posterior wall of the urinary bladder, and that these openings receive a color. (2) Color the anterior relations of the kidneys. The kidneys are shown as underlying, shaded silhouettes and receive no color of their own. (3) Color gray the arrows marking sites of potential obstruction by "stones."

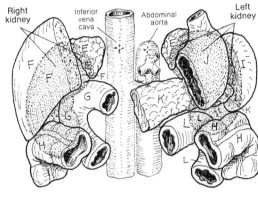

ANTERIOR RELATIONS OF THE KIDNEYS

Right kidney

Inferior vena cava

Abdominal aorta

Left kidney

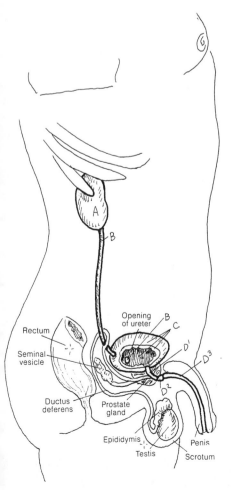

Rectum

Seminal vesicle

Opening of ureter

Ductus deferens

Prostate gland

Epididymis

Testis

Penis

Scrotum

The urinary tract consists of paired *kidneys* and *ureters* in the retroperitoneum, a single *urinary bladder*, and a *urethra*. The urinary tract represents a pathway for the elimination of metabolic by-products and toxic and other nonessential molecules, all dissolved in a small volume of water (urine). The *kidneys* are not simply instruments of excretion; they function in the conservation of water and maintenance of acid-base balance in the blood. The process is a dynamic one, and what is excreted as waste in one second may be retained as precious in the next.

The ureters are fibromuscular tubes, lined by transitional epithelium. Three areas of the ureters are relatively narrow and are prone to being obstructed by mineralized concretions ("stones") from the kidney (see arrows).

The fibromuscular urinary bladder lies in the true pelvis, its superior surface covered with peritoneum. The mucosa is lined with transitional epithelium. The bladder can contain as little as 50 ml of urine and can hold as much as 700–1000 ml without injury; as it distends, it rises into the abdominal cavity and bulges posteriorly. The mucosal area between the two ureteral orifices and the urethral orifice is called the trigone.

The fibromuscular, glandular urethra, lined with transitional epithelium except near the skin, is larger in males (20 cm) than females (4 cm). Hence, urethritis is more common in men, cystitis is more common in women.

The urethra is described in three parts in the male (*prostatic, membranous,* and *spongy*). The membranous urethra is vulnerable to rupture in the urogenital diaphragm with trauma to the low anterior pelvis.

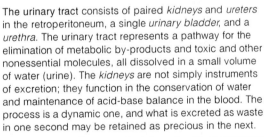

12th rib

Ovary

Uterine tube

Opening of ureter

Uterus

Vagina

Rectum

KIDNEYS & RELATED RETROPERITONEAL STRUCTURES

CN: Use red for B, blue for L, and a very light color for X (use a color, not gray). (1) Color the various structures in the abdominal cavity. Part of the peritoneum (X), whose title is among the upper diagrams, is shown covering much of the right side. (2) At the upper right, note the relationship of the retroperitoneum to the parietal peritoneum.

KIDNEY A
URETER A'
URINARY BLADDER A²

AORTA B & BRANCHES ⫶
CELIAC A. & BRANCHES C
SUPRARENAL A. D
SUP. MESENTERIC A. E
RENAL A. F
TESTICULAR A. G
INF. MESENTERIC A. H
COMMON ILIAC A. I
INTERNAL ILIAC A. J
EXTERNAL ILIAC A. K

INFERIOR VENA CAVA L
& TRIBUTARIES ⫶
INTERNAL ILIAC V. M
EXTERNAL ILIAC V. N
COMMON ILIAC V. O
TESTICULAR V. P
RENAL V. Q
SUPRARENAL V. R
HEPATIC VS. S

ORGANS & DUCTS ⫶
ESOPHAGUS T
SUPRARENAL GLAND U
RECTUM V
DUCTUS (VAS) DEFERENS W

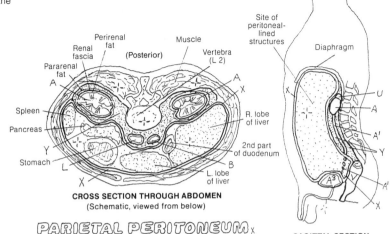

CROSS SECTION THROUGH ABDOMEN
(Schematic, viewed from below)

PARIETAL PERITONEUM X
RETROPERITONEUM Y

**SAGITTAL SECTION
THROUGH TRUNK**
(Schematic)

The paired kidneys and ureters lie posterior to the *parietal peritoneum* of the abdominal cavity; they are, therefore, in the *retroperitoneum*. During fetal development, some abdominal structures arise in the retroperitoneum (e.g., kidneys), and some become retroperitoneal as a result of movement of visceral organs (e.g., ascending/descending colon, pancreas). The abdominal *aorta and its immediate branches* and the *inferior vena cava and its immediate tributaries* all are retroperitoneal. Arteries and veins travel between layers of peritoneum to reach the organs they supply/drain. Lymph nodes, lumbar trunks, and the cysternal chyli (not shown) all are retroperitoneal. The ureters descend in the retroperitoneum and under the parietal peritoneum to reach the posterior and inferior aspect of the bladder. Pelvic viscera and vessels lie deep to the parietal peritoneum.

The kidneys are encapsulated in perirenal fat, secured by an outer, stronger layer (renal fascia). Each kidney and its fascia are packed in pararenal fat. These compartments do not communicate between left and right. Such a support system permits kidney movement during respiration but secures them against impact forces.

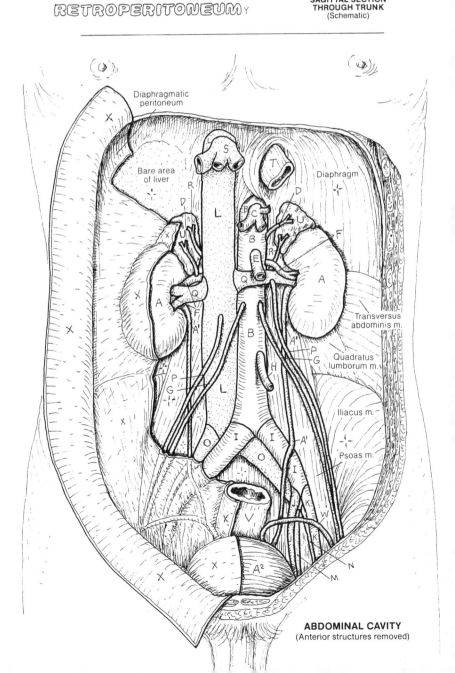

ABDOMINAL CAVITY
(Anterior structures removed)

KIDNEY & URETER

CN: Use red for J, blue for K, yellow for P, and very light colors for B, F, G, H, and I. (1) Begin with the large illustration and note that the thickness of the renal capsule (A) has been greatly exaggerated for coloring purposes. Color the cut edges of blood vessels in the cortex (B). Also color the titles and arrows reflecting blood and urine flow. (2) Color the large arrow (E) pointing to the concavity of the bean-shaped kidney, the renal hilum.

KIDNEY:

- RENAL CAPSULE A
- RENAL CORTEX B
- RENAL MEDULLA (PYRAMID) C
    - RENAL PAPILLA D
- RENAL HILUM E
    - MINOR CALYX F
    - MAJOR CALYX G
    - RENAL PELVIS H
    - RENAL SINUS I

RENAL ARTERY J
   OXYGEN-RICH BLOOD J'
RENAL VEIN K
   OXYGEN-POOR BLOOD K'

URETER L
  MUCOSA: TRANSITIONAL
    EPITHELIA M
  MUCOSA: LAMINA PROPIA N
  MUSCULARIS: CIRC. O / LONG. O'
  ADVENTITIA L'

URINE P

1300 mL/min J'
(Into both kidneys)

1299 mL/min K'
(Out of both kidneys)

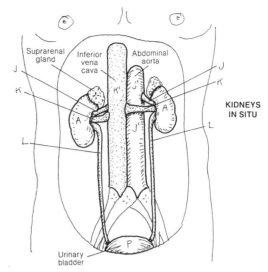

Suprarenal gland — Inferior vena cava — Abdominal aorta

KIDNEYS IN SITU

Urinary bladder

KIDNEY
(schematic coronal section)

Renal column

Blood vessels

Minor and major calyces are cut open

0.7 mL/min P

Circular muscle

URETER
(cross section)

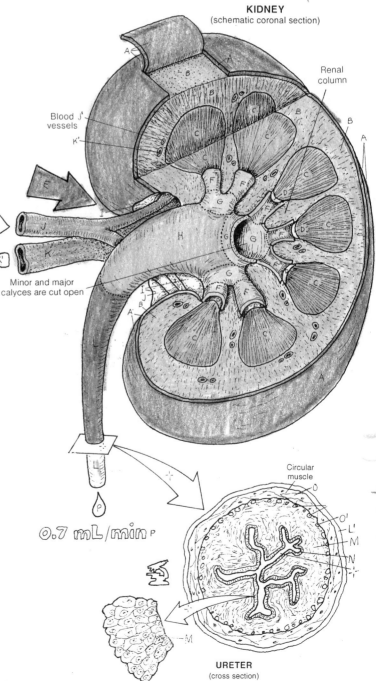

The **kidney** consists of filtering capsules, tubules, and blood vessels tightly pressed together into what is called the parenchyma. The parenchyma of the kidney consists of an outer *cortex* covered on its surface by a thin fibrous *capsule*, and an inner *medulla* consisting of pyramids of straight tubules. The cortex reaches down between the pyramids (renal columns). The cortex consists of convoluted tubules and filtering capsules. The apex of each medullary pyramid forms a *papilla* that fits into the small cup-shaped funnel called the *minor calyx*. These funnels, numbering 8–18, open into three much larger *major calyces,* all of which open into the cavity called the *renal pelvis*. In the concavity of the kidney (the hilum), in an area called the *renal sinus*, the renal pelvis narrows to form the proximal *ureter,* sharing the area with the renal artery and vein.

Renal blood flow (the amount of blood flowing through the kidneys) is about 1300 mL per minute (both kidneys). About 125–130 mL of plasma is filtered into the renal tubular systems each minute. Less than 1% of that filtered plasma (about 0.7 mL) is actually excreted as *urine*. Clearly, the kidney is in the water conservation business!

The structure of the ureter is a continuation of the renal pelvis. The epithelial layer is *transitional*, a variably stratified layer dependent upon the volume of urine in the ureter. The fibrous *lamina propria* supports the epithelia and blood vessels as well as nerves. The *muscular coat* consists of both longitudinal (inner) and circular layers. The outer layer of the urethra (*adventitia*) is fibrous.

RENAL TUBULE:-

CN: Use red for G, yellow for M, and a very light color for H. Some colors will be used for the same structures on Plate 150.
(1) Begin with kidney regions. (2) Color the two types of nephrons. (3) Color the detailed view of the cortical nephron. (4) At the bottom of the page, the capillary-like vessels of the glomerulus (G^1) are largely covered by podocytes (visceral layer, H^2); but not entirely. Color both; see smaller drawing where podocytes have been removed to reveal a porous capillary. The capsular space (H^3) is left uncolored.

RENAL TUBULE:-
NEPHRON:-
CORTICAL NEPHRON E
JUXTAMEDULLARY NEPH. F
RENAL CORPUSCLE:-
AFFERENT ARTERIOLE G
GLOMERULUS G¹
GLOMERULAR CAPSULE H
####### PARIETAL LAYER H¹
####### VISCERAL LAYER H²
####### CAPSULAR SPACE H³:-
EFFERENT ARTERIOLE G²
PROXIMAL CONVOLUTED TUBULE I
LOOP OF HENLE J
DISTAL CONVOLUTED TUBULE K
COLLECTING TUBULE L
COLLECTING DUCT L¹
PAPILLARY DUCT L²
URINE M

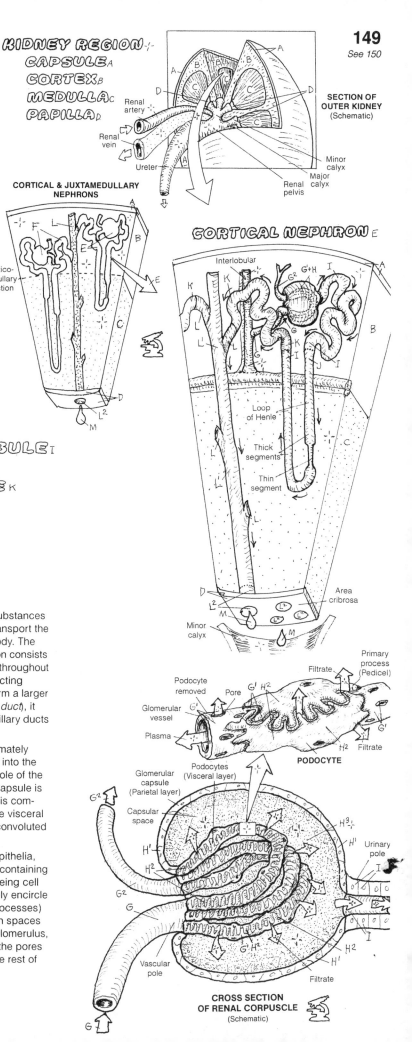

KIDNEY REGION:-
CAPSULE A
CORTEX B
MEDULLA C
PAPILLA D

Renal artery / Renal vein / Ureter

SECTION OF OUTER KIDNEY (Schematic)

Minor calyx / Major calyx / Renal pelvis

See 150

CORTICAL & JUXTAMEDULLARY NEPHRONS

Cortico-medullary junction

CORTICAL NEPHRON E

Interlobular

Loop of Henle / Thick segments / Thin segment

Area cribrosa

Minor calyx

Primary process (Pedicel)

Filtrate

Podocyte removed / Glomerular vessel / Plasma / Pore

PODOCYTE

Podocytes (Visceral layer)

Glomerular capsule (Parietal layer) / Capsular space

Urinary pole

Vascular pole

Filtrate

CROSS SECTION OF RENAL CORPUSCLE (Schematic)

The function of the renal (uriniferous) tubule is to (1) extract certain substances from the blood, (2) return certain substances to the blood, and (3) transport the leftovers to the urinary bladder for storage and expulsion from the body. The renal tubule consists of a *nephron* and a *collecting tubule*. The nephron consists of a renal corpuscle and a tubule that is both convoluted and straight throughout its course. At its termination, the nephron is continuous with the collecting tubule. Each collecting tubule joins with other collecting tubules to form a larger *collecting duct*. Passing through the papilla of the medulla (*papillary duct*), it empties into the minor calyx. Each papilla has multiple orifices of papillary ducts (area cribrosa).

The renal corpuscle consists of a *glomerular (Bowman's) capsule* intimately related to a tuft of capillaries (*glomerulus*). An *afferent arteriole* leads into the glomerulus and an *efferent arteriole* departs it, both at the vascular pole of the corpuscle (see next plate). The inner, indented wall of the epithelial capsule is its visceral layer; the outer wall is its parietal layer. The visceral layer is complexly interwoven with the glomerular vessels. The space between the visceral and parietal layers is the capsular space; it opens into the proximal convoluted tubule at the urinary pole of the renal corpuscle.

The cells of the visceral layer are highly modified simple squamous epithelia, called *podocytes*. Each has the shape of a centipede, with its "body" containing the nucleus and its multiple "legs" (primary processes or pedicels) being cell membrane–lined cytoplasmic extensions. These pedicels incompletely encircle a vessel, creating filtration slits. The "legs" have "feet" (secondary processes) that attach to the porous vascular wall and reduce further the filtration spaces "between the toes" (not shown). As the blood circulates through the glomerulus, plasma and its noncellular solutes (less proteins) are driven through the pores and slits to enter the capsular space. The filtrate's journey through the rest of the renal tubule is shown and described in the following plate.

TUBULAR FUNCTION & RENAL CIRCULATION

CN: Use the same colors as were used on the preceding plate for A, B, C, M, and F. Use red for D, blue for G, and purple for E. (1) Begin with the major blood vessels of the kidney. (2) Color the entire enlarged nephron the color (F) from the preceding plate. Note that the afferent and efferent arterioles, although part of the dotted nephron, receive arterial colors (D^5, D^6). When coloring the inset of the juxtaglomerular apparatus, leave the relevant cells blank, for easier identification.

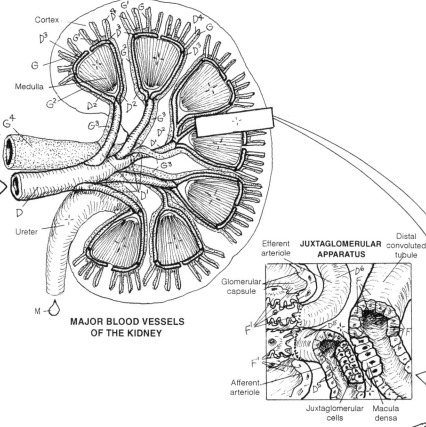

MAJOR BLOOD VESSELS OF THE KIDNEY

Cortex · Medulla · G^4 · D · Ureter · M

JUXTAGLOMERULAR APPARATUS

Efferent arteriole · Glomerular capsule · Afferent arteriole · Juxtaglomerular cells · Macula densa · Distal convoluted tubule

ARTERIES :·
RENAL A. D
SEGMENTAL A. D^1
INTERLOBAR A. D^2
ARCUATE A. D^3
INTERLOBULAR A. D^4

AFFERENT ARTERIOLE D^5
GLOMERULUS F^1
EFFERENT ARTERIOLE D^6
PERITUBULAR CAPIL. PLEXUS E
VASA RECTA E^1

VEINS :·
INTERLOBULAR V. G
ARCUATE V. G^1
INTERLOBAR V. G^2
SEGMENTAL V. G^3
RENAL V. G^4

JUXTAMEDULLARY NEPHRON F

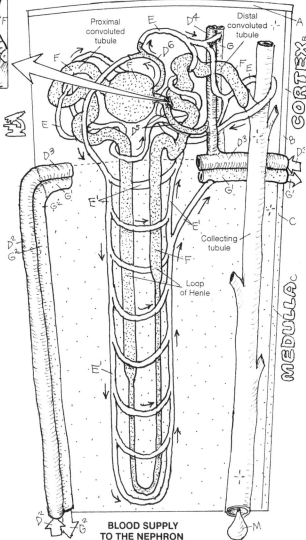

Proximal convoluted tubule · Distal convoluted tubule · CORTEX · MEDULLA · Collecting tubule · Loop of Henle

BLOOD SUPPLY TO THE NEPHRON

The **juxtamedullary nephron** shown here lies near the cortico-medullary border. Unlike nephrons of the deeper cortex, its descending tubules reach into the deep medulla, and its vascular relations are different. Of the 0.4–1 million or so nephrons in each kidney, about 30% are juxtamedullary in location.

In juxtamedullary nephrons, the efferent arteriole leads directly to the *vasa recta* (straight vessels) arranged close to the long tubules in the medulla. In the cortical nephrons, the efferent arteriole leads directly to the *peritubular capillary plexus*, which embraces the entire neuron. In some nephrons, both a peritubular capillary plexus and vasa recta may exist, as shown here. The venous side of the vasa recta flows into veins that may merge with the arcuate veins or continue up into the lower cortex to join the interlobular vein. The venous side of the peritubular capillary plexus joins the interlobular vein. It is the close relationship of tubule to blood vessel that makes possible the preservation of what is needed at any one moment and the rapid ejection of what is not needed, e.g., toxic substances.

The renal corpuscle filters the blood plasma. The resulting filtrate is discharged into the proximal tubule, where tubular reabsorption and tubular secretion begin. Water, sodium (Na^+), glucose, and amino acids are rapidly reabsorbed by the tubular cells. The descending thin segment of the *loop of Henle*, composed of simple squamous epithelia, also reabsorbs water and electrolytes, principally by simple diffusion. The thin and thick segments of the ascending limb of the loop, however, are largely impermeable to water but actively reabsorb sodium, chloride, and other ions, leaving very dilute water going into the *distal tubule*.

The first part of the distal tube comes into contact with the afferent arteriole of the glomerulus of origin. Modified smooth muscle, renin-secreting, juxtaglomerular (JG) cells in the arteriole are sensitive to arteriolar blood pressure. Modified epithelial cells in the adjacent distal tubule (macula densa) sense the solute content in the filtrate at that site. By their interaction, the glomerular filtration rate and blood pressure can be modulated. The distal tubule is impermeable to water while reabsorbing electrolytes. The dilute tubular fluid entering the collecting tubule is reabsorbed under the influence of anti-diuretic hormone (ADH). Fluid is continuously reabsorbed until the residual is discharged as urine at the papilla. For every 100 ml of filtrate formed, 1 ml of it will find the minor calyx.

INTRODUCTION

C, I: Use a very light color for C and a darker one for D (actually located on posterior surface of thyroid). (1) After coloring endocrine glands and tissues, color the scheme at lower left.

ENDOCRINE GLANDS
HYPOPHYSIS (PITUITARY) A
PINEAL B
THYROID C
PARATHYROID (4) D
THYMUS E
ADRENAL (SUPRARENAL) (2) F
PANCREAS G
OVARY (2) H
TESTIS (2) I

ENDOCRINE TISSUES
HYPOTHALAMUS J
HEART (ATRIA) K
KIDNEY (2) L
GASTROINTESTINAL TRACT M
PLACENTA N

Endocrine glands and tissues are discrete masses of secretory cells and their supporting tissues in close proximity to blood capillaries, into which the cells secrete their hormones. The glands and tissues are ductless. Hormones are chemical agents usually effective among cells (target organs) located some distance from their source. Hormonal secretion results in negative or positive feedback control mechanisms. In the broader scope, hormonal activity results in growth, reproduction, and related activity as well as metabolic stability in the internal environment. Stability of the internal environment is called homeostasis.

The classical endocrine glands listed and shown here are presented in the following plates, with the exception of the pineal gland (see Plate 75) and the thymus (see Plate 124). Also listed here are just a few of the myriad tissues/cells that secrete chemical agents influential in cellular activities. The role of the hypothalamus can be colored in Plates 152 and 153. The atria of the heart secrete atrial natriuretic peptide (ANP) during periods of weak myocardial contraction, resulting in increased excretion of sodium and water. The juxtaglomerular cells of the kidney (Plate 150) secrete renin, an enzyme that converts angiotensinogen to angiotensin I and indirectly induces increased blood pressure and conservation of body fluids, such as during hemorrhage. Numerous endocrine factors secreted by cells of the gastrointestinal tract influence intestinal motility and enzyme secretion. The placenta secretes, among many hormones, human chorionic gonadotropin, which contributes to the support of embryonic growth during the first 90 days post-fertilization by stimulating the growth of the corpus luteum (Plates 161, 163, 165, 166).

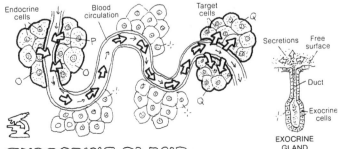

ENDOCRINE FUNCTION

Endocrine cells
Blood circulation
Target cells

Secretions
Free surface
Duct
Exocrine cells

EXOCRINE GLAND

ENDOCRINE GLAND O
HORMONAL SECRETION P
TARGET ORGAN Q

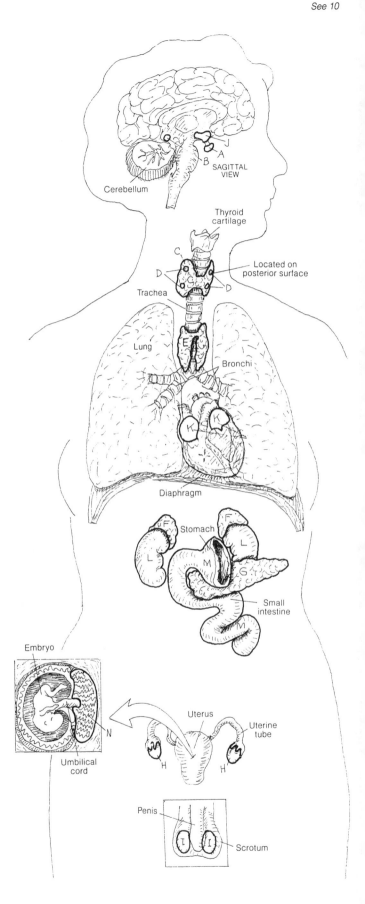

SAGITTAL VIEW
Cerebellum
Thyroid cartilage
Located on posterior surface
Trachea
Lung
Bronchi
Diaphragm
Stomach
Small intestine
Embryo
Umbilical cord
Uterus
Uterine tube
Penis
Scrotum

PITUITARY GLAND & HYPOTHALAMUS

See 153

CN: Use red for H, blue for K, purple for I, and a very light color for J.
(1) Begin with the enlarged view of the hypophysis and hypothalamus.

PITUITARY GLAND (HYPOPHYSIS):
ADENOHYPOPHYSIS:
PARS TUBERALIS A
PARS DISTALIS (ANT. LOBE) B
PARS INTERMEDIUS C
NEUROHYPOPHYSIS:
MEDIAN EMINENCE D
INFUNDIBULAR STEM E
PARS NERVOSA (POST. LOBE) F

HYPOTHALAMUS G

The pituitary gland (hypophysis) essentially consists of an anterior and a posterior lobe. It is suspended from the hypothalamus of the brain and fits into a recess of the sphenoid bone called the sella turcica. The pituitary is about the size of four peas (thank you, Dr. Marian Diamond). The three parts of the *adenohypophysis* are derived from an upward extension of the developing roof of the mouth; indeed, the gland was once thought to form mucus ("pituita") that was secreted into the nose. The posterior lobe is a downward migration from the floor of the hypothalamus. That floor, inferior to the third ventricle, consists of the hollow infundibulum (stem, stalk) surrounded by the *median eminence*. The lowest part of the infundibulum (below the median eminence, still part of the hypothalamic floor, but no longer hollow) is continuous with the posterior lobe. The three (infundibulum, median eminence, and posterior lobe) are often considered as the "*neurohypophysis.*" Note how the pars tuberalis of the adenohypophysis embraces the *infundibular stem* and median eminence. The pars intermedius is rudimentary and appears to secrete no significant levels of hormone.

ADENOHYPOPHYSIS:
HYPOTHAL. SECR. NEURON/HORMONE G'
SUP. HYPOPHYSEAL ARTERY H
HYPOPHYSEAL PORTAL SYSTEM I
CAPILLARY I' /PORTAL V. I² /SINUSOID I³
SECRETORY CELL J /HORMONES J'
INF. HYPOPHYSEAL VEIN K

The pars distalis contains a variety of cells that secrete several hormones (see next plate). There are three types that can be differentiated by staining: the acidophils, basophils (both chromophillic cells), and chromophobes. The secretory activity of these cells is stimulated by neurons of the hypothalamus. These neurons secrete releasing *hormones* into the vascular *hypophyseal portal system* at the level of the median eminence (next plate). Capillaries there drain into portal veins that reach the *sinusoids* of the pars distalis. Secretions from the cells of the pars distalis enter the sinusoids, which are drained by the *inferior hypophyseal* vein.

NEUROHYPOPHYSIS:
HYPOTHAL. SECRETORY NEURONS:
SUPRAOPTIC NUCL./HORMONE G²
PARAVENTRIC. NUCL./HORMONE G³
CAPILLARY PLEXUS L
HYPOPHYSEAL VEIN K'

The pars nervosa of the neurohypophysis has no secretory cells of its own. Axons of secretory neurons in the *supraoptic* and *paraventricular* nuclei of the hypothalamus extend down through the infundibulum to *capillary networks* in the posterior lobe. There these axon terminals release oxytocin and antidiuretic *hormones* into the circulation (see next plate).

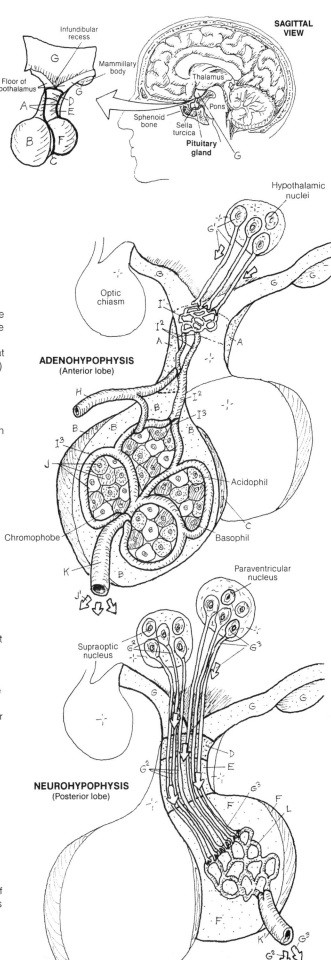

SAGITTAL VIEW

Infundibular recess

Floor of hypothalamus

Mammillary body

Thalamus

Sphenoid bone

Sella turcica

Pituitary gland

Pons

Hypothalamic nuclei

Optic chiasm

ADENOHYPOPHYSIS (Anterior lobe)

Acidophil

Chromophobe

Basophil

Paraventricular nucleus

Supraoptic nucleus

NEUROHYPOPHYSIS (Posterior lobe)

PITUITARY GLAND & TARGET ORGANS

CN: Use the color from the previous plate for hypothalamic hormones (A) and secretions (A¹). Color the major headings distributed throughout the illustration. (1) Begin with the arrows and circles representing those hormones and secretions, including the penetration of the cells of the anterior lobe. (2) Color the pituitary hormones. (3) Color the arrows representing the target organ hormones performing their feedback function.

PITUITARY GLAND HORMONES / SOURCE CELL

PARS DISTALIS

FOLLICLE-STIM. H. (FSH)ʙ /BASOPHIL c
LUTEINIZING H. (LH)ᴅ /BASOPHIL c
THYROID STIM. H. (TSH)ᴇ /BASOPHIL c
ADRENO CORTICOTROPIC H. (ACTH)ꜰ /BASOPHIL c
GROWTH H. (GH)ɢ /ACIDOPHIL ʜ
PROLACTIN ɪ /ACIDOPHIL ʜ

PARS NERVOSA

OXYTOCIN ᴊ
ANTIDIURETIC H. (ADH)ᴋ

TARGET ORGAN HORMONES ᴍ'

ESTROGEN ɴ
PROGESTERONE ᴏ
TESTOSTERONE ᴘ
THYROXINE ᴏ
ADRENAL CORTICAL H.ᴏ ʀ

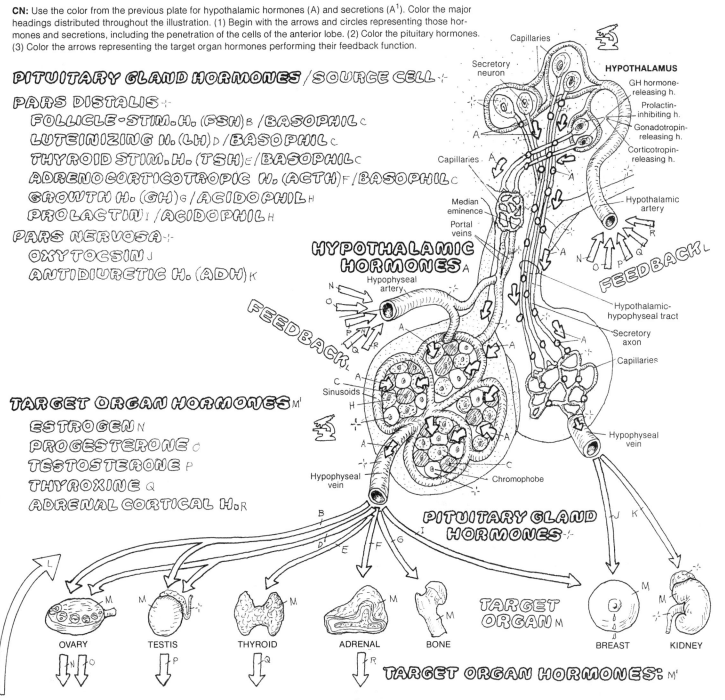

HYPOTHALAMIC HORMONES ᴀ

PITUITARY GLAND HORMONES

TARGET ORGAN ᴍ

OVARY   TESTIS   THYROID   ADRENAL   BONE   BREAST   KIDNEY

TARGET ORGAN HORMONES: ᴍ'

STRUCTURAL/FUNCTIONAL EFFECT/FEEDBACK ʟ

Hypothalamic releasing or inhibiting hormones act on the anterior lobe of the pituitary. These hormones stimulate/inhibit the target cells in the anterior lobe to increase/decrease their secretion of hormone. Inhibition of pituitary hormone secretion is most often controlled by negative feedback. For example, the hypothalamus is sensitive to the concentration of estrogen in the hypothalamic circulation (via the hypothalamic artery). As estrogen levels diminish, certain hypothalamic nuclei sense this and increase their secretion of gonadotropic-releasing hormone (GRH). GRH is released from secretory nerve endings into the hypophyseal portal system in the median eminence. GRH reaches the sinusoids of the anterior lobe and stimulates certain basophils there to secrete follicle-stimulating hormone. FSH is released into the circulation and has a stimulatory influence on the growth of ovarian follicles (as well as spermatogenesis in the male). Significantly increased levels of estrogen are sensed by the hypothalamus (feedback), and the hypothalamus turns off its secretion of GRH (negative feedback).

LH stimulates testosterone secretion, ovulation, development of the corpus luteum, and estrogen/progesterone secretion (Plate 163). TSH induces secretion of the thyroid hormone thyroxin (Plate 154). ACTH stimulates the release of adrenal cortical hormones (e.g., cortisol); it also has melanocyte-stimulating properties, dispersing pigment in the skin (Plate 155). GH stimulates body growth, especially bone. Prolactin mediates milk secretion (Plate 164) and is inhibited by prolactin-inhibiting hormone in the hypothalamus.
Oxytocin and antidiuretic hormone (ADH, vasopressin) are products of secretory neurons in the supraoptic and paraventricular nuclei of the hypothalamus; the secretory material is transported down long axons (hypothalamo-hypophyseal tract) to capillaries in the posterior lobe, where they are released into the general circulation via the hypophyseal vein. Oxytocin induces ejection of milk (Plate 164) and stimulates uterine contractions. ADH (plate 155) causes retention of body water by the kidneys. Its secretion is induced by osmoreceptors in the hypothalamus. ADH is also a powerful vasoconstrictor.

THYROID & PARATHYROID GLANDS

CN: Use red for H, blue for I, light colors for E, F, G, and the same colors as on Plate 151 for A and D. (1) Color the three upper views simultaneously, taking note of the arteries and veins that penetrate the thyroid. (2) Color the microscopic sections of hypoactive and hyperactive thyroid follicles; normal tissue lies between the two extremes. (3) Color the diagram of thyroid and parathyroid function.

THYROID A

THYROID FOLLICLE
FOLLICLE CELL B
COLLOID C
THYROXIN A'

PARATHYROID (4) D
PARATHORMONE D'

RELATED STRUCTURES
TRACHEA E
PHARYNX F
ESOPHAGUS G
ARTERIES H
VEINS I

The thyroid gland, covering the anterior surfaces of the 2nd to 4th *tracheal rings*, is bound by a fibrous capsule whose posterior layer encloses the four *parathyroid glands*. The thyroid gland, composed of right and left lobes connected by an isthmus, consists of clusters of *follicles* (like grapes) supported by loose fibrous tissue rich in blood vessels. A microscopic section through a follicle reveals a single layer of cuboidal epithelial *cells* forming the follicular wall. The follicle contains *colloid*, a glycoprotein (thyroglobulin) produced by the follicle cells. These cells take up thyroglobulin and dismantle it to form a number of hormones, primarily *thyroxin* (T4, tetraiodothyronine). Thyroxin is then secreted into the adjacent capillaries. Thyroid hormones contain iodide (a reduced form of iodine), which is absorbed by the follicle cells from the blood. Thyroxin formation and secretion is encouraged by thyroid-stimulating hormone (TSH) from the hypophysis. The relationship operates on a negative feedback mechanism: increased secretions of thyroxin inhibit further secretion of TSH. Thyroxin increases oxygen consumption in practically all tissues, and thus maintains the metabolic rate. It is involved at many levels in growth and development. Excessive secretion of thyroxin generally results in weight loss, extreme nervousness, and an elevated basal metabolic rate. Hypothyroidism results in diminished mental activity, voice changes, reduced metabolic activity, and the accumulation of mucus-like material under the skin (myxedema), giving a puffy appearance. Like all endocrine glands, the thyroid is highly vascular. The vessels shown here warrant very careful attention when considering undertaking an emergency tracheostomy or cricothyrotomy. Especially note the inferior thyroid veins on the anterior surface of the trachea. The pattern of these vessels is not always predictable.

The parathyroids consist of small buttons of highly vascular tissue containing two cell types, one of which (chief cells) secretes *parathormone* Parathormone maintains plasma calcium levels by inducing osteoclastic activity (bone breakdown), freeing calcium ions. Normal muscle activity and blood clotting depend on normal calcium levels in the plasma. Reduced parathyroid function lowers calcium levels and below certain levels causes muscle stiffness, cramps, spasms, and convulsions (tetany).

Hyoid bone
Thyrohyoid membrane
Thyroid cartilage
Cricothyroid muscle
Inferior thyroid a.
Costocervical trunk
R. subclavian a.
Brachiocephalic a.
R. brachiocephalic v.
Superior vena cava

Superior thyroid a.
Superior thyroid v.
Common carotid a.
Lateral lobe
Isthmus
Middle thyroid v.
Inferior thyroid a.
Internal jugular v.
Common inferior thyroid v.
L. brachiocephalic v.

ANTERIOR VIEW

Aortic arch

LATERAL VIEW

Internal carotid a.
Superior thyroid a.
Superior parathyroid
Inferior parathyroid

POSTERIOR VIEW

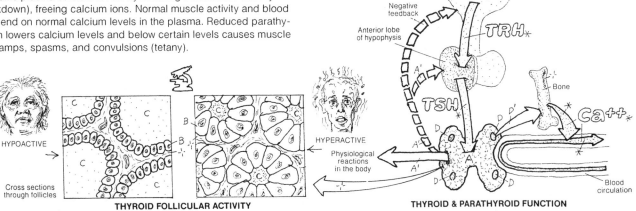

HYPOACTIVE

Cross sections through follicles

THYROID FOLLICULAR ACTIVITY

HYPERACTIVE

Hypothalamus
Negative feedback
Anterior lobe of hypophysis
TRH
TSH
Bone
Ca++
Physiological reactions in the body
Blood circulation

THYROID & PARATHYROID FUNCTION

ADRENAL (SUPRARENAL) GLANDS

CN: Use red for F, blue for G, yellow for H, and a very light color for E. (1) In the upper view, only those vessels with subscripts are to be colored. (2) Color the cross section through the adrenal, and related arrows and hormones. (3) Color the various organs associated with the "fight or flight" reaction, noting the listed effects.

ADRENAL GLAND $_A$
CAPSULE $_{A'}$
CORTEX:
ZONA GLOMERULOSA $_B$
ZONA FASCICULATA $_C$
ZONA RETICULARIS $_D$
MEDULLA $_E$

ARTERIES $_F$ -
SUPERIOR SUPRARENAL A. $_{F'}$
MIDDLE SUPRARENAL A. $_{F^2}$
INFERIOR SUPRARENAL A. $_{F^3}$

VEINS $_G$ -
R. $_{G'}$ & L. SUPRARENAL V. $_{G^2}$

SUPRARENAL PLEXUS $_H$
GREATER SPLANCHNIC N. $_{H'}$
CELIAC GANGLION $_{H^2}$
PLEXUS $_{H^3}$

MINERALOCORTICOIDS $_B$
(INCLUDING ALDOSTERONE)
GLUCOCORTICOIDS $_C$
(INCLUDING CORTISOL)
SEX STEROIDS $_D$
(ESTROGENS, PROGESTERONE, ANDROGENS)

HORMONES OF THE CORTEX

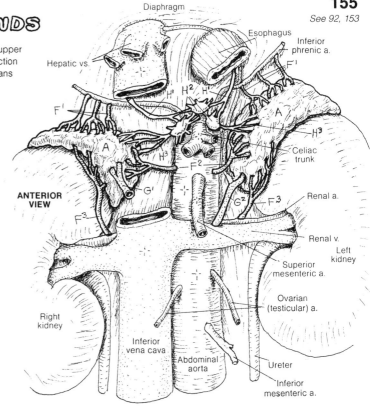

ANTERIOR VIEW

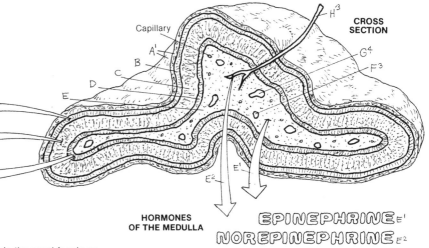

CROSS SECTION

HORMONES OF THE MEDULLA

EPINEPHRINE $_{E'}$
NOREPINEPHRINE $_{E^2}$

The adrenal (suprarenal) glands lie in the retroperitoneum within the renal fascia on the superior and medial aspects of each kidney (T11–T12 vertebral levels). Like other endocrine glands, the adrenals are abundantly vascularized. The adrenals are two different glands encapsulated as one: the outer *cortex* and the inner *medulla*.

The adrenal cortex is organized into three regions: the outer zona glomerulosa, the middle zona fasciculata, and the inner zona reticularis. When presented with a decrease in fluid volume, as with hemorrhage, the cells of the zona glomerulosa synthesize and secrete hormones called mineralocorticoids. The most well known of these is aldosterone. Mineralocorticoids act primarily on the distal tubules of the kidney, the sweat glands, and the gastrointestinal tract; they encourage the absorption of sodium (and water) and the secretion of potassium. Cells of the zona fasciculata, mediated by ACTH, secrete glucocorticoids. These hormones (primarily cortisol and secondarily corticosterone) stimulate the formation of glucose in the liver. Cells of the zona reticularis secrete the androgen dehydroepiandrosterone (DHEA) in small amounts. DHEA can convert to testosterone. The female sex hormones (estrogen and progesterone) are also secreted in small amounts. These adrenal androgens and estrogens have a limited effect during a lifetime.

The medulla consists of cords of secretory cells supported by reticular fibers, and an abundant collection of capillaries. Fibers of the greater splanchnic nerve pass through the celiac ganglia without synapsing to enter the adrenal gland. These fibers terminate on and stimulate the medullary secretory cells, 80% of which produce and release epinephrine; the rest secrete norepinephrine. These secretory cells are, in fact, modified post-ganglionic neurons. Their secretions elicit the "fight or flight" reaction in response to life-threatening situations, as diagrammatically represented at right.

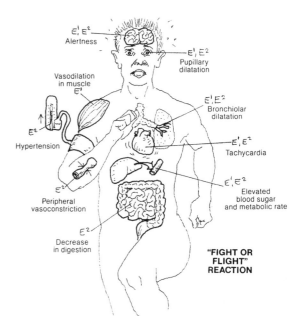

"FIGHT OR FLIGHT" REACTION

PANCREATIC ISLETS

CN: Use purple for N, and light colors for K and L. (1) In coloring the upper drawing, include the broken lines representing arteries within or on the posterior surface of the pancreas. (2) Color the microscopic section of the pancreas and the enlarged view of an islet. Color the arrows and the diagram reflecting the role of glycogen and insulin receptors in liver cells with respect to glucose and glycogen.

ARTERIES TO THE PANCREAS.

GASTRODUODENAL & BRS. A
ANT. PANCREATICO - DUOD. B
POST. PANCREATICO - DUOD. C

SPLENIC & BRS. D
DORSAL PANCREATIC E
INFERIOR PANCREATIC F

GREAT PANCREATIC G
SUPERIOR MESENTERIC H
INF. PANCREATICO - DUOD. I

The pancreas is supplied by numerous arteries from sources springing from the celiac and superior mesenteric arteries. The extensive capillary networks of the pancreas are drained by tributaries of the hepatic portal vein, which conducts the secreted hormones of the pancreatic islets to the liver and beyond for general circulation.

BLOOD SUPPLY TO PANCREAS

PANCREATIC ISLET (ENDOCRINE) J
ALPHA CELL K
GLUCAGON K' / RECEPTOR K²
BETA CELL L
INSULIN L' / RECEPTOR L²
DELTA CELL M
BLOOD CAPILLARY N

GLYCOGEN O   GLUCOSE O'

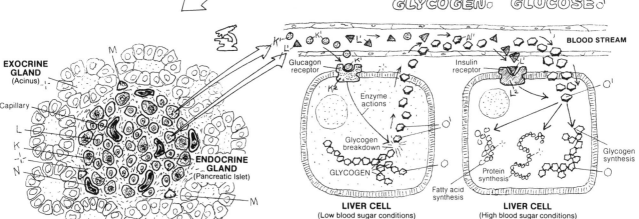

LIVER CELL (Low blood sugar conditions)

LIVER CELL (High blood sugar conditions)

The islands (islets) of endocrine tissue (and their *capillaries*) in the pancreas are surrounded by masses of grape-like clusters/follicles of exocrine gland cells. The secretions of these cells enter ducts that are tributaries of the pancreatic duct(s) opening into the duodenum.

The islets are characterized by three or four different cell types. *Alpha(A) cells*, generally located in the periphery of the islet, secrete *glucagon*, a polypeptide hormone that binds to glycogen *receptors* on liver cell membranes. Glucagon induces the enzymatic breakdown of glycogen to glucose, a process called glycolysis. Glucagon also facilitates the formation of glucose from amino acids in the liver, a process called gluconeogenesis. As a result of these processes, blood glucose levels increase.

Beta (B) cells, constituting 70% of the islet cell population, occupy the central part of the islet and secrete *insulin*, a polypeptide, primarily in response to increased plasma levels of *glucose*. Most insulin is taken up by the liver

and kidney, but almost all cells can metabolize insulin. Insulin expedites the removal of glucose from the circulation by increasing the number of proteins that transport glucose across cell membranes (glucose carriers; not shown) in muscle cells, fat cells, leukocytes, and certain other cells (not including liver cells). Insulin increases the synthesis of *glycogen* from glucose in liver cells. Uptake of insulin is facilitated by *insulin receptors* (proteins) on the external and internal surfaces of many—but not all—cell membranes. Decreased insulin secretion or decreased numbers or activity of insulin receptors leads to glucose intolerance and/or diabetes mellitus. The effects of insulin activity are far-reaching: mediating electrolyte transport and the storage of nutrients (carbohydrates, proteins, fats), facilitating cellular growth, and enhancing liver, muscle, and adipose tissue metabolism. *Delta (D) cells* occupy the periphery of the islet and make up about 5% of the islet cell population. They secrete somatostatin and inhibit the Alpha cells' secretion of glucagon and the Beta cells' secretion of insulin.

MALE REPRODUCTIVE SYSTEM

CN: Use red for L, blue for M, and very light colors for A, J, and K.
(1) Color the upper views simultaneously. In the sagittal view, only the urethra is shown in the median plane. (2) The coverings of the spermatic cord in the illustration below actually consist of several layers (recall Plate 51). Color the parts of K and L seen deep to the pampiniform plexus (M).

SCROTUM A
TESTIS B
EPIDIDYMIS C
DUCTUS DEFERENS D
SEMINAL VESICLE E
EJACULATORY DUCT F
URETHRA G
BULBOURETHRAL GLAND H
PROSTATE GLAND I
PENIS J

The male reproductive system consists of the primary organs, the *testes* (testicles), suspended within a sac of skin and thin fibromuscular tissue (the *scrotum*); a series of ducts; and a number of glands. Development of the male germinating cells (sperm) in the testes requires a temperature slightly lower than that of the body (about 35° C or 95° F); this is achieved by their separation from the warmer body cavities. The temperature within the scrotum can be adjusted slightly by the contraction/relaxation of smooth muscle (dartos muscle) in the scrotal wall, tightening or loosening the tension of the wall about the testes. Mature sperm are stored in the *epididymis*; with stimulus, sperm cells are induced to move into and through the *ductus* (vas) *deferens* by rhythmic contractions of the smooth muscle in the ductal wall. Within the ductus deferens, the sperm pass through the abdominal wall (via the inguinal canal) and pelvic cavity to enter the prostatic *urethra* via the pencil-point-shaped *ejaculatory duct*. Here the nutrient-rich secretions of the *prostate gland* and *seminal vesicles* are added to the population of sperm in the prostatic urethra, forming semen. Prior to the release of the semen (ejaculation), the *bulbourethral glands* add secretions to the urethra. The *penis* and scrotum constitute the external genital organs.

Enlargement of the prostate is common (prostatic hypertrophy/hyperplasia) in men 50 years and older. The glands and connective tissues surrounding the urethra are subject to thickening and blocking urine flow (benign prostatic hypertrophy). Neoplastic growth (prostatic carcinoma) is less common (5–15% of men with prostatic hypertrophy) and occurs in the more peripheral tissues of the prostate.

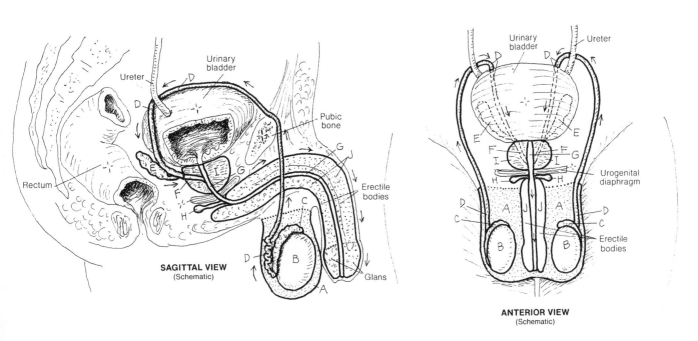

SAGITTAL VIEW
(Schematic)

ANTERIOR VIEW
(Schematic)

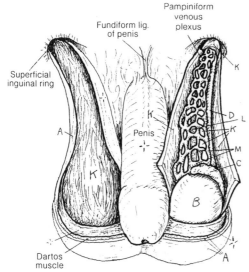

ANTERIOR VIEW OF SCROTUM
(With cord dissected)

SPERMATIC CORD :-
COVERINGS K
CONSTITUENTS :-
DUCTUS DEFERENS D
TESTICULAR ARTERY L
TESTICULAR VEIN M

The testicular artery and vein, and some nerves and lymphatics, join the *ductus deferens* just before entering the deep ring (intra-abdominal orifice) of the inguinal canal. The collection of these form the constituents of the spermatic cord. Passing through the inguinal canal, they become invested by a representative layer from each of the abdominal wall layers (less rectus); these are the coverings of the spermatic cord and testes (here represented as one layer; see Plate 51). In a vasectomy procedure, the ductus deferens is identified within the cord and it alone is divided. A number of techniques (ligatures, cauterization, folding and burial, and so on) are used to prevent the natural tendency of the transected duct sections to recanalize.

TESTIS

CN: Use the colors employed for the testis, epididymis, and ductus deferens on the previous plate with those same structures here (A, E, and F). Use red for U and light colors for G, H, I, S, and T. (1) Note that the spermatogenic epithelium is colored gray in the cross section through the tubules above and that the tubular lumen is not to be colored.

TUNICA ALBUGINEA A
  SEPTUM A'
SEMINIFEROUS TUBULE B
RETE TESTIS C
EFFERENT DUCT D
EPIDIDYMIS HEAD E BODY E' TAIL E²
DUCTUS DEFERENS F

SPERMATOGENIC EPITHELIUM *
  SPERMATOGONIUM G
  PRIMARY SPERMATOCYTE H
  SECONDARY SPERMATOCYTE I
  SPERMATID J
  SPERMATOZOON K
      HEAD ⋰
          ACROSOME L
          NUCLEUS M
      TAIL ⋰
          NECK N
          MIDDLE PIECE O
              MITOCHONDRION P
          PRINCIPAL PIECE Q
          END PIECE R
  SERTOLI (SUPPORTING) CELL S
  BASEMENT MEMBRANE B'
  INTERSTITIAL CELL (OF LEYDIG) T
  BLOOD VESSEL U

SECTIONAL VIEW OF A TESTIS
(Schematic)

Connective tissue
Lobules
Testosterone
Capillary
Lumen

Fibroblast
LINING OF SEMINIFEROUS TUBULE
Lumen

Head
Body
Tail
SPERMATOZOON
(Sperm cell)

The testes (testicles) arise on the posterior abdominal wall during fetal development; as the developing body lengthens, they appear to "descend" into outpocketings of the anterior abdominal wall (scrotum). The testes have two principal functions: development of male germ cells (sperm or *spermatozoa*) and secretion of testosterone, the male sex hormone.

Each testis has a dense, fibrous, outer capsule (tunica albuginea) from which *septa* are directed centrally to compartmentalize the testis into lobules. One to four highly coiled *seminiferous tubules* exist in each lobule. These tubules converge toward the posterior side of the testis, straighten (tubuli recti), and join a network of epithelial-lined spaces (*rete testis*). *Efferent ducts* leave the rete to form the head of the *epididymis*. The convoluted epididymal duct (*head, body, tail*) is lined with pseudostratified columnar epithelium, one type of which contains long, immobile cilia (stereocilia). At the lower portion of the epididymis, the tubule turns upward to form the *ductus deferens*. The wall of the *ductus deferens*, lined with pseudostratified columnar epithelium with stereocilia, contains significant smooth muscle, the rhythmic contractions of which drive sperm toward the prostate gland during emission.

Each seminiferous tubule consists of a lumen with walls of compact, organized masses of cells (*spermatogenic epithelia* and *supporting/(Sertoli) cells*) encapsulated by a thin, fibrous basement membrane. The most immature of the sperm-developing cells are the *spermatogonia*. These divide, and the daughter cells are pushed out toward the lumen of the tubule. These cells differentiate into *primary spermatocytes*, the largest of the developing germ cells. When they divide to become *secondary spermatocytes*, the chromosome number is reduced from 46 to 23 (meiosis). Each pair of newly formed secondary spermatocytes rapidly divides again to form four *spermatids*. These small cells mature by developing tails, condensing their nuclei and cytoplasm, and developing acrosomal caps (with enzymes to break down the wall of the ovum and permit penetration). The mature sperm cell (*spermatozoon*) consists of a *head* of 23 chromosomes (nucleus) including the acrosome, a *middle piece* containing mitochondria to power cell movement, and the rest of the *tail* (fibers containing microtubules; the end piece is essentially a single flagellum), whose flagellations provide the cell's motive force. However, early spermatozoa are essentially immotile and incapable of fertilizing ova. They are swept into the epididymis from the rete testis via the efferent ducts by ciliary action and fluid flow; there the they mature into potent and motile sperm cells.

The interstitial cells dispersed in the vascular loose connective tissue around the tubules include fibroblasts as well as the secretory cells (*of Leydig*), which are known to produce and secrete testosterone into adjacent capillaries. This male sex hormone stimulates the development of ducts and glands of the reproductive tract at puberty (generally between 11 and 14 years of age) as well as secondary sex characteristics.

MALE UROGENITAL STRUCTURES

CN: Use blue for I, red for J, yellow for K, and very light colors for D, E, and G. (1) Color the two upper views simultaneously, noting that the superficial and deep fascia (G, H) have been omitted from the coronal view. (2) Color the structural view and the cross section.

URETHRA:
PROSTATIC U. A
MEMBRANOUS U. B
SPONGY U. C

PENIS:
CORPUS CAVERNOSUM D
CRUS OF PENIS D'
CORPUS SPONGIOSUM E
BULB OF PENIS E'
GLANS PENIS E²
PREPUCE (FORESKIN) F

RELATED STRUCTURES:
SUPERFICIAL FASCIA G
DEEP FASCIA H
VEIN I ARTERY J NERVE K
SUSPENSORY LIG. L
LEVATOR ANI (PELVIC DIAPHRAGM) M
UROGENITAL DIAPHRAGM N
BULBOURETHRAL GLAND O

The urethra in the male has an extensive (20 cm or so) course from the neck of the bladder to the external urethral orifice at the end of the penis. The *prostatic urethra* receives urine from the urinary bladder, sperm from the ejaculatory ducts, seminal fluid from the seminal vesicles, and secretions from the prostate via several ducts. Reflex contraction of the bladder neck muscles prevents voiding of urine during the expulsion of semen. The urethra continues through the pelvic diaphragm and into the thin, fibromuscular urogenital diaphragm as the *membranous urethra*. The *spongy urethra* passes through the penis. Numerous mucus glands exist in the urethral mucosa.

The penis consists of three bodies of erectile tissue, ensheathed in two layers of fasciae. The *corpora cavernosa* (the two lateral bodies) arise from the ascending rami of the pubic bones; the central *corpus spongiosum* arises as a *bulb* suspended from the inferior fascia of the *urogenital diaphragm* (perineal membrane). Each body consists of erectile tissue with a fibrous capsule (tunica albuginea); the corpus spongiosum contains the urethra as well. The three bodies are bound together in a dense stocking of *deep perineal fascia* and hang as a unit suspended by the deep *suspensory* and more superficial fundiform *ligaments*. Deep to the skin of the penis is a

layer of *superficial fascia*. The erectile tissue consists of lakes of veins (cavernous sinuses) bound by fibroelastic tissue and smooth muscle. These sinuses are fed by arteries in the erectile bodies. During sexual activity, these *arteries* dilate secondary to increased autonomic motor activity, and the volume of blood entering the sinuses increases, expanding the erectile tissue. As a result, the *veins* at the periphery of the erectile bodies deep to the tunica albuginea are pressed against the capsule (unable to drain blood) and the penis enlarges and becomes rigid (erection). The glans remains non-rigid.

SAGITTAL VIEW (Schematic)

CORONAL VIEW (Schematic)

STRUCTURAL VIEW

CROSS SECTION OF PENIS (Through the middle)

FEMALE REPRODUCTIVE SYSTEM

CN: (1) Color the two (upper) views of the internal reproductive structures simultaneously. In the sagittal view, color the double line representing the peritoneum in gray. (2) In the lower drawings, color the vestibule (N) gray after coloring the other structures located in that area (L–P). (3) In the dissected view of the external structures, take note of the surrounding musculature, none of which is colored.

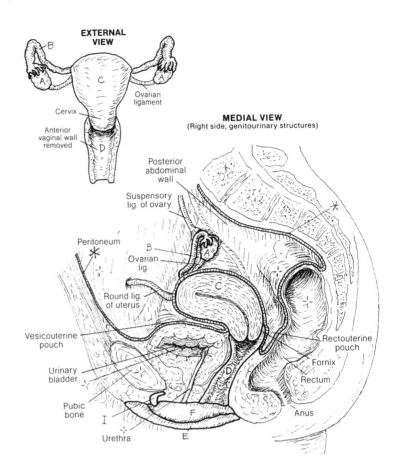

EXTERNAL VIEW

MEDIAL VIEW
(Right side, genitourinary structures)

INTERNAL STRUCTURES

OVARY A
UTERINE (FALLOPIAN) TUBE B
UTERUS C
VAGINA D

The primary organ of the female reproductive system is the *ovary,* which produces the female germ cells (ova) and secretes the hormones estrogen and progesterone. Each ovary, like the testis, arises on the posterior abdominal wall (adjacent to the kidneys) during early fetal development. It also descends along that wall, like the testis, but is interrupted early in its journey by a ligament and is retained in the pelvis. The *uterus* serves as a site for implantation and nourishment of the embryo/fetus. The *uterine tubes* provide a conduit for the freshly fertilized or unfertilized ovum enroute to the uterus. The *vagina,* a fibromuscular sheath, receives the semen from the penis and transmits it to the uterus and acts as a birth canal from the uterus to the outside for the newborn.

EXTERNAL STRUCTURES

LABIUM MAJUS E
LABIUM MINUS F
FRENULUM G
PREPUCE H
CLITORIS I
GLANS I'
BODY J
CRUS K

BULB OF THE VESTIBULE L
VESTIBULAR GLAND/DUCT M
VESTIBULE N*
URETHRAL ORIFICE O
VAGINAL ORIFICE D' /HYMEN P

The female external genitals collectively constitute the vulva. They are located within the perineum. The *labia majora* are fat-filled folds of skin largely obscuring the cavity/space between them (vestibule) that contains the *urethral* and *vaginal orifices.* Medial to the labia majora are thin folds of skin (*labia minora*) that approach the *clitoris* anteriorly and split around it, forming the *frenulum* and *prepuce* of the clitoris. Like the penis, the clitoris has a *crus* (pl. crura) arising from each ischiopubic ramus; the two crura join in the midline to form the *body* or corpus. The body is capped by a skin-covered, vascular, sensitive *glans.* These clitoral components contain erectile tissue (less in the glans) enclosed in fascial coverings; their erection or rigidity is accomplished by the same mechanism operative in the penis. The clitoris, unlike the penis, does not incorporate the urethra. The bulbs of the *vestibule* are homologous to the bulb of the penis, but separated into two erectile bodies. They are covered by the bulbospongiosus muscle and protrude into the vagina during sexual stimulation. The *vaginal orifice* is completely or incompletely covered or surrounded by a rim of mucosa called the *hymen.* Remnants of it (as shown) are often retained in the sexually active female.

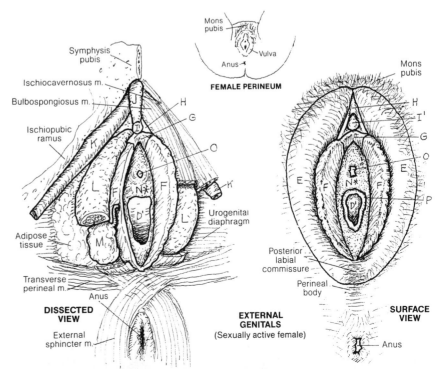

FEMALE PERINEUM

DISSECTED VIEW

EXTERNAL GENITALS
(Sexually active female)

SURFACE VIEW

OVARY

CN: Use the colors from the preceding plate for the ovary (A) and uterine tube (M). Use red for K and R, yellow for L, blue for S, and very light colors for C–J, M, O, and P. (1) Color the development of the female germ cell in both upper and lower views of the sectioned ovary. The oocyte (C) is colored through ovulation. In the large illustration, color the background stroma (B) gray; do not color the blood vessels in the stroma.

OVARIAN STRUCTURES:
EPITHELIUM / TUNICA ALBUGINEA A
CONNECTIVE TISSUE STROMA B*
OOCYTE DEVELOPMENT:
OOCYTE / OVUM C
PRIMORDIAL FOLLICLE D
PRIMARY FOL. E
SECONDARY FOL. F
MATURING FOL. G
MATURE (GRAAFIAN) FOL. H
RUPTURED FOL. I
DISCHARGED OVUM C'
ATRETIC FOL. J
CORPUS HEMORRHAGICUM K
YOUNG CORPUS LUTEUM L
MATURE CORPUS LUTEUM L'
CORPUS ALBICANS L²

RELATED STRUCTURES:
UTERINE TUBE M / FIMBRIAE M'
BROAD LIGAMENT N
MESOSALPINX O
MESOVARIUM P
SUSP. LIG. OF OVARY Q
OVARIAN A. R / V. S
UTERINE A. R' / V. S'
OVARIAN LIG. T

LUTEAL (SECRETORY) PHASE

Corona radiata
Ovulation
Zona pellucida
Corona radiata

FOLLICULAR (PROLIFERATIVE) PHASE
OVARIAN CYCLE

Uterus

UTERUS AND OVARIES
(Posterior view)

Tubal branch of uterine artery and vein

Branches of ovarian a. and v.

Cut edge of mesosalpinx

Zona pellucida

POSTERIOR VIEW
(Schematic)

Cut edge of broad lig.

Cut edges of mesovarium

Development of female germ cells and the secretion of the hormones estrogen and progesterone are the functions of the *ovary*. Confined by the thin but dense fibrous *tunica albuginea*, lined with epithelium, many ovarian follicles in various stages of development can be seen in the *connective tissue stroma*. A follicle consists of an immature epithelial germ cell (*oocyte*) surrounded by one or more layers of non-germinating cells. These germ cells were seeded in the ovary early in development—over 400,000 of them. Of these, only 500 or so will mature, the rest stopping short in their development and degenerating with their follicular cells (*atretic follicles*). Development of an ovum starts with the *primordial follicle*—an oocyte with one layer of follicular cells. The oocyte increases in size and maturity as the follicle cells increase in number. In *secondary follicles*, a small cavity (antrum) filled with follicular fluid appears. This antrum continues to increase/expand at the expense of the follicle cells, which are pushed away from the oocyte

except for a layer of cells (*mature follicle*). Those cells in the outermost part of the follicle secrete estrogen during the proliferative phase of the reproductive cycle. On about the 14th day of that cycle, the ovum (surrounded by a glycoprotein coat—the zona pellucida—and a "corona radiata" of follicular cells) bursts from the follicle into the waiting fingers (*fimbriae*) of the *uterine tube*. The *ruptured follicle* involutes, and some bleeding and clotting goes on (*corpus hemorrhagicum*) as the follicle cells transition, characterized by accumulating large amounts of lipid. This newly formed structure (*corpus luteum*) secretes estrogen and progesterone during the secretory phase of the cycle; in the event of pregnancy, it will support the developing embryo/fetus for up to three months with these secretions. Should pregnancy not ensue, the corpus luteum will involute and degenerate as the corpus *albicans*. Follicles or corpora albicans/lutea collectively relating to two or three different but sequential cycles can be seen in the ovary at one time.

UTERUS, UTERINE TUBES & VAGINA

CN: Use red for N, blue for O, and light colors for D, E, and Q. (1) Begin with the left half of the large illustration. Only parts of the ovarian and uterine veins are shown. Nerves and lymph vessels that may accompany arteries and veins are not shown. (2) Color the two views of the anteflexed and the retroflexed uterus.

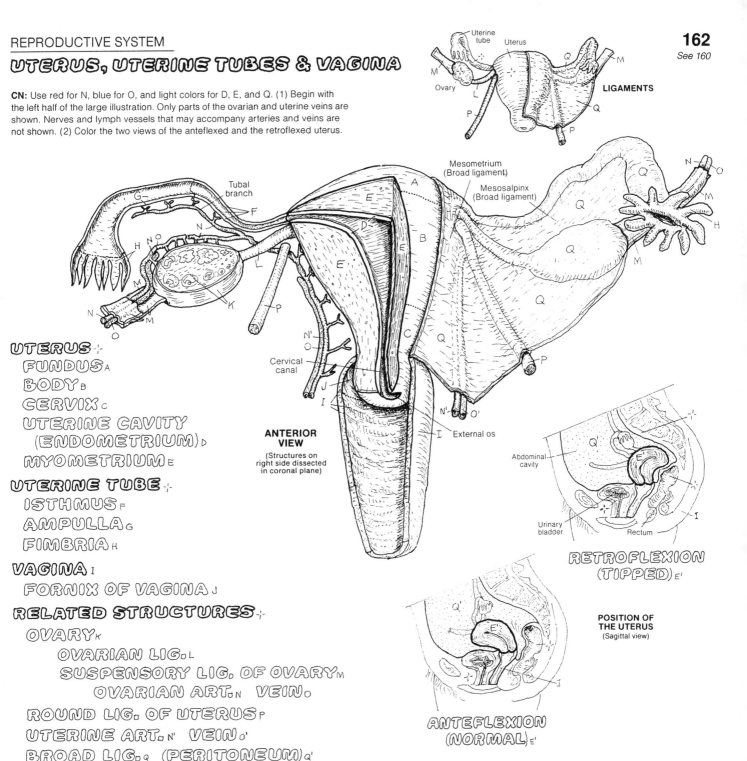

LIGAMENTS

ANTERIOR VIEW
(Structures on right side dissected in coronal plane)

External os

RETROFLEXION (TIPPED) E'

POSITION OF THE UTERUS
(Sagittal view)

ANTEFLEXION (NORMAL) E'

UTERUS -:-
 FUNDUS A
 BODY B
 CERVIX C
 UTERINE CAVITY
  (ENDOMETRIUM) D
 MYOMETRIUM E
UTERINE TUBE -:-
 ISTHMUS F
 AMPULLA G
 FIMBRIA H
VAGINA I
 FORNIX OF VAGINA J
RELATED STRUCTURES -:-
 OVARY K
  OVARIAN LIG. L
  SUSPENSORY LIG. OF OVARY M
   OVARIAN ART. N VEIN O
  ROUND LIG. OF UTERUS P
  UTERINE ART. N' VEIN O'
  BROAD LIG. Q (PERITONEUM) Q'

The ovaries, uterus, uterine tubes, and vagina make up the internal organs of reproduction in the female. The ovaries are suspended on the posterior layer of the *broad ligament* by a peritoneal extension (mesovarium), and supported by the *suspensory ligament of the ovary* (a lateral extension of the broad ligament and mesovarium), the *ovarian ligament*, and the *round ligament* (from the lateral wall of the uterus to the medial wall of the ovary). In this view, the ovaries have been brought to the horizontal to better clarify their relationship to the uterine tubes. The uterine tubes, suspended in a part of the broad ligament (mesosalpinx), are lateral extensions of the uterus, lined with ciliated columnar epithelium supported by connective tissue and smooth muscle. The rhythmic contractions of this muscle aid the ovum in its trek from the *fimbriae* to the *uterine cavity,* and the lining cells support it nutritionally. The tube shows three rather distinct parts: the distal *fimbriae* (finger-like projections), which "catch" the discharged ovum and whisk it into the tubular lumen; the *ampulla* or widest part of the tube; and the *isthmus,* whose lumen narrows as it enters the uterine wall/cavity.

The uterus is a pear-shaped structure whose neck (*cervix*) fits into the upper part of the vagina and whose body (*fundus*) is bent (*anteflexed*) and tilted (*anteverted*) anteriorly over the bladder. Backward bending/tilting

(*retroflexion/retroversion*) of the uterus is not uncommon, particularly in women who have given birth. The retroflexed uterus predisposes to mild slipping into the vagina (prolapse) when the uterus is more in the axis of the cervix/vagina. Such an event is generally resisted by the pelvic and urogenital diaphragms, the perineal body, and numerous fibrous ligaments (broad ligament, and condensations of the pelvic fasciae, not shown) mooring the uterus and its tubes to the pelvic wall and sacrum. The wall of the uterus is largely smooth muscle (*myometrium*) lined with a glandular surface layer of variable thickness (*endometrium*) that is extremely sensitive to the hormones estrogen and progesterone.

The vagina is an elastic, fibromuscular tube with a mucosal lining of stratified squamous epithelium. The anterior and posterior mucosal surfaces are normally in contact. The anterior vaginal wall incorporates the short (4 cm) urethra. Remarkably, the mucosa of the vagina lacks glands; secretory activity during sexual stimulation is derived from a transudate of plasma from the local capillaries and from glands of the cervix. The vaginal lining reveals few sensory receptors. Where the cervix fits into the vagina, a circular moat or trough is formed around it (*fornix*, fornices). The fibroelastic posterior fornix is capable of significant expansion during intercourse.

MENSTRUAL CYCLE

CN: Use yellow for B, red for G, and a very light color for A. (1) Color the time bar of the menstrual cycle at the bottom of the main diagram. Color the arrows C and D in the drawing on hormonal influences above. Then color the hormonal curves C and D in the main diagram, followed by the different follicular stages of the ovarian cycle (A, B), noting how these hormones influence the follicular changes. (2) Color the arrows E and F and the endometrium in the diagram above. Color the curves E and F in the main diagram, followed by the uterine structures in the menstrual cycle, noting how these hormones influence endometrial growth and menstruation. Color only the epithelial surface, glands, and vessels of the endometrium. (3) The days indicated are approximate. The hormonal curves reflect relative plasma hormone levels and are not absolute values.

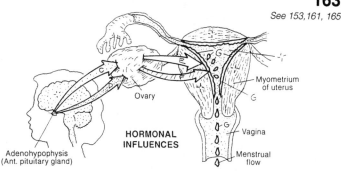

HORMONAL INFLUENCES

Adenohypophysis (Ant. pituitary gland)
Ovary
Myometrium of uterus
Vagina
Menstrual flow

OVARIAN CYCLE
PRIMORDIAL FOL.ₐ
PRIMARY FOL.ₐ'
SECONDARY FOL.ₐ²
MATURE FOL.ₐ³
OVULATION ₐ⁴
CORPUS LUTEUM ᵦ,ᵦ'
CORPUS ALBICANS ᵦ²

HORMONAL CYCLE
HYPOPHYSEAL HORMONES
FSH c
LH d
OVARIAN HORMONES
ESTROGEN e
PROGESTERONE f

MENSTRUAL CYCLE
PHASES
MENSTRUATION g
PROLIFERATIVE h
SECRETORY i
ENDOMETRIUM
EPITHELIUM j
GLAND i'
SPIRAL ARTERY g'

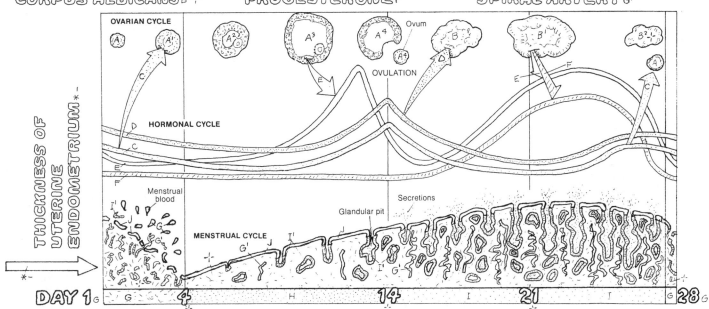

OVARIAN CYCLE

Ovum

OVULATION

HORMONAL CYCLE

THICKNESS OF UTERINE ENDOMETRIUM

Menstrual blood

Glandular pit

Secretions

MENSTRUAL CYCLE

DAY 1   4   14   21   28

The 28-day human female reproductive cycle, initiated and maintained by hormones, involves significant alterations in ovarian (follicular) and uterine (endometrial) structure. The cycle is characterized by periods of endometrial breakdown and discharge (*menstruation*), which begin at about 12 years of age (menarche) and end at about 45 years of age (menopause). The progressive changes that occur in the ovary and uterus during each cycle serve to develop and release the female germ cell for possible fertilization by the male germ cell and to prepare the *endometrium* for implantation of the fertilized ovum.

The menstrual period constitutes the first five days of the cycle. Note the loss of endometrial tissue and attendant bleeding during this time. Endometrial regrowth begins on about the 5th day of the menstrual cycle and is precipitated by hormones from the ovarian follicles. The ovarian cycle is regulated by hormones from the adenohypophysis (anterior pituitary gland), specifically *follicle stimulating hormone (FSH)* and *luteinizing hormone (LH)*. During the last few days of the previous cycle and the first several days of the next, these hormones stimulate follicular development.

As the selected follicle develops, it begins to produce *estrogen* on about the 7th day. Estrogen enters the circulation and influences endometrial growth (*proliferative phase*). On about the 14th day of the menstrual cycle, the combined "spikes" of increased concentrations of FSH, LH, and estrogen induce *ovulation*: bursting of the mature follicle and release of the immature ovum into the fimbriae of the uterine tube. After ovulation, the burst follicle undergoes significant reconstruction (*corpus luteum*)

influenced by LH. On about the 21st day, this body secretes *progesterone* as well as estrogen, a combination with remarkable influence on endometrial glandular development (*secretory phase*). This phase is characterized by the development of numerous secretory cells in the *epithelium*, a connective tissue stroma edematous with secretions from developing *glands*, and *spiral arteries* taking a tortuous course about the many glands—a condition conducive to nutritional support for an implanted fertilized ovum. If fertilization occurs (on about day 16), the corpus luteum becomes the principal source of hormones supporting development of the embryo and will continue as such for the next 90 days or so or until the placenta is capable of producing its own hormones.

In the absence of fertilization, on about day 26 the corpus luteum begins to involute (forming a *corpus albicans*), and estrogen/progesterone levels drop. Lacking hormonal stimulation, the endometrium experiences reduced glandular secretion in the presence of continued fluid absorption by the local veins, and the tissues collapse. The spiral arteries are flexed by these events, rupture, and hemorrhage with considerable hydraulic force; epithelial lining, glands, and fibrous tissues are disrupted and the structural integrity of the endometrium is largely destroyed. The vessels rapidly constrict, and bleeding is generally limited. The broken tissue (menstruum, mostly glandular tissue and secretions), blood, and one or more unfertilized ova gravitate toward the vagina. After 3–5 days of menstruation, only about 1 mm (in height) of endometrium is left for regeneration. Within the next two weeks, it will regenerate 500% to a height of about 5 mm!

BREAST (MAMMARY GLAND)

CN: Use yellow for E, pinks, tans, or browns for J and K, and light colors for A, D, E, and G. (1) Color the two illustrations of the breast and underlying breast structures simultaneously. (2) Color the arrows indicating the direction of lymph flow and the lymph nodes of the chest. If you wish, you may color over the network of lymph vessels. (3) Color the diagrams of breast development. (4) Color the enlargement of glands and ducts in the lower right corner.

RIB ᴀ CLAVICLE ᴀ'
INTERCOSTAL MUSCLE ʙ
PECTORALIS MAJOR M. ᴄ
DEEP FASCIA ᴅ
SUPERFICIAL FASCIA (FAT) ᴇ
SUSPENSORY LIGAMENT ꜰ
GLANDULAR LOBE ɢ
LACTIFEROUS DUCT ʜ
LACTIFEROUS SINUS ɪ
NIPPLE ᴊ
AREOLA ᴋ

LYMPHATIC DRAINAGE ʟ

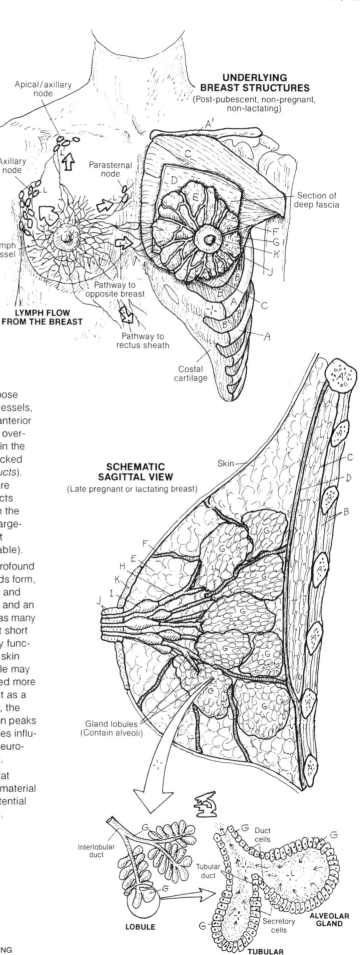

UNDERLYING BREAST STRUCTURES
(Post-pubescent, non-pregnant, non-lactating)

Apical/axillary node

Axillary node

Parasternal node

Lymph vessel

Section of deep fascia

Pathway to opposite breast

LYMPH FLOW FROM THE BREAST

Pathway to rectus sheath

Costal cartilage

SCHEMATIC SAGITTAL VIEW
(Late pregnant or lactating breast)

Skin

Gland lobules (Contain alveoli)

Interlobular duct

Tubular duct

Duct cells

Secretory cells

LOBULE

TUBULAR GLAND

ALVEOLAR GLAND

The breast (in both males and females) is an area of fatty (adipose and loose areolar) fibrous tissue, with associated nerves and blood and lymphatic vessels, in the subcutaneous fascia overlying the pectoralis major muscle on the anterior chest wall. The fatty tissue is supported by extensions of the deep fascia overlying the muscle (*suspensory ligaments*) and functions most prominently in the young, well-developed, post-pubescent (after puberty) female breast. Packed within the adipose tissue is a collection of branching ducts (*lactiferous ducts*). In the male and in the non-pregnant (non-lactating) female, these ducts are undeveloped. There are few or no glands (alveoli) associated with the ducts in those populations. At puberty, the increased secretion of estrogen from the ovaries (and perhaps the adrenal glands) in the female influences an enlargement of the *nipple* and *areola* and a generally marked increase in local fat proliferation. As a result, the breast enlarges to some degree (highly variable).

In the early stages of pregnancy, the lactiferous duct system undergoes profound proliferation, and small, inactive tubular and alveolar (*tubuloalveolar*) glands form, opening into alveolar ducts. A *lobule* consists of a number of these ducts and glands; a *lobe* (of which there are 15–20) consists of a number of lobules and an interconnecting *interlobular duct*. The interlobular ducts converge to form as many as 20 lactiferous ducts. These ducts dilate to form *lactiferous sinuses* just short of the nipple, then narrow again within the nipple. These sinuses probably function as milk reservoirs during lactation. The nipple consists of pigmented skin with some smooth muscle fibers set in fibrous tissue. Erection of the nipple may enhance flow of milk through the ducts. The circular areola, also pigmented more highly than the surrounding skin, contains sebaceous glands that may act as a skin lubricant during periods of nursing. In the latter stages of pregnancy, the alveolar glands undergo maturation and begin to form milk. Milk production peaks after delivery of the newborn as the result of the action of several hormones influencing the gland cells. The letdown and excretion of milk results from a neuroendocrine reflex mechanism initiated by the baby's sucking on the nipple.

The lymphatic vessels are an important part of the breast: they drain the fat portion of the milk produced during lactation. They also transfer infected material or neoplastic (cancer) cells from the breast to more distant parts. The potential lymphatic avenues for metastasis or spread of infection are shown above.

BREAST DEVELOPMENT

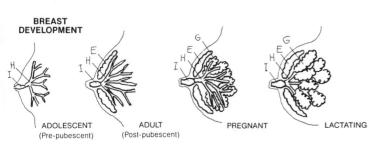

ADOLESCENT
(Pre-pubescent)

ADULT
(Post-pubescent)

PREGNANT

LACTATING

DEVELOPMENT OF EMBRYO (1)

CN: Use light colors throughout. (1) Follow the events from fertilization to implantation. The number of days cited in this and the following two plates are days after (post-) fertilization. Fertilization occurs about 14 days after the last day of menstruation; physicians date fetal age by time since last menstrual period (LMP). Thus, fetal age according to LMP is 14 days earlier than true (post-fertilization) age.

ZONA PELLUCIDA A

FERTILIZATION :-
(1ST STAGE) :-
FEMALE PRONUCLEUS B
HEAD OF SPERM C
MALE PRONUCLEUS C'

BLASTOMERE
(CLEAVAGE) STAGE :-
2-CELL D
4-CELL E
MORULA F

BLASTOCYST :-
TROPHOBLAST G
INNER CELL MASS H
BLASTOCOELE I

PATHWAYS OF SPERM

Uterine cavity

Uterine tube

Ovary

SPERM C²

Vagina

Penis

Uterine tube

Ampulla

Fimbria

Ovulation

Ovary

Myometrium

Uterine cavity

Endometrium

DAY 2

DAY 4

ZYGOTE (First cell)

Fusion of nuclei

B+C'

DAY 1

Enlarging sperm head

Degenerating tail

OVUM (Secondary oocyte)

Corona radiata

Plasma membrane

FERTILIZATION

Degenerating

EARLY BLASTOCYST

DAY 7

LATE BLASTOCYST

Developing cavity

Syncytiotrophoblast

Epithelium of endometrium

0.1mm

IMPLANTATION

Following ovulation, the ovum enters the uterine tube and proceeds toward the uterus. It reaches the ampulla of the tube in about 30 minutes. If sperm-laden semen has been deposited in the fornices of the vagina in the preceding several minutes to 24 hours, a few hundred of the original 50 million or more sperms will succeed in reaching the ampulla. Over a period of several hours, the sperms become activated, and with the aid of sperm-produced enzymes, one of the sperms will penetrate the corona radiata (retained follicular cells) and *zona pellucida* of the ovum, fuse with the plasma membrane (leaving its cell membrane attached to the ovum's plasma membrane), and enter the ovum. This event is called fertilization. As the tail breaks down and disappears, the *head of the sperm* enlarges and forms the *male pronucleus*. The nucleus of the ovum is the *female pronucleus*. The two pronuclei approach each other, fuse nuclear membranes, and form a single nucleus. The male and female chromo-

somes join up in the metaphase stage of the first mitotic division of the fertilized ovum. The first cell of the new individual is called a zygote.

The zygote undergoes division (cleavage stage) to form two *blastomeres*. Over the next two days or so post-fertilization, within the restraints of the zona pellucida, the cells divide to form a *four-cell blastomere* and again to form eight cells, and so on, until a ball of cells (*morula*) is formed. After about five days, the cells within the morula disperse enough to accommodate progressively enlarging fluid-filled cavities. Some cells are pushed aside to form a peripheral rim of cells (*trophoblast*) enclosing a large single cavity (*blastocoele*); some cells form an *inner cell mass* within the blastocoele. This multicellular structure is called a blastocyst. The blastocyst enters the uterus and implants in the endometrium on about the 7th day post-fertilization.

DEVELOPMENT OF EMBRYO (2)

CN: Use the same color as on the previous plate for trophoblast (C) and note that the syncytiotrophoblast (D) is now given a separate color. Use yellow for F. Complete each drawing before proceeding to the next.

2 LAYER EMBRYONIC DISC ⫶
EPIBLAST A
HYPOBLAST B

TROPHOBLAST C
SYNCYTIOTROPHOBLAST D
AMNION E /AMNIOTIC CAVITY E'
YOLK SAC: PRIMARY F, SECONDARY F'
EXOCOELOMIC MEMBRANE G
EXTRAEMBRYONIC MEMBRANE H
CONNECTING STALK H'
EXTRAEMBRYONIC COELOM I

3 LAYER EMBRYONIC DISC ⫶
ECTODERM A'
MESODERM J
ENDODERM K

On day 11 post-fertilization, the inner cell mass gives rise to a flat embryonic disc, consisting of a layer of columnar cells (*epiblast*) and an adjacent layer of cuboidal cells (*hypoblast*). The epiblast will develop almost entirely into the embryo. The *amniotic cavity* develops among the *trophoblast* cells adjacent to the epiblast; the roof of the cavity is called the *amnion*. The embryo and subsequent fetus will develop within this cavity. The trophoblast also gives rise to the *primary yolk sac*; the cells lining this sac are continuous with those of the hypoblast. Though it has no yolk, the sac probably has a nutritional function for the embryonic disc. Cells of the trophoblast form an *extraembryonic mesoderm* tissue (*membrane*) that largely fills the cavity once known as the blastocyst.

By day 14, the primary yolk sac diminishes in size, replaced by a *secondary yolk sac*. Cavities within the extra-embryonic membrane form a single cavity (it looks paired, but the connection between yolk sacs does not create two cavities). This cavity (I) surrounds the amnion/amniotic cavity and the yolk sac except where the amnion retains a *connecting stalk* to the trophoblast layer.

By day 16, the epiblast undergoes significant changes. The primary yolk sac is gone. Cells emerge from the epiblast and migrate into the area between the epiblast and hypoblast and into the hypoblast itself. The cells between are embryonic *mesoderm* cells; the cells migrating into the hypoblast layer form embryonic *endoderm*. The remaining epiblast cells become embryonic *ectoderm*. The earlier two-layered embryonic disc has formed into a three-layered embryonic disc. These three layers are called germ layers, and they give rise to the cells and tissues of the body. From ectoderm form the skin and related glands, nervous system, hypophysis, lens of the eye, and inner ear. From mesoderm form bones, muscle and connective tissues, lymphoid organs, blood, the urogenital system, and serous membranes. From endoderm form the epithelial part of the gastrointestinal system and respiratory system as well as the epithelia of the pharynx and thyroid.

By day 24 post-fertilization, the once flat embryonic disc has rounded to form within the amniotic cavity an embryo with a definitive head end and tail end, secured to the chorion (C, D, H) by the connecting stalk. As the lateral folds of embryonic mesoderm encircle the ventral (anterior) part of the embryo to form the anterolateral abdominal walls, the yolk sac is pinched off and formation of the primitive gut begins. By the end of three weeks post-fertilization, the gastrointestinal tract, brain, and heart have begun their development.

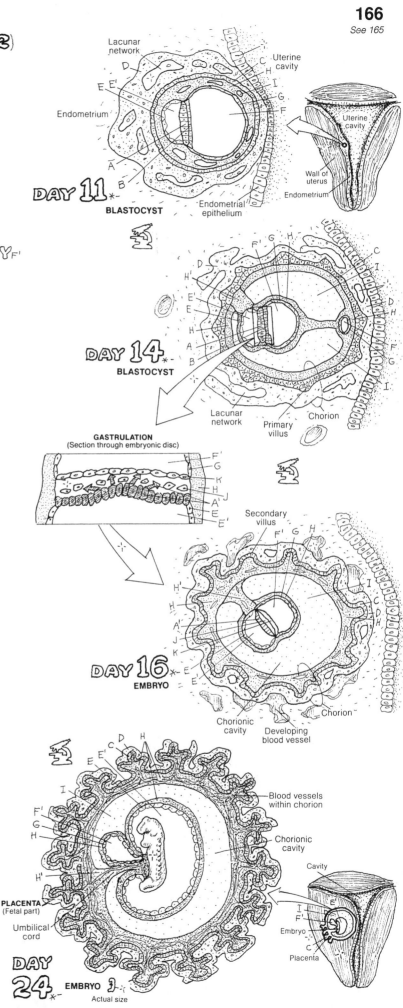

EMBRYO/FETUS COVERINGS

CN: Use the same colors for structures B and C that were given to them on the preceding plate. Use the color given to "connecting stalk" for the umbilical cord (A), and use the color given to "trophoblast" for chorion (D). (1) Color the embryonic coverings. The uterine cavity is colored gray, though it is actually lined with the decidua capsularis (E). Note that the *amniotic cavity* (C^1), *chorionic cavity* (D^1), and the embryo/fetus are left uncolored. (2) The umbilical cord is composed of different blood vessels but receives one color (A). The band representing the uterine wall (below) is colored with both G and H.

EMBRYO
UMBILICAL CORD A
YOLK SAC B
AMNION C & CAVITY C'
CHORION D /CAVITY D' /VILLI D²

UTERUS
ENDOMETRIAL DECIDUA
D. CAPSULARIS E
D. BASALIS F
D. PARIETALIS G
MYOMETRIUM H
CERVIX I

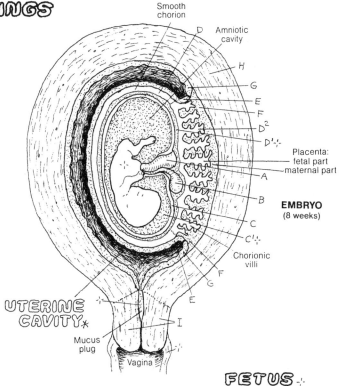

EMBRYO
(8 weeks)

The developing embryo (called a fetus after eight weeks of development) lies within and is supported, nurtured, and protected by membranes and sacs. These coverings have both maternal and fetal origins.

Those of fetal origin include the *amnion* and *chorion/chorionic villi*, the *umbilical cord*, and the *yolk sac*. The chorion forms a sac around the early embryo; the cavity of the sac is the *chorionic cavity* (recall Plate 166). As the embryo grows, the chorionic sac is obliterated and the amnion and chorion fuse (*amniochorionic membrane*). The chorion exhibits villi circumferentially early on (e.g., 24-day embryo); in time, most of the villi are absorbed except for those in the developing placenta (8-week embryo), creating a smooth chorion around the amnion and a bushy one (the villi and an underlying chorionic plate) in the future placenta. This is the situation of the fetal membranes at term (40 weeks).

The coverings of maternal origin (the *decidua*), are thickened, fairly distinct layers of the uterine mucosa (endometrium) in which the blastocyst implanted. In the 8-week embryo (above), the *decidua basalis* is integrated with the fetal villi to form the placenta. The *decidua capsularis* surrounds the embryo and its membranes. The *decidua parietalis* lines the uterine cavity, superficial to the *myometrium*.

The parietalis is continuous with the capsularis, as shown. This is the situation with the maternal membranes at eight weeks.

When the fetus grows to the point of pushing the decidua capsularis against the parietalis, the uterine cavity is obliterated. The capsularis soon degenerates, leaving only the parietalis (lower illustration). This layer will be retained after birth as the basal endometrium. The decidua basalis and chorionic villi (placenta) will be discharged after birth.

The fetus develops within the fluid-filled amniotic cavity. The plasma-like fluid gives freedom to the embryo to develop its form without mechanical pressure. It also acts as a water cushion, absorbing shock forces. Just prior to birth, the amniochorionic membrane surrounding the fetus bursts, sending a half liter or more fluid into the *vagina* and to the outside (breaking the "bag of waters"). Parturition (childbirth) generally occurs about 280 days (40 weeks) after fertilization.

FETUS
UMBILICAL CORD A
AMNION C /CAVITY C'
CHORION D /VILLI D²

Liver
Stomach
Intestine
Diaphragm
Spleen
Aorta
Uterine wall
Umbilical vessels
Urinary bladder
FETUS AT TERM
(40 weeks)
Mucus plug
Vagina
Rectum

ENDOCHONDRAL OSSIFICATION

CN: Use the same colors as used on the previous plate for hyaline cartilage (A), periosteal bone (B), which was compact bone on Plate 7, and endochondral bone (E), which was spongy bone. Use red for D. Complete each stage before going on to the next. Do not color the periosteum, which appears adjacent to periosteal bone in step 3 and continues to the end. Color the small shapes (E) that appear in the epiphyses and, to a lesser extent, the diaphyses (views 5–8). They represent spongy (cancellous) bone of endochondral origin.

Bone development occurs by intramembranous and/or endo-chondral ossification. Here we show longitudinal sections of developing long bone, demonstrating both forms of ossification but emphasizing endochondral bone growth.

The endochondral process begins at about 5 weeks of post-fertilization age with formation of cartilage models (bone proto-types) from embryonic connective tissue. Subsequently (over the next 16–25 years), the cartilage is largely replaced by bone. The rate and duration of this process largely determines a person's standing height. Intramembranous bone development begins in embryonic connective tissue (membrane) and does not involve replacement of cartilage. The flat cranial bones, the clavicle, and the bone collar surrounding the shaft of cartilage models develop in this fashion.

Endochondral ossification begins with a *hyaline cartilage* model (1). As the cartilage structure grows, its central part dehydrates. The cartilage cells there begin to degenerate: they enlarge, die, and calcify (2). At the same time, *blood vessels* bring bone-forming cells to the waist of the cartilage model, and a collar of bone is formed around the cartilage shaft (2) within the membranous perichondrium (intramembranous ossification). This vascular, cellular, fibrous membrane around the bone collar is now called periosteum. The new bone collar (*periosteal bone*) becomes a supporting tubular shaft for the cartilage model, with a core of degenerating, calcifying cartilage (3).

Blood vessels from the fibrous periosteum penetrate the bone collar, enter the cartilage model (periosteal bud), and proliferate, conducting periosteal osteoblasts into the cartilage model (4). Starting at about 8 weeks post-fertilization, these bone-forming cells line up along peninsulas of *calcified cartilage* at the extremes of the shaft (*diaphysis*) and secrete new bone (5). The calcified cartilage degenerates and is absorbed into the blood; endochondral bone has now replaced the cartilage. The two sites of this activity are called primary centers of ossification (5). The direction of growth at these sites is toward the ends of the developing bone. The calcified cartilage and some endo-chondral bone of the diaphysis are subsequently absorbed, forming the medullary or marrow cavity (5). This cavity of the developing tubular bone shaft becomes filled with gelatinous red marrow in the fetus. Productive primary (diaphyseal) centers of ossification are well established at birth.

Beginning in the first few years after birth, secondary centers of ossification begin at the ends or *epiphyses* as blood vessels penetrate the cartilage there (6). The healthy cartilage between the epiphyseal and diaphyseal centers of ossification becomes the *epiphyseal plate* (7). It is the growth of this cartilage that is responsible for bone lengthening; it is the gradual replacement of this cartilage by bone cells in the metaphysis (7) that thins this plate and ultimately permits fusion of the epiphyseal and diaphyseal ossification centers (8), ending longitudinal bone growth (at 12–20 years of age). Dense areas of bone at the fusion site may remain into maturity (*epiphyseal line*). Epiphyseal bone is less structured (irregular beams) than that of the diaphysis (organized columns or osteons), and in maturity it is called spongy or cancellous bone (recall Plate 7).

Intramembranous ossification of the diaphyseal shaft (bone collar to compact bone) is responsible for the widening of developing long bone. The ossification process is regulated by growth hormone (from the pituitary gland) and the sex hormones.

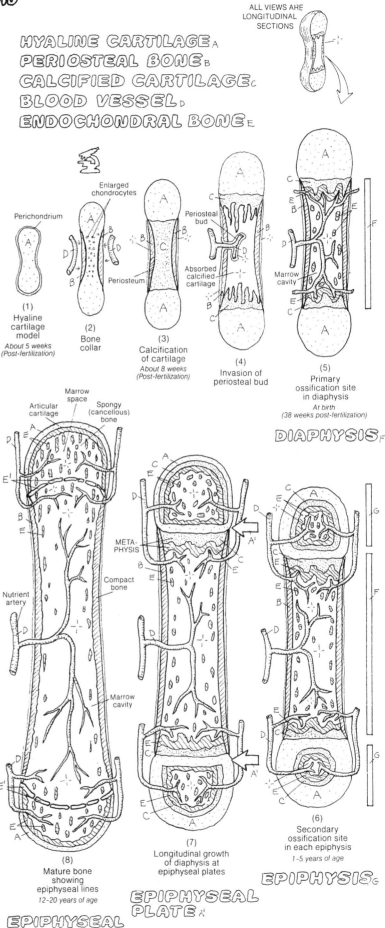

ALL VIEWS ARE LONGITUDINAL SECTIONS

HYALINE CARTILAGE A
PERIOSTEAL BONE B
CALCIFIED CARTILAGE C
BLOOD VESSEL D
ENDOCHONDRAL BONE E

Perichondrium

(1) Hyaline cartilage model
About 5 weeks (Post-fertilization)

Enlarged chondrocytes

Periosteum

(2) Bone collar

Periosteal bud

Calcification of cartilage

Absorbed calcified cartilage

(3) Calcification of cartilage
About 8 weeks (Post-fertilization)

(4) Invasion of periosteal bud

Marrow cavity

(5) Primary ossification site in diaphysis
At birth (38 weeks post-fertilization)

DIAPHYSIS F

Marrow space

Articular cartilage

Spongy (cancellous) bone

Nutrient artery

Compact bone

META-PHYSIS

Marrow cavity

(8) Mature bone showing epiphyseal lines
12-20 years of age

(7) Longitudinal growth of diaphysis at epiphyseal plates

(6) Secondary ossification site in each epiphysis
1-5 years of age

EPIPHYSIS G

EPIPHYSEAL PLATE A

EPIPHYSEAL LINE E'

DEVELOPMENT OF CENTRAL NERVOUS SYSTEM

CN: Use light colors for A and C. (1) Begin with the two dorsal views of the 20-day-old embryo. Color as well the large arrows pointing to the surface locations. Simultaneously color the diagrammatic cross section to its right. Follow the same procedure for the later views of the growing embryo. (2) Color the stages of brain development in the head end of the neural tube.

NEURAL PLATE_A
FOLD_A'
TUBE_A²
NEURAL GROOVE_B
NEURAL CREST_C

The nervous system develops from the dorsal surface of the ectodermal germ layer (future skin) of the embryo. In the 20- to 21-day embryo, a longitudinal groove (*neural groove*) begins to form on this thickened layer (*neural plate*). In the central part of the plate, the groove deepens, forming *neural folds* on either side. Deepening of the neural groove proceeds toward the head and tail ends of the embryo. By 22 days, the dorsal part of the folds fuse in the central part of the groove, forming a *neural tube*. During this process, the neural tube separates from the ectoderm. By 24 days, formation of the neural tube has progressed to the extreme ends of the embryo. Much of the neural tube will form the spinal cord; the head end of the tube will form the brain. The *neural crest* cells, formed from the neural folds, will develop into certain nerve cells of the peripheral nervous system and Schwann cells. The surrounding mesoderm will form the cranium and the vertebral column and related muscles. The notochord (a primitive supporting rod for the embryo) will be absorbed by the developing vertebral column, and remnants of it will remain as the core of the intervertebral discs (nucleus pulposus). The endoderm will contribute to the development of the digestive tract.

FOREBRAIN_D
TELENCEPHALON_E
DIENCEPHALON_F
MIDBRAIN_G
(MESENCEPHALON)_G
HINDBRAIN_H
METENCEPHALON_I
MYELENCEPHALON_J
SPINAL CORD_K

By the end of three weeks of embryonic development, three regions of the developing brain are apparent: *forebrain*, *midbrain*, and *hindbrain*. With further growth, the forebrain expands to form the massive *telencephalon* (endbrain; future cerebral hemispheres) and the more central *diencephalon* ("between" brain; future top of the brain stem). The midbrain retains its largely tubular shape as the *mesencephalon* (midbrain; future upper brain stem). The hindbrain differentiates into the upper *metencephalon* ("change" brain; future middle brain stem) with a large dorsal outpocketing (future cerebellum), and the lower *myelencephalon* (spinal brain; lowest part of the future brain stem). The brain stem narrows to become the *spinal cord* at the level of the foramen magnum of the skull.

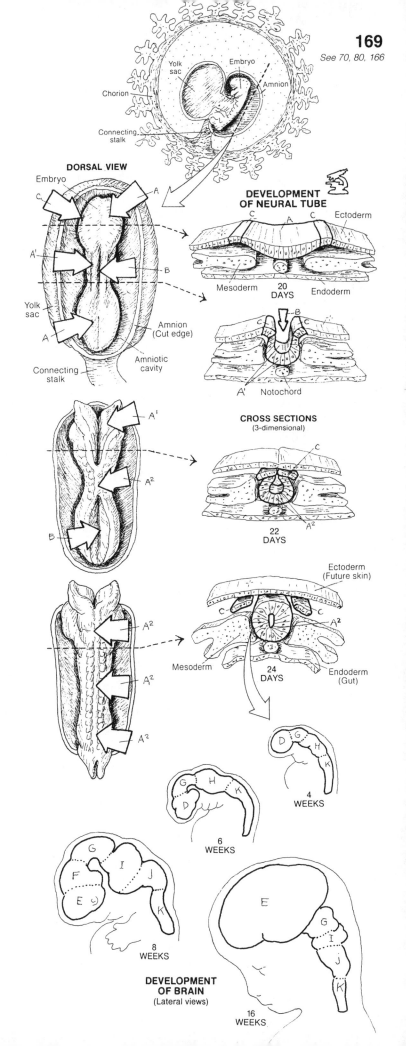

DORSAL VIEW

DEVELOPMENT OF NEURAL TUBE

CROSS SECTIONS
(3-dimensional)

DEVELOPMENT OF BRAIN
(Lateral views)

FETAL CIRCULATION

PLACENTA A
OXYGEN-RICH BLOOD ⟹ B
UMBILICAL VEIN B'
DUCTUS VENOSUS C
OXYGEN-POOR BLOOD ⟹ D
MIXED BLOOD ⟹ E
FORAMEN OVALE F
DUCTUS ARTERIOSUS G
UMBILICAL ARTERY E'

CN: Use red for B (oxygenated blood, represented by a dotted arrow) and B¹ (umbilical vein). Use blue for D (deoxygenated blood, represented by a light-lined arrow). Use purple for E (mixed deoxygenated and oxygenated blood, represented by a dark-lined arrow) and E¹ (umbilical artery). Use bright colors for C, F, and G. (1) Color the placenta and the large number 1, as well as the enlarged rectangular portion of the placenta magnified to show capillary exchange between fetal and maternal vessels. (2) Color the large numbers while coloring related structures and blood flow arrows. (3) Color the placenta and components of the umbilical cord in the uterus at lower left.

The fetus in the uterus does not breathe air; its lungs are deflated. This plate reveals how the fetus gets oxygen-rich blood to its system (in the absence of breathing air) and gets oxygen-poor blood out of the body.

The placenta (numeral 1) is an organ in the uterus of a pregnant woman that provides gaseous and nutritional support for the fetus. The placenta communicates with the fetus by an umbilical cord (2). The vessel taking oxygen-rich blood from the placenta to the fetus is the *umbilical vein* (2) which runs to the underside of the liver (3) to join the portal vein. Here the oxygen-rich blood of the former is mixed with the oxygen-poor blood of the latter. A vein existing only in the fetus (*ductus venosus*) diverts the blood directly to the hepatic vein, bypassing the liver sinusoids. The mixed blood then enters the inferior vena cava to the right heart. The blood is directed to the left (systemic) side of the heart by two means: an opening in the interatrial wall (*foramen ovale*; 4) and a short vessel between the pulmonary trunk and the descending part of the aortic arch (*ductus arteriosus*; 5). Only a fraction of mixed blood gets pumped to the non- functioning (but living) lungs. The mixed blood leaves the heart via the aorta (6) to reach the body tissues. The oxygen-carrying capacity of fetal hemoglobin is particularly great in comparison with that of the adult; the fetal tissues get sufficient oxygenation from mixed blood to permit remarkably rapid growth.

Paired umbilical arteries, arising from the internal iliacs, return the oxygen-poor blood from the fetus to the umbilical cord and placenta. After birth, because of altered hemodynamic patterns associated with breathing, the circulation in the fetal umbilical vessels and ducts of the newborn diminishes significantly and the vessels soon thrombose. The umbilical vein atrophies to become the ligamentum teres in the falciform ligament (Pl. 106); the umbilical arteries become the medial umbilical ligaments (Pl. 75); the ductus venosus becomes the ligamentum venosum; revised circulation to the lungs induces closure of the foramen ovale; flow through the ductus arteriosus trickles down and the vessel closes and becomes a ligamentous strand (ligamentum arteriosum; Pl. 66).

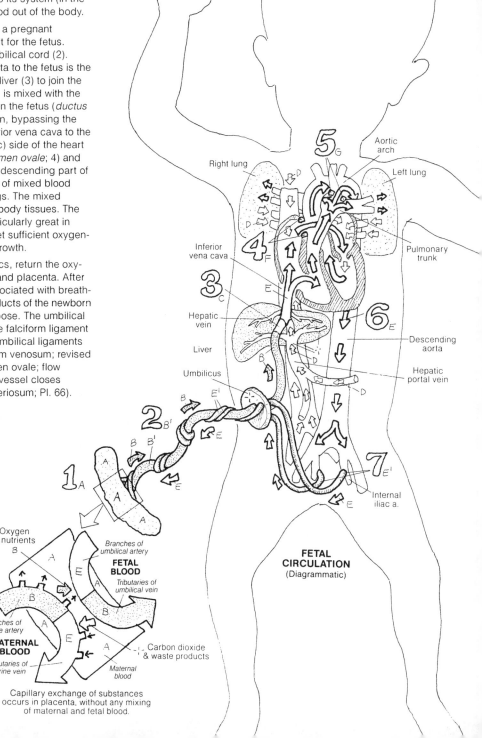

Right lung
Aortic arch
Left lung
Inferior vena cava
Pulmonary trunk
Hepatic vein
Liver
Umbilicus
Descending aorta
Hepatic portal vein
Internal iliac a.

FETAL CIRCULATION
(Diagrammatic)

Uterine wall
Umbilical cord

Oxygen & nutrients
Branches of umbilical artery
FETAL BLOOD
Tributaries of umbilical vein
Branches of uterine artery
MATERNAL BLOOD
Tributaries of uterine vein
Carbon dioxide & waste products
Maternal blood

Capillary exchange of substances occurs in placenta, without any mixing of maternal and fetal blood.

BIBLIOGRAPHY AND REFERENCES

Agur, A. (ed), *Grant's Atlas of Anatomy*, 10th ed., Williams & Wilkins, Baltimore, 1999

Alcamo, I. Edw., L.M. Elson, *The Microbiology Coloring Book*, HarperCollins, New York, 1996

Basmajian, J., *Muscles Alive*, 5th ed., Williams & Wilkins, Baltimore, 1985

Bergman, R. S. Thompson, A. Affifi, F. Saadeh, *Compendium of Human Anatomic Variation*, Urban and Schwarzenberg, Baltimore, 1988

Dawson, D., M. Mallett, L. Millender, *Entrapment Neuropathies*, 2nd ed., Little Brown and Co., Boston, 1990

Diamond, M.C., A.B. Scheibel, L.M. Elson, *The Human Brain Coloring Book*, HarperCollins, New York, 1985.

DuBrul, L., *Sicher's Oral Anatomy*, 7th ed., C.V. Mosby, St. Louis, 1980

Guyton, A.C., J.E. Hall, *Textbook of Medical Physiology*, 10th ed., W.B. Saunders, Philadelphia, 2000

Haymaker, W., B. Woodhall, *Peripheral Nerve Injuries: Principles of Diagnosis*, 2nd ed., W.B. Saunders, Philadelphia, 1953

Hoppenfeld, S. *Physical Examination of the Spine and Extremities*, Appleton-Century-Crofts, New York, 1976.

Junqueira, L.C., J. Carneiro, R. Kelley, *Basic Histology*, 9th ed., Appleton & Lange, Connecticut, 1998

Kapit, W., R. Macey, E. Meisami, *The Physiology Coloring Book*, Benjamin Cummings, San Francisco, 2000

Levinson, W., E. Jawetz, *Medical Microbiology & Immunology*, Lange Medical Books/McGraw-Hill, New York 2000

Lockhart, R.D., G.F. Hamilton, F.W. Fyfe, *Anatomy of the Human Body*, 2nd ed., J.B. Lippincott Co., Philadelphia, 1959

Marieb, E.N., J. Mallatt, *Human Anatomy*, 3rd ed., Benjamin Cummings, San Francisco, 2001

Moore, K. *The Developing Human: Clinically Oriented Embryology*, 6th ed., W.B. Saunders, Philadelphia, 1998

Netter, F., S. Colacino (ed.), *Atlas of Human Anatomy*, Ciba-Geigy Corp., Summit, N.J., 1998

_____ , *Nomina Anatomica*, 6th edition, Churchill Livingstone, New York, 1989

Roberts, M., J. Hanaway, *Atlas of the Human Brain in Section*, 2nd ed., Lea and Febiger, Philadelphia, 1986

Rothman, S., W. Glenn, *Multiplanar CT of the Spine*, University Park Press, Baltimore, 1985

Skinner, H., *The Origin of Medical Terms*, 2nd ed., Williams & Wilkins, Baltimore, 1961

Warfel, J., *The Head, Neck, and Trunk: Muscles and Motor Points*, 6th ed., Lea and Febiger, Philadelphia,1993

Warfel, J., *The Extremities*, 6th ed., Lea and Febiger, Philadelphia, 1993

White, A., M. Panjabi, *Clinical Biomechanics of the Spine*, 2nd ed., J.B. Lippincott, Philadelphia,1990.

Williams, P.L. (ed.), *Gray's Anatomy*, 38th ed., Churchill Livingstone, New York, 1995

ANSWER KEYS

PLATE 36
Bones of the upper limb
A   Clavicle
B   Scapula
C   Humerus
D   Ulna
E   Radius
F   Carpals
G   Metacarpals
H   Phalanges

Joints of the upper limb
1   Acromioclavicular joint
2   Glenohumeral
    (Scapulohumeral) joint
3   Sternoclavicular joint
4   Humeroulnar joint
5   Radiohumeral joint
6   Proximal Radioulnar joint
7   Distal Radioulnar joint
8   Radiocarpal joint
9   Intercarpal joint
10 Carpometacarpal joint
11 Intermetacarpal joint
12 Metacarpophalangeal joint
13 Interphalangeal joints

PLATE 43
Bones of the lower limb
A   Hip
B   Femur
C   Patella
D   Tibia
E   Fibula
F   Tarsal
G   Metatarsals
H   Phalanges

Bones of the upper limb
A^1 Scapula
B^1 Humerus
D^1 Ulna
E^1 Radius
F^1 Carpals
G^1 Metacarpals
H^1 Phalanges

Joints of the lower limb
1   Sacroiliac joint
2   Hip joint
3   Patellofemoral joint
4   Tibiofemoral joint
5   Proximal tibiofibular joint
6   Distal tibiofibular joint
7   Ankle joint
8   Intertarsal joint
9   Tarsometatarsal joint
10 Intermetatarsal joint
11 Metatarsophalangeal joint
12 Interphalangeal joints

PLATE 60
Muscles of the upper limb
MUSCLES ACTING PRIMARILY ON THE SCAPULA
A   Trapezius
A^1 Rhomboids
A^2 Serratus anterior

MUSCLES MOVING THE SHOULDER JOINT
B   Deltoid
B^1 Pectoralis major
B^2 Latissimus Dorsi
B^3 Infraspinatus
B^4 Teres Minor
B^5 Teres Major
B^6 Coracobrachialis

MUSCLES MOVING ELBOW & RADIOULNAR JOINTS
C   Biceps brachii
C^1 Brachialis
C^2 Triceps brachii
C^3 Anconeus
C^4 Brachioradialis
C^5 Pronator teres

MUSCLES MOVING WRIST & HAND JOINTS
D   Flexor carpi radialis
D^1 Palmar longus
D^2 Flexor carpi ulnaris
D^3 Extensor carpi radialis longus
D^4 Extensor carpi radialis brevis
D^5 Extensor digitorum
D^6 Extensor digiti minimi
D^7 Extensor carpi ulnaris

FOREARM MUSCLES MOVING THE THUMB
E   Abductor pollicis
E^1 Extensor pollicis longus
E^2 Extensor pollicis brevis

THENAR MUSCLES MOVING THE THUMB
F   Opponens pollicis
F^1 Abductor pollicis brevis
F^2 Flexor pollicis brevis

HYPOTHENAR MUSCLES MOVING THE 5TH DIGIT
G   Opponens digiti minimi
G^1 Abductor digiti minimi
G^2 Flexor digiti minimi brevis

OTHER MUSCLES ACTING ON THE THUMB & FINGERS
H   Adductor pollicis
H^1 Lumbricals
H^2 Dorsal interosseous

ANSWER KEYS

PLATE 68
Muscles of the lower limb

MUSCLES ACTING PRIMARILY ON THE HIP JOINT
A   Obturator internus
A^1 Iliopsoas
A^2 Gluteus medius
A^3 Tensor fasciae latae
A^4 Gluteus maximus
A^5 Pectineus
A^6 Adductor longus
A^7 Adductor magnus

MUSCLES ACTING PRIMARILY ON THE KNEE JOINT
B   Rectus femoris
B^1 Vastus lateralis
B^2 Vastus medialis
B^3 Sartorius
B^4 Gracilis
B^5 Biceps femoris
B^6 Semitendinosus
B^7 Semimembranosus

MUSCLES ACTING PRIMARILY ON THE ANKLE JOINTS
C   Gastrocnemius
C^1 Plantaris
C^2 Soleus
C^3 Flexor digitorum longus
C^4 Flexor hallucis longus
C^5 Tibialis anterior
C^6 Extensor digitorum longus
C^7 Extensor hallucis longus
C^8 Fibularis (Peroneus) tertius

MUSCLES ACTING PRIMARILY ON THE SUBANKLE JOINTS
D   Fibularis (Peroneus) longus
D^1 Fibularis (Peroneus) brevis

MUSCLES ACTING ON DIGITS OF THE FOOT
E   Abductor hallucis
E^1 Abductor digiti minimi
E^2 Extensor digitorum brevis

PLATE 115
Review of major arteries

A Aortic Arch

ARTERIES OF THE UPPER LIMB
B Brachiocephalic
C Subclavian
D Axillary
E Brachial
F Radial
G Ulnar
H Deep palmar arch
I Superficial palmar arch
J Palmar digital

ARTERIES OF THE HEAD AND NECK
K Common carotid
L Internal carotid
M External carotid

ARTERIES OF THE CHEST
A   Aortic arch
A^1 Thoracic aorta
N   Intercostal
0   Internal thoracic
P   Musculophrenic
Q   Superior epigastric
R   Pulmonary trunk
S   Pulmonary

ARTERIES OF THE ABDOMEN AND PELVIS
A^2 Abdominal aorta
T   Celiac
U   Superior mesenteric
V   Inferior mesenteric
W   Renal
X   Testicular/Ovarian
Y   Common iliac
Z   Internal iliac
1   External iliac
2   Inferior epigastric

ARTERIES OF THE LOWER LIMB
3   Femoral
4   Popliteal
5   Anterior tibial
6   Dorsalis pedis
7   Arcuate
8   Dorsal metatarsal
9   Dorsal digital
10 Posterior tibial
11 Fibular
12 Medial plantar
13 Lateral plantar
14 Plantar arch

PLATE 120
Review of major veins

VEINS OF THE UPPER LIMB
A Dorsal digital
B Dorsal digital network
C Basilic
D Cephalic
E Brachial
F Axillary
G Subclavian
H Brachiocephalic
I Superior vena cava
J Digital
K Superficial palmar arch
L Deep palmar arch
M Radial
N Ulnar

VEINS OF THE HEAD AND NECK
O Internal jugular
P External jugular

VEINS OF THE CHEST
Q Pulmonary
R Intercostal
S Azygos
T Thoracoepigastric

VEINS OF THE LOWER LIMB
U Dorsal digital
V Dorsal metatarsal
W Dorsal venous arch
X Great saphenous
Y Lesser saphenous
Z Plantar digital
1   Plantar metatarsal
2   Deep plantar venous arch
3   Medial plantar
4   Lateral plantar
5   Posterior tibial
6   Dorsal
7   Anterior tibial
8   Popliteal
9   Femoral

VEINS OF THE PELVIS AND ABDOMEN
10 External iliac
11 Internal iliac
12 Common iliac
13 Testicular/Ovarian
14 Renal
15 Inferior mesenteric
16 Splenic
17 Superior mesenteric
18 Gastric
19 Hepatic portal
20 Hepatic
21 Inferior vena cava

The spinal cord segments and spinal nerve roots give origin to spinal nerves whose axons are distributed among the peripheral nerves. Cranial nerves are not included here, but can be reviewed in Plate 83. The motor axons of these nerves innervate skeletal muscle. The loss of a nerve supply threatens the life of a skeletal muscle. When a muscle is denervated, its functional loss is usually characterized by sensory loss, reduction of a related deep tendon reflex, and muscle atrophy/weakness. Nerve roots shown in boldface are particularly likely to reveal deficits (radiculopathy) if injured. Here are listed the skeletal muscles, the nerves that supply them, and the related spinal nerve roots and spinal cord segments. A nerve in parentheses indicates small contribution.

SPINAL INNERVATION OF SKELETAL MUSCLES

SKELETAL MUSCLE	NERVE SUPPLY	SPINAL CORD SEGMENT/ NERVE ROOT
NECK		
Sternocleidomastoid	Spinal Accessory/C2, 3, 4	C1 – C5
Geniohyoid	C1 by way of Hypoglossal	C1
Sternohyoid	Ansa Cervicalis	C1 – C3
Sternothyroid	Ansa Cervicalis	C1 – C3
Thyrohyoid	C1 by way of Hypoglossal	C1
Omohyoid	Ansa Cervicalis	C1 – C3
Longus Colli/Capitis	Muscular Brs.	C2 – C6
Rectus Capitis	Muscular Brs.	C1 – C2
Scalenus Anterior	Muscular Brs.	C4 – C6
Scalenus Medius/Posterior	Muscular Brs.	C3 – C8

SPINAL INNERVATION OF SKELETAL MUSCLES

SKELETAL MUSCLE	NERVE SUPPLY	SPINAL CORD SEGMENT/ NERVE ROOT
UPPER LIMB		
Trapezius	Spinal Accessory	C1 – C5
Rhomboids Major/Minor	Dorsal Scapular	C4 – C5
Levator Scapulae	Dors. Scap. (C5); Musc. Br.	C3 – C5
Serratus Anterior	Long Thoracic	C5 – C7
Pectoralis Minor	Med./Lat. Pectoral	C5 – T1
Subclavius	Nerve to Subclavius	C5 – C6
Supraspinatus	Suprascapular	C5 – C6
Infraspinatus	Suprascapular	C5 – C6
Subscapular	Upper/Lower Subscapular	C5 – C6
Teres Minor	Axillary	C5 – C6
Deltoid	Axillary	**C5** – C6
Pectoralis Major	Medial/Lateral Pectoral	C5 – T1
Latissimus Dorsi	Thoracodorsal	C6 – C8
Teres Major	Lower Subscapular	C5 – C7
Biceps Brachii	Musculocutaneous	**C5** – C6
Brachialis	Musculocutan./(Radial)	C5 – (C7)
Coracobrachialis	Musculocutaneous	C5 – C7
Brachioradialis	Radial	C5 – **C6**
Triceps Brachii	Radial	C6, **C7**, C8
Anconeus	Radial	C6 – C8
Supinator	Radial	**C6** – C7
Pronator Teres	Median	C6 – C7
Pronator Quadratus	Median	C7 – C8
Palmaris Longus	Median	C7 – T1
Palmaris Brevis	Ulnar	C8 – T1
Flexor Carpi Radialis	Median	C6 – C7
Flexor Carpi Ulnaris	Ulnar	C7, **C8**, T1
Flexor Digit. Superficialis	Median	C8, T1
Flexor Digit. Profundus	Median/Ulnar	C8, T1
Flexor Pollicis Longus	Median	C7, C8
Thenar Muscles	Median	C8, T1
Hypothenar Muscles	Ulnar	C8, T1
Hand Intrinsic Muscles	Ulnar	C8, T1
Interossei	Ulnar	C8, T1
Lumbricales 1, 2	Median	C8, T1
Lumbricales 3, 4	Ulnar	C8, T1
Wrist Extensors	Radial	C6 – C8
Digit Extensors	Radial	**C7**, C8

SPINAL INNERVATION OF SKELETAL MUSCLES

SKELETAL MUSCLE	NERVE SUPPLY	SPINAL CORD SEGMENT/ NERVE ROOT
LOWER LIMB		
Psoas Major	Lumbar Musc. Brs.	L1 – L3
Psoas Minor	Lumbar Musc. Br.	L1
Iliacus	Femoral	L2 – L3
Adductors of the Hip	Obturator	L2, L3, (L4)
Adduct. Magnus	Obturator/Sciatic	L2, L3, (L4)
Pectineus	Femoral/Obturator	L2, L3
Quadriceps Femoris	Femoral	L2 – L4
Sartorius	Femoral	L2 – L3
Tensor Fasciae Latae	Superior Gluteal	L4 – S1
Gluteus Maximus	Inferior Gluteal	L5, S1, (S2)
Gluteus Medius/Minimus	Superior Gluteal	L4 – S1
Hamstrings (Post. Thigh)	Sciatic	L5 – S2
Lateral Hip Rotators	Sacral Plexus	L5 – S2
Piriformis	Nerve to Piriformis	L5 – S2
Obturator Internus	Nerve to Obturator Int.	L5 – S1
Obturator Externus	Obturator (Posterior Br.)	L3 – L4
Gemelli Sup. /Inf.	N. to Obt. Int./N. to Quad. F.	L5 – S1
Quadratus Femoris	N. to Quadratus Femoris	L5 –S1
Tibialis Anterior	Deep Fibular (Peroneal)	**L4** – L5
Extensor Hallucis Longus	Deep Fibular (Peroneal)	L5
Extensor Digitorum Long.	Deep Fibular (Peroneal)	**L5**, S1
Fibularis (Peron.) Tertius	Deep Fibular (Peroneal)	L5 – S1
Fibularis Longus/Brevis	Superficial Fibular (Peroneal)	L5 – S1
Gastrocnemius/Soleus	Tibial	S1 – S2
Plantaris	Tibial	S1 – S2
Tibialis Posterior	Tibial	L4 – L5
Flexor Hallucis Longus	Tibial	L5, **S1**, **S2**
Flexor Digitorum Longus	Tibial	L5 – **S2**
Foot Intrinsic Muscles	Tibial/Plantar	L5 – S3

SPINAL INNERVATION OF SKELETAL MUSCLES

SKELETAL MUSCLE	NERVE SUPPLY	SPINAL CORD SEGMENT/ NERVE ROOT
THORACIC WALL		
Ext./Int./In. Intercostals	Intercostal	T1 – T11
Thoracic Diaphragm	Phrenic	C3 – C5
Serratus Post. Superior	Thoracic Posterior Rami	T1 – T3
Serratus Post. Inferior	Thoracic Posterior Rami	T9 – T12
Subcostalis/Transv. Thor.	Intercostal	T12/T1 – T11
ABDOMINAL WALL		
External/Internal Oblique	Thor./Lumbar Post. Rami	T6 – T12, L1
Cremaster (from Int. Obl.)	Genito-Fem. N./Gen. Br.	L1 – L2
Transversus Abdominis	Thor./Lumbar Post. Rami	T6 – T12, L1
Rectus Abdominis	Thoracic Post. Rami	T5 – T12
Pyramidalis	Subcostal	T12
Quadratus Lumborum	Thor./Lumbar Ant. Rami	T12, L1 – L3
DEEP BACK		
Suboccipital Muscles	Cervical Posterior Ramus	C1
Erector Spinae Muscles	Post. Rami all Spinal Ns.	C2 – S1
Splenius Capitis/Cervicis	Post. Rami of Cerv. Sp. Ns.	C3 – C8
Transversospinalis Ms.	Post. Rami of Spinal Ns.	C2 – S1
PELVIS/PERINEUM		
Levator Ani	Pudendal/Sacr. Ant. Rami	S2 – S3
Coccygeus	Ant. Rami of Sacral Ns.	S3 – S4; Co 1 (?)
Perineal Muscles	Pudendal/Sacral Sp. Ns./ Pelvic Splanchnic Ns.	S2 – S4
Urethral Sphincter Ms.	Pudendal/Sacral Sp. Ns./ Pelvic Splanchnic Ns.	S2 – S4
Anal Sphincter Ms.	Perineal/Rectal Branches of Pudendal / S4 Sp. N.	S2 – S4

GLOSSARY

Anatomical terms as set forth and revised by the International Anatomical Nomenclature Committee of the International Congress of Anatomists, published in the 6th Edition of the *Nomina Anatomica* (1989), are included herein. For further inquiry, consult a standard medical dictionary. The terms here are compatible with those listed in *Dorland's Illustrated Medical Dictionary*, 27th Edition. Pronunciation of terms is given phonetically (as they sound, not by standard dictionary symbols). The primary accent (emphasis) is indicated by capitalized letters, e.g., ah-NAT-oh-mee, included with the definitions. The plural form is in parentheses following the term defined, e.g., alveolus(i) or alveoli. Pl. = plural.

A

A-, an-, without.

Ab-, away from the midline.

AB, antibody.

Abdomen, the region between the thoracic diaphragm and the pelvis.

Abscess (AB-sess), a cavity in disintegrating tissue, characterized by the presence of pus and infective agents.

Achilles, in Greek mythology, one of the sons of Peleus, a young king, and Thetis, one of the immortal goddesses of the sea. Not wanting Achilles to be mortal like his father, Thetis dipped him into the River Styx, holding him by the heel cord (tendocalcaneus), making him invulnerable to harm except at that spot. Achilles later became a great Greek warrior. In the many wars between Greece and Troy, Achilles was invulnerable to harm. At last, a Trojan, aided by the god Apollo, slew Achilles with an arrow into the vulnerable heel cord. The term "Achilles heel" refers to one's vulnerabilities; the Achilles tendon is the tendocalcaneus.

Acinus(i) (ASS-ee-nus), a saclike gland.

Actin, a protein of muscle, associated with the contraction/relaxation of muscle cells.

Ad-, toward the midline.

Adeno- (ADD-eh-no), gland.

Afferent, leading to a center.

Ag, antigen.

AIDS, acquired immunodeficiency syndrome.

-algia, pain.

Alimentary canal, the digestive tract from mouth to anus.

Alveolus(i) (al-VEE-oh-lus), grape-shaped cavity, rounded or oblong. Refers to the shape of exocrine glands, air spaces within the lungs, and the bony sockets for teeth.

Amino acid (ah-MEEN-oh), a two-carbon molecule with a side chain that contains either nitrogen (in the form of NH_2) or a carboxyl group (-COOH).

Amorphous (ay-MORF-us), without apparent structure at some given level of observation. What appears amorphous at 1000X magnification may be quite structured at 500,000X.

Amphi-, double, about, around, both sides.

Amphiarthrosis(es) (AM-fee-ar-THRO-sis), see joint classification, functional.

Ampulla(e), dilatation of a tubular structure.

Anastomosis(es) (ah-NASS-toh-moh-sis), connection between two vessels.

Anatomy (ah-NAT-oh-mee), *ana* = up, *tome* = to cut; the study of structure.

Anemia (ah-NEE-mee-ah), a condition of inadequate numbers of red blood corpuscles.

Angina (an-JYNE-ah), pain, especially cardiac pain.

Angio-, a vessel.

Angle, the point of junction of two intersecting lines, as in the inferior angle of the scapula between the vertebral (medial) and axillary (lateral) borders of that bone.

Angulus(i), an angle.

Ankle, the tarsus. The region between the leg and foot.

Annulus(i) (AN-new-lus), a ringlike or circular structure.

Ano-, anus.

Anomaly (ah-NOM-ah-lee), an abnormality, especially in relation to congenital or developmental variations from the normal.

A.N.S. or ANS, autonomic nervous system.

Ansa, loop.

Anserine, like a goose. *Pes anserinus*, goose foot.

Ante- (AN-tee), forward.

Antebrachium, forearm.

Antecubital, in front of the elbow (cubitus).

Anti-, against.

Antibody, a complex protein (immunoglobulin). A product of activated B lymphocytes and plasma cells, it is synthesized as part of an immune response to the presence of a specific antigen.

Antigen, any substance that is capable of inciting an immune response and reacting with the products of that response. Antigens may be in solution (toxins) or may be solid structures (microorganisms, cell fragments, and so on). Particulate matter that is phagocytosed but does not incite an immune response does not constitute antigen. Specific antibodies formed by cloning (monoclonal antibodies) may react with certain surface molecules on a cell membrane; those surface molecules constitute antigens.

Antigenic determinant, the specific part of an antigen that reacts with the product of an immune response (antibody, complement).

Aperture (AP-er-chur), an opening.

Apical, an apex or pointed extremity.

Aponeurosis(es), a flat tendon.

Apophyseal (app-oh-FIZZ-ee-al), refers to apophysis.

Apophysis(es) (ah-POFF-ee-sis), an outgrowth; a process.

Arborization (ar-bor-eye-ZAY-shun), branching of terminal dendrites.

GLOSSARY

Areolar (ah-REE-oh-lar), filled with spaces.

Arm, that part of the upper limb between the shoulder and elbow joints.

Arrhythmia(s) (a-RITH-mee-ah), a variation from the normal rhythm of the heartbeat; the absence of rhythm.

Arterio-, artery.

Arthr- (AR-thr), joint.

Arthritis(ides) (ar-THRI-tiss), inflammation of a joint.

Articular, joint.

Articular process, an outgrowth of bone on which there is a cartilaginous surface for articulation with another similar surface.

Articulation, a joint or connection of bones, movable or not; occlusion between teeth; enunciation of words.

Aspera, rough.

Aster, a ray, as in rays of light; in the cell, rays of microtubules projecting from centrioles.

Atherosclerosis, a form of arteriosclerosis or hardening of the arteries; specifically, characterized by yellowish plaques of cholesterol and lipid in the tunica intima of medium and large arteries.

ATP, adenosine triphosphate, a nucleotide compound containing three high-energy phosphate bonds attached to a phosphate group; energy is released when the ATP is hydrolyzed to adenosine diphosphate and a phosphate group.

Atrophy (AT-troh-fee), usually associated with decrease in size, as in muscle atrophy.

Avascular (ay-VASS-kew-lur), without blood vessels or, in some cases, blood.

Avulsion, tearing a part away from the whole, as in tearing a tendon from its attachment to bone.

B

Back, the region making up the posteriormost wall of the thorax and abdomen, supported by the thoracic and lumbar vertebrae. Strictly defined, it excludes the neck and sacrum/coccyx (pelvis).

Basal lamina(e), a thin layer of interwoven collagen fibrils interfacing epithelial cells (and certain other nonepithelial cells) and connective tissue. Seen only with an electron microscope.

Basement membrane, basal lamina and a contiguous layer of collagenous tissue. Seen with a light microscope, it controls diffusion and transport into/out of the cell.

Basilar, at the base or bottom.

Benign, nonmalignant; often used to mean mild or of lesser significance.

Bi-, two.

Bicipital, two-headed.

Bicuspid, a structure, e.g., a tooth or valve, with two cusps.

Bifurcate (BY-fur-cate), to branch.

Bilateral, both sides (left and right).

-blast, formative cell; immature form.

Blephar-, eyelid.

Blood-borne, refers to some structure carried by the blood.

Blood–brain barrier, a state in the CNS in which substances toxic or harmful to the brain are physically prevented from getting to the brain; it is represented by tight endothelial junctions in capillaries of the brain, tight layers of pia mater around vessels, and the presence of neuroglial endfeet surrounding vessels.

Bolus, a mass of food; any discrete mass.

Bone, immature, see bone, woven.

Bone, lamellar, mature bone characterized by organized layers or lamellae of bone.

Bone, mature, see bone, lamellar.

Bone, primary, see bone, woven.

Bone, secondary, see bone, lamellar.

Bone, woven, immature bone characterized by random arrangements of collagen tissue and without the typical lamellar organization seen in more mature bone.

Brachi-, arm.

Bronch-, referring to bronchi or bronchioles of the respiratory tract.

Bursa(e), synovial-lined sac between tendons and bone or muscle and muscle, or any other site in which movement of structure tends to irritate or injure adjacent structure. It contains synovial fluid and is lined externally by fibrous connective tissue.

Bursitis, inflammation of a bursa.

C

CD4, CD, "clusters of differentiation." The abbreviation refers to a collection of cell surface molecules with specific structural characteristics (markers) reflecting a common lineage. The identification of these markers is made by purebred (monoclonal) antibodies, which react only with surface markers of cells of a common lineage. Cells exhibiting cell surface markers of a common lineage belong to a cluster (of differentiation), identified by number, e.g., 4. Most helper T lymphocytes have markers of three different clusters—CD3, CD4, and CD8. Cytotoxic T lymphocytes are CD3, CD4, and CD8.

Cadaver (ka-DA-ver), a dead body.

Canaliculus(i), a small canal.

Cancellous (KAN-sell-us), having a lattice-like or spongy structure with visible holes.

Cancer, a condition in which certain cells undergo uncontrolled mitoses with invasiveness and metastasis (migrating from the point of origin to other sites, usually by way of the lymphatic and/or blood vascular systems). There are two broad divisions: carcinoma, cancer of epithelial cells; sarcoma, cancer of the connective tissues.

Capillary attraction, the force that attracts fluid to a surface, such as water flowing along the undersurface of a pouring tube.

Capitulum, a rounded process of bone, usually covered with articular cartilage. Synonym: capitellum.

Caput medusae, the head of the mythical Medusa, whose golden hair entranced Neptune. The jealous Minerva turned the hair of Medusa into a mass of snakes. The term "caput medusae" is used for the snake-like appearance of the dilated, interwoven mass of subcutaneous veins surrounding the umbilicus in the condition of portal vein obstruction.

Cardio-, heart

Carpus, carpo-, wrist.

Cauda equina (horse's tail), the vertically oriented bundle of nerve roots within the vertebral canal below the level of the first lumbar vertebra (L1). Includes nerve roots for spinal nerves L2 through the Co2, bilaterally.

Cauda equina syndrome, irritation/compression of the cauda equina, resulting in bilateral symptoms and signs that may include bladder and bowel incontinence, weakness in the lower limb muscles, sensory impairment from the perineum to the toes, and reflex changes.

Cauterization, destruction of tissue by heat, as with an electrocauterizing instrument.

Cavity, potential, a space between membranes that can enlarge with fluid accumulation, as in the peritoneal cavity (ascites) or pericardial cavity (cardiac tamponade).

Cell body, the main, largest single mass of a neuron, containing the nucleus surrounded by organelles in the cytoplasm.

-centesis, puncture.

Central, at or toward the center.

Ceph-, head.

Cerebro-, brain; specifically, cerebral hemisphere.

Cerumen (sur-ROO-men), the wax secretion of the external ear.

Cerv-, neck.

Cheil- (KY-el), lip.

Chest, the thorax.

Chir- (kir), hand.

Choana(e) (KOH-ah-nah), referring to a funnel, as in the nasal passageways or apertures.

Chol- (koll), bile.

Chondro- (KOND-row), cartilage.

Chromosome (KRO-moh-sohm), "colored body."

Circulare(s), circle

-clast (klast), disruption, breaking up.

Clearing, the process of clearing water or solvent out of a specimen in preparation for microscopic study.

Cleavage, division into distinct parts.

Clinical, the setting in which a person is examined for evidence of injury or disease.

Clot, coagulated blood; a reticular framework of fibrin, platelets, and other blood cells. Associated fluid is serum.

cm, centimeter.

C.N.S. or CNS, central nervous system, consisting of the brain and spinal cord.

Co-, con-, together.

Coagulation, the clotting of the blood.

Coelom (SE-lom), the embryonic body cavity.

Collagen (KOLL-ah-jen), the protein of connective tissue fibers. Several different types are found in fasciae, tendons, ligaments, cartilage, bone, vessels, organs, scar tissue, and wherever support or binding is needed. Formed by fibroblasts, endothelia, muscle cells, and Schwann cells.

Collateral circulation, alternate circulatory routes; vessels between two or more points that exist in addition to the primary vessels between those points. Such circulation exists by virtue of anastamoses among a number of vessels.

Colli-, neck.

Colo-, colon.

Complement, a group of proteins in the blood whose activation causes their cleavage and fragmentation. The fragments have several biologic functions, of which one is combining with antibody/antigen complexes, enhancing the destruction of antigen.

Concentric contraction, a type of muscle contraction in which the internal contracting force of a muscle is greater than the external load imposed on it (positive work), so that the muscle shortens.

Conch (kawnk), a large spiral shell.

Concha(e) (KAWNK-ah, or KAWN-cha; pl. KAWNK-ee or KAWN-chee), a structure shaped like a conch shell.

Concretion, an inorganic or mineralized mass, usually in a cavity or tissue.

Condylar, condyloid, referring to a rounded process, as in a joint surface

Condyle, a rounded projection of bone; usually a joint surface, covered with articular cartilage.

Contiguous (kon-TIG-yu-us), adjoining and being in contact. The basement membrane is contiguous with the basal surfaces of certain epithelial cells.

Contra-, against.

Contraction, shortening.

Cornu(a) (KOR-new), a horn-shaped process.

Corona, crown.

Corona radiata, radiating crown. The term refers to the appearance of the subcortical white matter and, specifically, the projection system.

Coronoid (KOR-oh-noid), crownlike or beak-shaped; refers to a bony process.

Corpus(ora), body.

Corpuscle (KOR-pus-il), any small body, not necessarily a cell. Red blood corpuscles lack nuclei and are not considered cells.

Costa, rib

Costochondritis, an inflammation surrounding the cartilage of a joint of a rib, usually involving the synovium and fibrous joint capsule and perhaps related ligaments.

Coxa(e), hip; the hip (coxal) joint. Deformities of the upper femur often include the term (such as coxa varus or coxa valgus). Preceded by the term "os," it refers to the coxal or hip bone.

Crani-, cranium.

Cranium, that part of the skull containing the brain.

Cribriform, perforated; like a sieve.

-crine (krin), separate off, referring to glands that separate from classical epithelial surfaces.

Cruciate, shaped like a cross.

Crus (crura), leg.

Crux, cross.

Cu., cubic.

Cubital, front (anterior aspect) of the elbow.

Cusp, a triangular structure characterized by a tapering projection.

Cutan-, cutaneous (kew-TANE-ee-us), referring to the skin.

Cystitis, inflammation of the urinary bladder.

Cysto-, bladder.

-cyte (site), cell.

Cytokine, a product of a cell that facilitates destruction of antigen by inducing or enhancing an immune response.

Cytolysis, the dissolution and destruction of a cell.

Cytotoxin, a product of a cell that acts to destroy another cell or has a toxic effect.

D

Dachry-, relating to tears.

Dactyl, finger, toe.

Decussation, crossing over.

Defecation, elimination of waste material through the anal canal/anus from the rectum.

Deglutition, swallowing.

Demi-, half.

Denervation (dee-nerv-AY-shun), a condition in which a muscle or area of the body is isolated from its nerve supply.

Dentin (DEN-tin), the hard portion of a tooth. It is more dense (harder) than bone, less dense (softer) than enamel.

Depolarization, neutralization of a polarity; in biological systems, it is an electrical change in stimulated excitable tissues (nerves, specialized cardiac muscle cells) from a baseline polarity (about –90 millivolts) toward neutral (0 millivolts). Such an event induces the conduction of an electrochemical wave (impulse) to move along an excitable tissue (e.g., nerve).

Derm-, skin.

-desis, fixation.

Desiccation (dess-ee-KAY-shun), drying out; without water.

Desmo-, fibrous.

Dexterity, skill with the hands.

Di-, twice.

Diaphragm(ae) (DIE-ah-fram), a partition separating two cavities. There are three significant fibromuscular diaphragms in the body: thoracic (separating thorax and abdomen), pelvic (separating pelvis and perineum), and urogenital (separating the anterior recesses of the ischiorectal fossa from the superficial perineal space).

Diarthrosis(es) (die-ar-THRO-sis), see joint classification, functional.

Differentiation, making something different; in the development of a cell, it is the structural and functional changes within that cell that make it different from other cells; an increase in heterogeneity and diversification.

Diffusion, spontaneous movement of molecules without the application of additional forces.

Digit, finger or toe.

Diploic, referring to the marrow layer between the inner and outer layers of compact bone in the flat bones of the skull.

Dis-, apart.

Disc, a wafer-shaped, rounded or oval fibrocartilaginous structure; if crescent-shaped, it is called a meniscus. It may interface the articular cartilage surface in a synovial joint (articular disc) or it may interface opposing cartilage endplates of vertebral bodies (intervertebral disc).

Discharge, to set off or release, to fire, to let go.

Dissect (dis-SECT), to cut up, to take apart. In gross anatomy laboratories, the human body is studied by an ordered dissection by regions.

Dys-, abnormal, painful, or difficult.

Dorsum, back. Refers to the posterior aspect of the hand and the "top" of the foot.

E

Ec-, out.

Eccentric contraction, a type of muscle contraction wherein contracted muscle is stretched and lengthened during the contraction, such as antigravity contractions by antagonists during movement directed toward gravity. Even though there is a load on the muscle, the muscle is stretched (negative work).

-ectasis(es), dilatation.

-ectomy, removal.

Efferent, leading away from a center (organ or structure).

Elbow, the region between the arm and forearm.

Electrochemical, referring to combined properties of electrical and chemical, such as the neuronal impulse.

Ellipsoid, a closed curve more oval than a perfect circle. Ellipsoid joints are reduced forms of ball-and-socket joints; broadly speaking, they include condylar-shaped joints.

Em-, in.

Embalm (em-BAHM), to treat a dead body with preservative chemicals to prevent structural breakdown by microorganisms.

-emia, blood.

Emissary vein, a vein that drains a dural venous sinus and passes through the skull bone by way of a foramen.

Emission, an involuntary release of semen; also, the movement of sperm from the epididymis to the prostate during sexual stimulation in the male.

En-, in.

Encapsulate, to surround with a capsule.

Encephalo-, brain.

Endo-, in.

Endochondral (en-do-KON-dral), *endo* = in, *chondral* = cartilage.

Endochondral ossification, see ossification.

Endocrine (EN-do-krin), *endo* = in, *crine* = separate. Glands that secrete their products into the tissue fluids or vascular system.

Endocytosis, the ingestion of matter into a cell by surrounding the material with the cell membrane and budding it off in the cytoplasm.

Endometr-, endometrium.

Endosteum(a), the lining of the medullary canal of long bones, consisting of a thin sheet of collagen fibers and large numbers of osteoprogenitor cells.

Endothelium(a) (en-do-THEE-lee-um), the epithelial lining of blood and lymph vessels and the heart cavities. Endothelia are of mesenchymal origin, not ectodermal, and have properties different from classical epithelia.

Entero-, referring to the intestines.

Enteroendocrine, refers to cells of the epithelial layer/ glands of the gastrointestinal mucosa, which secrete hormones that stimulate/inhibit (regulate) intestinal/ pancreatic gland secretion and/or motility of smooth muscle.

Enzyme, a protein molecule that facilitates a reaction without becoming involved (changed or destroyed) in the reaction. Enzymes are identified by the suffix *-ase*.

Epi-, upon, at.

Epicondyle, an elevation of bone above a condyle.

Epidid-, epididymis.

Epidural, outside the dura, between the dura and the skull.

Epithelium(a) (ep-ee-THEE-lee-um), *epi* = upon, *thelia* = nipple.

Erg, a unit of work.

Ergo-, a combining form meaning "work."

Ex-, exo-, out.

Excretion (ex-CREE-shun), the discharging of or elimination of materials, such as waste matter. If the material excreted has some useful in-body function or use outside the body (e.g., semen), it has probably been secreted, not excreted, although there is no universal agreement on this. See secretion.

Exocrine (EX-oh-krin), *exo* = out, *crine* = separate off; referring to glands that separate from classical epithelial surfaces.

Exocytosis, removal of matter from a cell.

Extracellular, outside of the cell, such as the fibrous tissue supporting cells, and vascular spaces.

Extrinsic, coming from the outside. With reference to a specific area (e.g., thumb, hand, foot), extrinsic muscles are those with origins outside of the specific area, but which insert in the area and have an effect on the specific area. See intrinsic.

F

Facet (FASS-et), a small plane or slightly concave surface. The flat cartilaginous surfaces of a joint may be called facets, as on the articular processes of vertebrae.

Facet joint, a joint between articular processes of adjacent vertebrae; also called *zygapophyseal joints*.

Facilitation, enhancement of or assistance in an event.

Falx inguinalis (conjoint tendon), a tendon composed of fibers from transversus abdominis and internal oblique that arcs over the spermatic cord and attaches to the pectineal line of the pubic bone. See Plate 51.

Fascia(e) (FASH-uh, pl. FASH-ee), a general term for a layer or layers of loose or dense, irregular, fibrous connective tissue. Superficial fascia, often infiltrated with adipose tissue, is just under the skin. Deep fascia envelops skeletal muscle and fills in spaces between superficial fascia and deeper structure, and between/among muscle bellies (myofascial structure). Extensions of deep fasciae form intermuscular

septa, support viscera (e.g., endopelvic fascia), act as fibrous bands, and support neurovascular bundles. Smaller, microscopic layers of fibrous tissue (e.g., perimysium, endomysium, vascular tunics) do not constitute deep fascia, even though they may be distant extensions of it. These fibrous connective tissue investments, integrated with tendons, ligaments, periosteum, and bone, blend into a unibody construction, resistant to all but the most traumatic of forces.

Fascia, thoracolumbar, strong layer enveloping the deep back or paravertebral muscles from the iliac crest and sacrum to the ribs/sternum. Plays an important role in limiting and moving motion segments of the back.

Fascicle(s) (FASS-ih-kul), a bundle.

Feedback, a communication relationship between two structures, e.g., wherein the output (secretion) of one substance induces an inhibition or facilitation of the secretion of another substance. Negative feedback reflects an inhibitory effect; positive feedback reflects a facilitating relationship.

Fibers, elongated lengths of tissue, e.g., living muscle fibers (cells or their parts), connective tissue fibers (nonliving cell products), living nerve fibers (extensions of cell bodies).

Fibril (FY-brill), en elongated structure smaller than and part of a fiber.

Fibrous (FY-brus), referring to a fiber or fiberlike quality.

Fibrosus (fy-BROHS-us), a fibrous stucture.

Filament, a small delicate fiber; in biology, a structure of some length, often smaller than a fibril, which is smaller than a fiber.

Filtration, movement of a fluid by the application of a force, such as pressure, vacuum, or gravity.

Fissure, a narrow crack or deep groove.

Fixation, a process in preparation of tissue for microscopic study. Treatment of fresh tissue with a fixative preserves structure, preventing autolysis and bacterial degradation.

Flaccid (FLA-sid or FLAK-sid), without tone; denervated; lax or soft.

Foot, the most distal part of the lower limb. The skeleton of the foot consists of the tarsus, metatarsal bones, and phalanges. It joins with the leg at the ankle (talotibiofibular joint).

Foramen(ina) (foh-RAY-men), opening or hole.

Forearm, that part of the upper limb between the elbow and wrist (radiocarpal) joints.

Forefoot, that part of the foot anterior to the transverse tarsal (talonavicular and calcaneocuboid) joints.

Fossa(e), a depressed or hollow area; a cavity.

Fusiform, spindle-shaped; shaped like a round rod tapered at the ends.

G

Gastro-, stomach.

Gastrointestinal, stomach and intestines.

Genia-, origin.

Genital(s), L., belonging to birth. Refers to reproductive structures; loosely, the term refers to the external genitals of either sex.

Glia, see neuroglia.

Glomerulus, a small cluster of vessels or nerve endings, as in the glomerulus of the kidney.

Glosso-, tongue.

Glyco-, sweet, pertaining to sugar or carbohydrate, e.g., glycogen (starch), glycoprotein (sugar-protein complex).

Glycoprotein, an organic compound consisting of carbohydrate and protein.

Glycosaminoglycan, a long chain of double sugars (disaccharides) connected with a nitrogen-containing group (amine); *glyco* = sugar, *glycan* = polysaccharide. Previously termed mucopolysaccharide. Proteins combined with glycans are temed proteoglycans.

Gomphosis(es) (gom-PHO-sis), bolting together. See joint classification, structural.

Gray matter, brain and spinal cord substance consisting largely of neuronal cell bodies, glia, and unmyelinated processes. Collections of gray matter are generally called nuclei or centers.

Groove, a linear depression in bone.

H

Hallucis, genitive form of hallux.

Hallux, great (first) toe.

Hand, the most distal part of the upper limb. The skeleton of the hand consists of the carpus, metacarpus, and phalanges. It joins with the forearm at the wrist (radiocarpal) joint.

Haustra(e), sacculations of the large intestine held in tension by longitudinal bands of smooth muscle (taeniae).

Haversian system, a cylindrical arrangement of bone cells and their lacunae, named after C. Havers, a 17th-century anatomist; the central tubular cavity, the Haversian canal, contains vessels. Seen in compact bone.

Head, that part of the body supported by the skull and superior to the first cervical vertebra.

Hem-, blood.

Hematocrit (he-MAT-oh-krit), the measurement of red blood cell volume in a tube of centrifuged blood; the tube itself is called a hematocrit tube.

Hematoma (hee-mah-TOE-ma), *hemat* = blood, *oma* = tumor or swelling. A collection of blood under the skin, fascia, or other extracellular membrane.

Hematopoiesis (hee-mah-toh-po-EE-sus), blood cell formation; occurs in the bone marrow and, in early life, in the liver and spleen; blood cells include red blood corpuscles and white blood cells.

Hemi-, half.

Hemopoiesis (hee-mo-po-EE-sus), see hematopoiesis.

Hemorrhage (HEM-or-ij), bleeding; escaping of blood from blood vessels into the adjacent tissues or onto a surface.

Hemorrhoid, a varicose dilatation of a vein that is a part of the superior/inferior rectal (hemorrhoidal) plexus of veins.

Hemosiderin (hee-mo-SID-er-in), storage form of iron.

Heparin, a glycoprotein present in many tissues that has anticoagulation ("blood thinning") properties.

Hepat-, liver.

Herniation, a protrusion through a wall or wall-like structure.

Heterogeneous, varied, as in a mixture of nonuniform elements.

Hg, mercury (chemical symbol).

Hiatus, an opening.

Hindfoot, that part of the foot posterior to the transverse tarsal (talonavicular and calcaneocuboid) joints.

Hip, the coxal bone; the region of the hip (coxal) joint.

Histamine, a nitrogenous molecule whose effects include contraction of smooth muscle and capillary dilatation.

HIV, human immunodeficiency virus.

Homogeneous, of uniform quality.

Hydroxyapatite, Ca (PO) (OH), a mineral or inorganic compound that makes up the mineral substance of bone and teeth. A very similar structure is found in nature outside the body.

Hyper, excessive.

Hyperplasia, increased number of normal cells.

Hypertonia, increased muscle tension; increased resistance to stretching of muscle.

Hypertrophy, increase in size of muscle.

Hypo, inadequate or reduced.

Hypoesthesia, reduced sensation.

Hyster-, uterus.

I

-iasis, condition, presence of.

Ileo-, ileum of the small intestin.

Ilio-, ilium of the coxal (hip) bone.

Immuno-, refers to the immune system or to some activity or part of that system.

Immunosuppression, suppression of immune (lymphoid) system activity; also called immunodepression.

Impinge, to have an effect on something; contact, irritate, strike.

Infarction (in-FARK-shun), an area of dead tissue caused by interruption of the blood supply to the tissue.

Infection, the invasion of body cells, tissues, or fluids by microorganisms, usually resulting in cell or tissue injury, inflammation, and immune response.

Inflammation, a vascular response to irritation, characterized by redness, heat, swelling, and pain; may be acute or subacute (lasting more than two weeks, or chronic).

Infra-, under.

Inhibition, restraint or restraining influence.

Injury, anatomic disruption at some level of body organization in response to an external force (e.g., blunt, penetrating, electrical, radiation, thermal).

Innate, inborn, congenital.

Innervation (in-nerv-AY-shun), provision of one or more nerves to a part of the body.

Innominate, unnamed. First applied to the coxa (hip bone) by Galen; first applied to the artery by Vesalius.

GLOSSARY

Integument, the skin.

Inter-, between; e.g., interscapular, between the scapulae.

Intercalated, inserted between.

Interface, surfaces facing one another; to face a surface.

Interstitial, **interstices**, **interstitium**, interspaces of a tissue; between two or more definitive structures.

Intima, innermost part.

Intra-, within; e.g., intracellular, within a cell.

Intramembranous ossification, see ossification.

Intravenous, within a vein.

Intrinsic, part of a specific area and not extending beyond that area (e.g. thumb, hand, foot). Muscles that arise (originate) and insert within the hand region are known as intrinsic muscles (of the hand).

Investing, surrounding or enclosing.

Isometric contraction, a contraction that involves muscle contraction without bone movement, so that the muscle maintains the same apparent length. Fibril shortening in such a contraction is offset by the inherent elasticity of the myofascial tissue.

-itis, inflammation. Term does not specify the cause of inflammation; therefore, it does not mean infection, but may refer to the inflammation induced by or associated with an infection.

J

Jejuno-, jejunum of small intestine.

Joint classification, functional; joints are classified according to the degree of movement, i.e., immovable, partly movable, freely movable. Immovable joints are called synarthroses, partly moveable joints are called amphiarthroses, and freely movable joints are called diarthroses. Immovable joints may be fibrous (sutures, gomphoses) or cartilaginous (synchondroses). Synovial joints are not normally immovable. Partly movable joints may be fibrous (syndesmoses) or cartilaginous (symphyses). Freely movable joints are always synovial. Synovial joints are limited in their motion by joint architecture and ligaments, but within those limitations, they are normally freely movable. See also syn-.

Joint classification, structural; joints are classified according to the material that makes the joint, i.e., fibrous, cartilaginous, bony, synovial. Fibrous joints are further classified as sutures (thin fibrous tissue between flat bones of the skull), syndesmoses (ligamentous sheets between the bones of the forearm and leg), and gomphoses (fibrous tissue between tooth and bony socket). Cartilaginous joints are further classified as synchondroses (hyaline cartilage between the end and shaft of developing bone) and symphyses (fibrocartilaginous discs between bones, as between vertebral bones and between the pubic bones). Bony joints are fibrous or cartilaginous joints that have ossified over time (synostoses). Classification of synovial joints can be seen in Plate 22.

Jugular (JUG-yoo-lar), referring to the neck or a neck-like structure. Specifically refers to the vein(s) of the neck so named.

K

Kary-, nuclear.

Keratin, a sclero-protein that is insoluble and fibrous. It is the principal constituent of the outer layer of stratified squamous epithelia in skin (stratum corneum; see Plate 19), hair, and tooth enamel (Plate 138).

Kerato-, outer skin.

-kine, movement.

Kinin (KY-nin), a polypeptide (short protein) that influences reactions, such as antigen-antibody complexes.

Knee, the region between the thigh and the leg.

Kyphosis (ky-PHO-sis), humpback. Anatomically, a curve of the vertebral column in which the convexity is directed posteriorly; in orthopaedics, it is an excessive curvature of the thoracic vertebrae.

L

Labium(i), lip, or any fleshy border.

Labyrinthine (laba-RINTH-een), interconnecting, winding, as in an interwoven series of passageways.

Lacerum (lahss-AYR-um), an irregular aperture or opening.

Lacuna(e), a cavity or lake-like pit.

Lacrimal, referring to tears.

Lamella(e), a thin, plate-like structure; may be circular, as seen in the Haversian system of bone.

Lamina(e), layer.

Laryngo-, larynx.

Latency, inactivity. Usually a period between moments of activity.

Latent, see latency.

Leg, that part of the lower limb between the knee joint and the ankle joint.

-lemma, covering or sheath.

Lepto-, slender.

Leptomeninges, pia mater and arachnoid combined.

Levator, a lifter; an elevator.

Lieno-, spleen.

Ligament, fibrous tissue connecting bone to bone; also a peritoneal attachment between organs.

Lip-, pertaining to lipids; fat; triglyceride (composed of glycerol and three fatty acids).

-listhesis, slip.

Lith-, stone.

Lithotomy, removal of a stone.

Lordosis, a curve of the back seen in the cervical and lumbar regions in which the convexity is directed anteriorly; anatomically, it refers to any curve of the back so described; orthopaedically, it is an excessive curve as described.

Lumen(ina) (LEWM-un), a cavity, space, or tunnel within an organ.

Lunar, referring to the moon. **Semi-lunar**, half-moon-shaped.

Lymphatic, refers to the system of vessels concerned with drainage of body fluids (lymph).

Lymphoid, refers to the tissue or system of organs (lymphoid or immune system) whose basic structure is lymphocytes and reticular tissue.

Lymphokine (LIM-fo-kine), a product of activated lymphocytes that enters into solution and influences immune responses, generally by enhancing destruction of antigen.

-lysis(es) (LYE-sis), destruction or dissolution.

M

Macro, large, as in macromolecule.

Magnum, great.

-malacia, softening, as in demineralization of bone; changes in matrix of a tissue resulting in a loss of turgor or fibrous quality.

Mamm-, breast.

Manual, referring to the hand.

Manus, hand.

Mastication (*masticate*, to chew), the act of chewing.

Mastoid, breast-shaped.

Matrix(ices) (MAY-trix), fluid or viscous background or ground substance, often apparently amorphous and homogeneous, often colorless. A variety of organic compounds and minerals may be dispersed within.

Meatus (mee-AYT-us), an opening or passageway.

Media, middle.

Mediastinum(a) (mee-dee-ahs-TY-num), middle partition; the partition or septum between the lungs in the thorax.

Mediate, influence.

Mediator, an influential substance; a substance that acts indirectly but influentially in a reaction or in inducing a reaction.

Medulla, inner part.

Medusa, the radiating, contorted, dilated venous network bulging out on the surface of the anterior abdominal wall of chronic sufferers of portal vein hypertension/obstruction has been given the name caput Medusae (head of the Medusa). In Greek mythology, Medusa was one of the Gorgon sisters, characterized as winged monsters with heads of snakes in place of hair. When a person looked at one of them, he was turned to stone. Medusa was the only mortal Gorgon. In offering service to his tyrant king, Perseus pursued Medusa and cut off her head (which, though detached, still had the power to turn onlookers into stone). Perseus presented the head to the vile king and his men, who, upon casting their eyes on the snake-covered head, promptly turned to stone. Perseus then became king.

Mega-, big, great, as in megakaryocyte.

-megaly, enlargement.

Menin-, refers to meninges.

Meninges, dura mater, arachnoid, and pia mater coverings of the spinal cord and brain, and the first part of cranial and spinal nerves.

Ment-, referring to the chin, as in mental foramen.

Mesenchyme (mesenchymal), embryonic connective tissue, often with plenipotentiary cells.

Mesothelium(a) (meezo-THEE-lee-um), the epithelium lining the great (closed) body cavities, e.g., pleura, peritoneum, and pericardium. It is of mesenchymal origin, not ectodermal, and has different properties from classical epithelia.

Meta-, change.

Metr-, uterus.

Micro, small, as in microtubule.

Microorganism, one of a group of organisms including bacteria, viruses, fungi, protozoans, and other microscopic life forms.

Micturition, urination; discharge of urine outside the body.

Mineralization, the process of mineral (calcium complexes) deposition, especially in bone formation and remodeling as well as formation of teeth.

mm, millimeter.

mm Hg, millimeters of mercury. A pressure-measuring system in which the open end of an evacuated (vacuum) graduated cylinder (tube) is placed in a container of liquid mercury. The pressure of the atmosphere or fluid pressing on the mercury will push the mercury up the cylinder. The distance the mercury moves up the tube is measured in mm Hg and reflects the pressure imposed.

Modulate, to induce a change.

Modulator, a controlling element or agency.

Mortise, a recess that receives a part, as the talus fits into the recesses of the tibia and fibula.

Motor, referring to movement; with respect to the nervous system, refers to that part concerned with movement.

Mucosa(e) (mew-KOS-ah), a lining tissue of internal cavities open to the outside. Epithelial/gland cells secrete a mucus onto the free surface of the lining, which consists of epithelial lining cells, glands, and underlying connective tissue and nerves/vessels; it may have a thin layer of muscle.

Mucous, referring to mucus.

Mucus, a secretion of certain glandular cells, composed largely of glycoproteins in water, forming a slime-gel consistency, thicker than serous fluid.

Multi-, many.

Muscularis (muss-kew-LAHR-is), a layer of muscle.

Musculoligamentous, consisting of muscle and ligament.

Musculoskeletal, consisting of muscle, bones, ligaments, tendons, fasciae, and joints.

Musculotendinous, consisting of muscle and tendon.

Myelin (MY-eh-lin), compressed cell membranes of Schwann cells in the PNS and oligodendrocytes in the CNS, arranged circumferentially, in layers, around axons. Composed of cholesterol, components of fatty acids, phospholipids, glycoproteins, and water.

Myelo-, marrow; usually refers to spinal cord.

Myelopathy, neurologic deficit resulting from spinal cord injury or disease.

Myo-, referring to muscle.

Myoepithelium(a), contractile epithelial cells. Usually located at the base of gland cells, with tentacle-like processes embracing secretory cells. Particularly prominent in sweat, mammary, lacrimal, and salivary glands.

Myofascia, skeletal muscle ensheathed by vascular and sensitive fibrous connective tissue.

Myoglobin, the oxygen-containing, pigment-containing protein molecule of muscle.

Myosin, the principal protein of muscle associated with

contraction and relaxation of muscle cells.

Myriad, a great number.

Myx-, mucus.

N

Naso-, nose, nasal.

Neck, that part of the body inferior to the head and superior to the first thoracic vertebra and confluent with the shoulders, upper back, and upper chest; cervical region.

Necrosis (neh-KRO-sis), a state of cellular or tissue death.

Nephro-, kidney.

Neuro- (NOO-roh), nervous, referring to nervous structure or the nervous system.

Neuroglia (noo-ROHG-lee-ah), nonconducting support cells of the nervous system, including the astrocytes, oligodendrocytes, ependyma, and microglia of the CNS, and Schwann cells and satellite cells of the PNS.

Neurologic (neurology), concerned with disorders of the nervous system. Also refers to nerve/neuronal disorders seen in a clinical setting.

Neuron (NOO-ron), nerve cell.

Neurovascular, refers to nerve(s) and vessel(s), as in neurovascular bundle.

Nociceptor (no-see-SEP-tur), a receptor for pain.

Nucha- (NOO-kaw), posterior neck.

O

Oculus(i), eye.

-oid, having similar form; -like.

-oma, tumor.

Omni-, all, universally; e.g., omnidirectional, in all directions.

Ooph-, ovary.

Ophth-, eye.

Optic, relating to the eye.

Or-, mouth.

Orb, sphere, round structure.

Orbicular, rounded, circular.

Orbit, the bony cavity containing the eyeball.

Orchi-, testis.

Organelle(s) (or-gan-ELL), small functional structures within the cell cytoplasm.

Os-, bone.

Oscilloscope, an instrument that permits visualization of baseline and waves of changes in electrical voltage.

-osis, condition or state of; e.g., arthrosis is a generic term for a condition of a joint.

Osseous, relating to bone.

Ossification, endochondral, formation of bone by replacement of cartilage/calcified cartilage.

Ossification, intramembranous, formation of bone directly from osteoprogenitor cells in embryonic connective tissue (mesenchyme) or in fibrous tissue adjacent to fractured bone. There is no intermediate stage of cartilage formation or replacement.

Ossification, primary center of, the principal center of bone formation in the diaphysis or center of developing bone.

Ossification, secondary center of, a satellite center of ossification, as in the epiphysis.

Osteo-, bone.

Osteoblastic, referring to bone-forming cells (osteoblasts).

Osteoclastic, referring to bone-destroying cells (osteoclasts).

Osteoid (OSS-tee-oyd), bonelike; nonmineralized bone.

Osteoprogenitor, a primitive cell that has the potential, when stimulated, to become a bone-forming cell (osteoblast).

-ostomy, operation that makes an artificial opening.

Ovale, oval.

Oxy-, oxygen.

P

Pachy-, thick.

Pachymeninx, dura mater.

Palpable (PAL-pah-bul), touchable; by touch.

Palpate, to touch or feel (a common clinical technique).

Palsy, weakness.

Para-, alongside.

Parenchyma (pah-REN-keh-ma), the functional substance of an organ.

Paresis, weakness caused by incomplete paralysis.

Parietal (pah-RY-et-all), referring to a wall or outer part.

-pathy, disease.

Ped-, foot.

Pedal, foot.

Pedicle, footlike process; narrow stalk.

Pedo-, child.

Peduncle, a narrow stalk, specifically, masses of white matter in the CNS.

Pelvic girdle, the two coxal (hip) bones.

Pelvis(es), the ring of bone consisting of the two coxal (hip) bones and the sacrum and coccyx.

-penia, deficiency or decrease.

Penicillar, resembling a painter's brush or pencil.

Pennate, feather-shaped.

Peri- around.

Perichondrium (paree-KOND-ree-um), the fibrous envelope of cartilaginous structures (except articular), containing blood vessels, fibroblasts, and chondroblasts (immature cartilage cells).

Perineal, referring to the region inferior to the pelvis.

Periodontal, around a tooth.

Periosteum (paree-OS-tee-um), the fibrous envelope surrounding bone, containing osteoprogenitor cells, osteoblasts, fibroblasts, and blood vessels, serving as the life support system of bone.

Peripheral, away from the center, near or toward the periphery.

Peristalsis (paree-STAHL-sis), waves of coordinated and rhythmic muscular contractions in the walls of a cavity or tubelike organ, induced by hormones or other secreted factors and by nerves of the autonomic nervous sustem.

Peroneal, the lateral (fibular) side of the leg.

Perpendicular, refers to a plane at right angles (90 degrees) to an adjoining plane.

Pes, foot.

GLOSSARY

Pes anserinus, goose's foot. Refers to the tendons (sartorius, gracilis, and semitendinous) that collectively insert on the medial proximal tibia.

Petrous (PEET-russ), rocky or like a rock.

-pexy, fixation or suspension.

Phagocyte, a cell that takes up cell fragments or other particulate matter into its cytoplasm by endocytosis. Phagocytes with a segmented nucleus are called polymorphonuclear leukocytes (neutrophils); mononuclear phagocytes (of the monocyte-macrophage lineage) are known by several names, depending on their location—e.g., macrophages, monocytes of the blood, histiocytes of the connective tissues, Kupffer cells of the liver, alveolar (dust) cells of the lung, microglia of the cental nervous system. Many cells that are phagocytic under certain circumstances are not called or considered phagocytes.

Phagocytosis (fago-site-OH-sus), the taking of fragments or other particulate matter into a cell.

Phlebo-, vein.

-physis(es), growing part.

-pial, referring to pia mater.

Pinocytosis, cellular ingestion of fluid.

Pituitary (archaic), referring to mucus.

-plasia, referring to development or growth.

Plasm-, referring to the substance of some structure, e.g., cytoplasm (cell substance).

-plasty, surgical correction.

Plenipotentiary, having the capacity to develop along a number of different cell lines. Undifferentiated mesenchymal cells, pericytes, and certain other cells have such capability.

Pneumo-, air.

P.N.S. or PNS, peripheral nervous system, consisting of cranial and spinal nerves and the autonomic nervous system.

Pole (polar), either extremity of an axis, as in south and north poles of the Earth. Also refers to processes of a neuron (e.g., unipolar).

Pollex, thumb. Pollicis, genitive form.

Poly-, many or multi-.

Polymodal, with many modalities; polymodal receptors are responsive to several different stimuli.

Portal circulation, veins that drain a capillary bed and terminate in a second capillary or sinusoid network, as in the hepatic portal vein and the portal system of the hypophysis.

Post-, back of, after, posterior to.

Pre-, in front of, anterior to.

Precursor, a forerunner, whose existence precedes something that is formed from it.

Pro-, in front of.

Procerus (pro-SE-russ), long, slender muscle.

Process, bony, a projection sticking out from a surface.

Process, neuronal, an extension of a neuron, containing cytoplasm/organelles and limited by a cell membrane. A neuronal process (dendrite or axon) is part of a living cell.

Procto-, rectum.

Prolapse, the sinking down or displacement of a structure, such as the sinking of the uterus into the vagina.

Propria (prohp-ree-ah), common.

Protein, a chain of amino acids of varying length.

Proteoglycan, chain of disaccharides (carbohydrates) connected to a core of protein; a binding material.

Proteolytic, causing digestion or breakdown of protein.

Protuberance, a projection, or something sticking out from a surface.

Proviral, refers to viral DNA that has been integrated into the DNA of the host cell.

Pseudo (SOO-doh), false. In anatomy or medicine, having the appearance of one structure or phenomenon but not, in fact, being such a structure or phenomenon.

Pterygoid (TAYR-ee-goid), winglike.

-ptosis, falling, drooping.

Pulp, a soft, spongy tissue, often vascular.

Pyel-, pelvis.

Pyo-, pus.

Q

Quad-, four.

Quadrant, one-quarter of a circle.

Quadrate, four-sided; rectangular, usually square.

R

Radi-, ray.

Radiculitis, inflammation/irritation of a nerve root.

Radiculopathy, nerve root deficit characterized by change in the deep tendon (stretch) reflex, sensory loss (objective numbness), and muscle weakness.

Radix, root.

Ramus (RAY-mus), a branch.

Ratio, a fixed relationship or proportion between two things; e.g., 1:4 means that there is 1 unit for every 4 other units.

Recto-, rectum. See also procto-.

Reflux, backward flow.

Renal, referring to the kidney. See also nephro-.

Repolarization, an electrical change in excitable tissue away from neutral polarity, e.g., increasing polarity from 0 millivolts to –90 millivolts.

Residue (REZ-ih-doo), the material left over after processing and extraction of other parts.

Reticulum(a), a small network.

Retro-, back, behind, posterior; opposite of antero-.

Retroperitoneum, the area posterior to the posterior layer of parietal peritoneum. It lies anterior to the muscles of the posterior abdominal wall and includes the kidneys, ureters, abdominal aorta and immediate branches, inferior vena cava and immediate tributaries, pancreas, and ascending and descending colon.

-rhaphy, suture.

Rotundum, round.

S

Salpingo-, referring to uterine tubes.

Salpinx, uterine tube.

Sarco-, flesh.

Scavenger cell, see phagocyte.

Schwann cell, cell of the peripheral nervous system that provides myelin for some and a membranous covering for all axons. A line of Schwann cells forms a tube for axonal regeneration after axonal injury.

Sciatica, pain in the buttock radiating to the foot via the posterior and/or lateral thigh and leg; it follows the distribution of the sciatic nerve, and therefore is assumed to be irritation of that nerve or its roots (radiculitis).

Scoliosis (sko-lee-OH-sis), any significant lateral curvature of the vertebral column. Some degree of lateral curvature is seen in most spines, probably related to use of the dominant hand.

-scopy, inspection or examination of.

Sebum, the oil lying on the surface of skin, secreted by sebaceous glands (see Plate 19).

Secondary sex characteristics, anatomic and physiologic changes occurring as result of increased sex hormone secretion (testosterone in the male, estrogen in the female); these characteristics develop at puberty (generally at 11–14 years of age). In the male, they include growth of body hair, change in voice due to change in laryngeal structure, increased skeletal growth, increase in size of external genitals, functional changes in internal genitals, and changes in mental attitude. In females, they include enlargement of breasts, change in body shape due to skeletal growth and distribution of body fat, and maturation of internal and external genital structures.

Secretion, elaboration of a product from a gland into a duct, vessel, or cavity. See excretion.

Sella, saddle.

Sellar, saddle-shaped.

Semi-, half or partly.

Sensitive, responsive to stimuli, eliciting an awareness of touch, pressure, temperature, and/or pain; innervated.

Sensory, referring to sensation (e.g., touch, perception of temperature, vision).

Septum(a), a wall or an extension of a wall; a structure that separates.

Serosa (sir-OH-sa), lining tissue of cavities closed to the outside, consisting of a layer of squamous or cuboidal cells and underlying connective tissue.

Serotonin, a nitrogenous molecule with many functions, including acting as a neurotransmitter, inhibitor of gastric secretion, and vasoconstrictor.

Serous, watery; see serum.

Serum, any clear fluid; also blood plasma less plasma (clotting) proteins.

Sesamoid, pea-shaped. Generally refers to small bones of the hand and foot. The largest sesamoid bone is the patella. These bones are formed within the tendons or ligaments at points of stress.

Sharpey's fibers, fibrous bands of ligaments, tendons, and/or periosteum inserting directly into bone.

Shoulder, the part of the body where the upper limb is joined to the trunk; specifically, the shoulder joint and surrounding area, including the upper lateral scapula and distal clavicle (acromioclavicular area).

-sial, referring to saliva.

Sinus(es) (SY-nuss), a cavity or channel. A venous sinus is a large channel, larger than an ordinary vein; an air sinus is a cavity.

Sinusoid, sinuslike; usually refers to thin-walled, porous vessels in glands. Generally slightly larger than capil-

laries, sinusoids vary in their structure depending on their location.

Soft tissue, any tissue not containing mineral, e.g., not bone, teeth. Generally refers to myofascial tissues.

Soma, the body; the body wall.

Somatic, referring to the body or body wall, e.g., the cell body of a neuron (soma); in organizational terms, contrasted with viscus or viscera (organs containing cavities).

Spasm, rapid, violent, involuntary muscle contraction, usually resulting in some contortion of the body part experiencing the spasm.

Spheno- (SPHEE-no), shaped like a wedge; refers to a triangular-shaped structure that comes to a thin edge on one side.

Sphincter, a concentric band of muscle surrounding a narrowed cavity or passage.

Spindle, a structure that is round and tapered.

Spinosum, spiny or spinelike.

Spleno-, spleen. See also lieno-.

Spondyl-, vertebra.

Squamous, platelike, thin. Generally refers to flat, thin epithelial cells.

Stenosis (sten-OH-sis), narrowing.

-stomy, hole or opening.

Stratified, layered; having more than one layer.

Stria, stripes or parallel markings.

Styloid (STYL-oyd), having the form of a pointed spike or pillar.

Sub-, under.

Subchondral, under cartilage; specifically, the bone adjacent to articular cartilage.

Subcutaneous (sub-kew-TANE-ee-us); under the skin.

Subdural, under the dura; between the dura and the brain or spinal cord.

Supra-, above.

Suture (SOO-chur), a type of fibrous or bony junction characterized by interlocking, V-shaped surfaces, as in the skull.

Swallowing, deglutition.

Sym-, see syn.

Symphysis(es) (SIM-fih-sis); see joint classification, structural.

Syn- (SIN), together, with, alongside.

Synarthrosis(es) (sin-arth-RO-sis); see joint classification, functional.

Synchondrosis (sin-kon-DRO-sis); see joint classification, structural.

Syndesmosis(es) (syn-des-MO-sis); see joint classification, structural.

Synostosis(es) (syn-os-TOH-sis); see joint classification, structural.

Synovial (sih-NOH-vee-ul), refers to a viscous fluid similar in consistency to uncooked egg white. This fluid and the membrane that secretes it line freely movable joints (synovial joints), bursae, and tendon sheaths.

Synthesis(es), formation of a structure from smaller parts; integration of parts.

T

Taenia(e) coli, strips of longitudinal muscle in the mus-

GLOSSARY

cularis externa of the large intestine (excluding the rectum and anal canal).

Tarsal, tarso-, the ankle.

Tendinitis, inflammation of a tendon.

Tendinous (TEN-dih-nuss), referring to tendon.

Tendon, fibrous tissue connecting skeletal muscle to bone or other muscle. May be cord-like or sheet-like (aponeurosis).

Thigh, that part of the lower limb between the hip joint and the knee joint.

Thorax, the region between the neck and the abdomen.

Thrombosis(es), a condition of clots or thrombi within a vessel or vessels.

Thrombus(i), a clot within a blood vessel, obstructing flow.

-tomy, incision.

Tone, normal tension in muscle, resistant to stretch.

Torso, the part of the body less the limbs and head; the trunk.

Transcriptase, an enzyme (polymerase) directed by DNA to facilitate synthesis of a single strand of RNA that is structurally complementary to a strand of DNA.

Transcriptase, reverse, a polymerase (enzyme) directed by RNA to facilitate synthesis of a single or double strand of DNA that is structurally complementary to a strand of RNA. In HIV infection of cells, RNA-directed reverse transcriptase makes possible the transcribing of viral RNA sequences into double-stranded DNA; this is then integrated into the host cell's DNA. The combined DNA is called proviral DNA.

Trauma, an anatomic or psychic response to injury.

Trochanter, a large process. Specifically, two processes of the upper femur.

Trochlea (TROHK-lee-ah), a pulley-shaped structure.

-trophic, relating to nutrition.

Truss, a collection of members (beams) put together in such a way as to create a supporting framework.

Tubercle (TOOB-er-kul), a rough, small bump on bone.

Tuberosity (toob-eh-ROSS-eh-tee), a bump of bone, generally larger than a tubercle, smaller than a process.

Tunica, referring to a coat or sheath; a layer.

Turcica (TUR-sih-kah), Turkish, as in sella turcica (Turkish saddle).

U

Uni-, one. A unicellular gland is a one-cell gland.

Unibody, of one body; a structure with parts integrated into one unit.

Unit, a single thing or quantity; the basic part of a complex of parts.

Urogenital, referring to structures of both the urinary and genital (reproductive) systems.

Urogenital diaphragm, a layer in the perineum consisting of the sphincter urethrae and deep transverse perineal muscles and their fasciae. Also called the deep perineal space.

V

Vacuolation (vac-u-oh-LAY-shun), formation of small cavities or holes; part of a degenerative process in cartilage during bone development.

Vacuum, a space devoid of air, with, therefore, no pressure. In the relative sense, decreased pressure in the thoracic cavity during inspiration represents a partial vacuum, drawing air from a space with air of higher pressure.

Varix(ces), an enlarged, tortuous (twisted) vessel.

Varicosity(ies), an enlarged and irregular-shaped, highly curved (tortuous) vein(s). Most often seen in superficial veins of the lower limbs and the testes/scrotum.

Vas(a), vessel.

Vasa vasorum, vessel that supplies a larger vessel.

Vascular, referring to blood or lymph vessels or to blood supply.

Vasorum, of the vessels.

Ventricle, a cavity.

Vessel, a tubelike channel for carrying fluid, such as blood or lymph.

Vestibule, an entranceway, cavity, or space.

Villus(i), a fingerlike projection of tissue, as in the intestinal tract or placenta.

Viral, relating to a virus.

Virion, a single virus, also called a virus particle, consisting of genetic material (DNA or RNA) and a protein shell (capsid).

Virus, one of a group of extremely small infectious agents, consisting of genetic material and a protein shell. A virus is not capable of metabolism, and thus requires a host to replicate. On attachment to a surface molecule on a cell membrane, a virus particle is enveloped by the cell membrane and brought into the cytoplasm, thus infecting the cell.

Viscous, a fluid or semi-fluid state wherein molecules experience significant friction during movement.

Viscus(era), an organ with a cavity in it.

Vomer, a plowshare-shaped structure.

W

White matter, a substance of the brain and spinal cord consisting of largely myelinated axons arranged in the form of bundles or tracts. It appears white in the living or preserved brain.

Wrist, the carpus, the region between the forearm and hand.

Wrist drop, a condition in which the extensors of the wrist are weak or paralyzed. The wrist cannot be extended and therefore the wrist "drops" when one attempts to hold the hand horizontally or vertically upward. This condition is usually the result of radial nerve denervation.

X

Xeno-, foreign.

Xero-, dry.

Z

Zygapophysis(es) (zi-gah-POFF-ee-sis), an articular process of a vertebra; also a joint between vertebrae (zygapophyseal articulation). Such synovial joints may be called facet joints. See also facet.

Zygo- (ZY-go), referring to a yoke or union; joined.

Words are indexed by generic terms (artery, ligament, foramen, process, and so on). For example, if you wish to locate a specific artery, look under "Artery." Principal reference among a number of cited plates is in bold (dark) type. Our reference medical dictionary is *Dorland's Illustrated Medical Dictionary.*

A

INDEX